Simulation and Modeling:
Current Technologies and Applications

Asim Abdel Rahman El Sheikh
The Arab Academy for Banking and Financial Sciences, Jordan

Abid Al Ajeeli
University of Bahrain, Bahrain

Evon M. O. Abu-Taieh
The Arab Academy for Banking and Financial Sciences, Jordan

IGI PUBLISHING

Hershey • New York

Acquisition Editor:	Kristin Klinger
Senior Managing Editor:	Jennifer Neidig
Managing Editor:	Sara Reed
Development Editor:	Kristin Roth
Copy Editor:	Alana Bubnis
Typesetter:	Amanda Appicello
Cover Design:	Lisa Tosheff
Printed at:	Yurchak Printing Inc.

Published in the United States of America by
       IGI Publishing (an imprint of IGI Global)
       701 E. Chocolate Avenue
       Hershey PA 17033
       Tel: 717-533-8845
       Fax: 717-533-8661
       E-mail: cust@igi-pub.com
       Web site: http://www.igi-pub.com

and in the United Kingdom by
       IGI Publishing (an imprint of IGI Global)
       3 Henrietta Street
       Covent Garden
       London WC2E 8LU
       Tel: 44 20 7240 0856
       Fax: 44 20 7379 0609
       Web site: http://www.eurospanonline.com

Library of Congress Cataloging-in-Publication Data

Simulation and modeling : current technologies and applications / Asim El Sheikh, Abid Thyab Al Ajeeli and Evon M. Abu-Taieh, editors.
    p. cm.
  Summary: "This book offers insight into the computer science aspect of simulation and modeling while integrating the business practices of SM. It includes current issues related to simulation, such as: Web-based simulation, virtual reality, augmented reality, and artificial intelligence, combining different methods, views, theories, and applications of simulations in one volume"--Provided by publisher.
  Includes bibliographical references and index.
  ISBN-13: 978-1-59904-198-8 (hc)
  ISBN-13: 978-1-59904-200-8 (ebook)
  1. Computer simulation. I. El Sheikh, Asim. II. Al Ajeeli, Abid Thyab. III. Abu-Taieh, Evon M.
  QA76.9.C65S528 2007
  003.3--dc22
                2007007288

British Cataloguing in Publication Data
A Cataloguing in Publication record for this book is available from the British Library.

Simulation and Modeling:
Current Technologies and Applications

Table of Contents

Preface

Introduction

Nowadays, technology advances in sync with the speed of light, noting that simulation is a practice, which extends to take into account factors concluded from scientific research emanating in the U.S. since the late 1940s, yet simulation is unanimously viewed as a science; therefore, the book tackles the current technologies and applications in simulation and modeling in a systematic, comprehensive manner.

Due to the high costs of network infrastructure and the constant rising of unanswered questions regarding technology, simulation has become a good choice for estimation of the performance of networks. Additionally, such methodology for requirements determination can be extended to serve as a blueprint for business (or management) simulation by providing an initial model for creating a business simulation to subsequently show how it can be incorporated into an application.

This allows one to capitalize on conceptual models in a business that have been created for requirements determination by extending them with the conceptual model of runtime management, thus covering the core decision-taking science, particularly in view of the well-known fundamental economic decision theory, to which individuals attribute rational choices from a range of alternatives.

The Challenges

The need to define what appropriate path to follow at any given crossroad triggers the concept behind this project, as any attempt to deal with a problem demands an adequate understanding of the challenges that exist. Such challenges can be further illustrated, as addressed in this book:

First, simulation of a system with limited data is challenging, as it calls for a certain degree of intelligence built in to the system.

Second, the overall employment of remotely controlled vehicles functioning in the ground, air, and marine domains requires investigating the critical issues in the command and control of such vehicles.

Third, a proper understanding of the simulation tools, underlying system algorithms, and user needs is challenging.

Fourth, healthcare systems pose many of the challenges, including difficulty in understanding the system being studied, uncertainty over which data to collect, and problems of communication between problem owners.

Searching for a Solution

Solutions to the problem of defining simulation technologies and application are tackled in this book: for instance, the book presents various based simulation methodologies that may be customized and used in the simulation of a wide variety of problems. Additionally, the book presents a model-based approach resulting in simulation architecture that integrates proven design concepts, such as the model-view-controller paradigm, distributed computing, Web-based simulations, cognitive model-based high-fidelity interfaces and object-based modeling methods.

Moreover, the book shows how simulation allows the identification of critical variables in the randomized clinical trial (RCT) by measuring their effects on the simulation model's "behaviour."

Organization of the Book

The book is organized into 15 chapters. A brief description of each of the chapters follows:

Chapter I provides a comprehensive explanatory platform of simulation background, reviewing simulation definitions, forms of models, the need for simulation, simulation approaches and modeling notations.

Chapter II offers an overview on the distributed simulation in industry, in view that, although the observance of a distinction between continuous and discrete simulations has long been a practice in the simulation community at large, human interactivity in simulation ("human-in-the-loop") *HLA* literature often uses a different terminology and refers to *time-stepped* and *event-driven* simulation.

In addition, Chapter III presents the object-oriented approach for the development of an optical burst switching (OBS) simulator, called OBSim, built in Java.

Subsequently, Chapter IV illustrates how natural language modeling (NLM), a conceptual modeling language, methodology for requirements determination can be extended to serve as a blueprint for business (or management) simulation by providing an initial model for creating a business simulation.

Consequently, Chapter V presents a suggested system development life cycle, "relay race methodology" (RRM). The RRM is based on the philosophy of a relay race, where each runner in the race must hand off the baton within a certain zone, usually marked by triangles on the track race.

On another note, Chapter VI sets forth a new model-based simulation methodology that may be customized and used in the simulation of a wide variety of problems involving multiple source-destination flows with intermediate agents. It explains the model based on a new class of neural networks, called differentially fed artificial neural networks, and the system level performance of the same.

Additionally, Chapter VII presents a model-based approach that the authors adopted for investigating the critical issues in the command and control of remotely operated vehicles (ROVs) through an interactive model-based architecture.

Furthermore, Chapter VIII reports on the use of simulation in supporting decision-making about what data to collect in a randomized clinical trial (RCT). The chapter shows how simulation also allows the identification of critical variables in the RCT by measuring their effects on the simulation model's "behavior."

In the same token, Chapter IX addresses the problem of modeling finished products and their associated sub-assemblies and/or raw materials. A production system is a set of policies that monitors and controls finished products and raw materials, as it determines how much of each item should be manufactured or be kept in warehouses, when low items should be replenished, and how many items should be assembled or ordered when replenishment is needed.

Chapter X illustrates the use of mathematical modelling and simulation to discover the reasons for data to behave in certain ways, as it suggests the use of simulation and modeling of knowledge-mining architecture by using recurrent hybrid nets; particularly in view that hybrid nets combine arithmetic and integrator elements to and from nodes for modeling the complex behavior of intelligent systems.

Likewise, Chapter XI demonstrates the development of a novel compromise linear programming having fuzzy resources (CLPFR) model as well as its simulation for a theory-of-constraints (TOC) product mix problem using MATLAB® v. 7.04 R.14 SP.2 software. The product mix problem considers multiple constraint resources. The developed CLPFR model helps in finding a robust solution with better profit and product mix solution in a non-bottleneck situation. The authors simulate the level-of-satisfaction of the decision maker (DM) as well as the degree of fuzziness of the solution found using the CLPFR model. Simulations have been carried out with MATLAB® v. 7.04 R.14 SP.2 software.

However, Chapter XII provides mainly an overview of the ongoing technology shift inside the vehicles and couples this to simulation possibilities and thereby introduces the business process simulator-based design (SBD). The perspective in this chapter is human-machine interaction (HMI) and therefore addresses human-in-the-loop simulators, keeping in mind the fact that simulation could and even must be used on other levels in order to optimize and verify more technical functions.

On another note, Chapter XIII tackles business aspects of simulation, amongst other things: describing the relationship between business process reengineering (BPR) and change management, the role of simulation in supporting BPR, notwithstanding the future challenges of business process simulation, along with an illustration of simulation technology limitations in reengineering business processes, characteristics of successful simulation and some simulation applications.

While Chapter XIV introduces virtual reality and augmented reality as a basis for simulation visualization, within this context, it shows how these technologies can support simulation visualization and gives important considerations about the use of simulation in virtual and augmented reality environments. Hardware and software features, as well as user interface and examples related to simulation, using and supporting virtual reality and augmented reality, are discussed, stressing their benefits and disadvantages. The chapter discusses virtual and augmented reality in the context of simulation, emphasizing the visualization of data and behavior of systems. The importance of simulation to give dynamic and realistic behaviors to virtual and augmented reality is also pointed out. The work indicates that understanding the integrated use of virtual reality and simulation should create better conditions for the development of innovative simulation environments as well as for the improvement of virtual and augmented reality environments.

In conclusion, Chapter XV aims to develop artificial mechanisms that can play the role emotion plays in natural life, in order to build agents with the mission to "to bring life" to several applications, amongst other things: information, transaction, education, tutoring, business, entertainment and e-commerce. In light of the fact that artificial emotions play an important role at the control level of agent architectures, emotion may lead to reactive or deliberative behaviors, it may intensify agent's motivations, it can create new goals (and then sub-goals) and it can set new criteria for the selection of the methods and the plans the agent uses to satisfy its motives. Since artificial emotion is a process that operates at the control level of agent architecture, the behavior of the agent will improve if the agent's emotion process improves.

Acknowledgment

The editors would like to extend their deepest appreciation for the efforts of all participants in the collation and review process of the book. Additionally, the editors would like to acknowledge their support, as this project could not have efficiently been completed without their significant participation. A further special note of thanks goes also to all the staff at IGI Global, especially Kristin Roth, Jan Travers, and Mehdi Khosrow-Pour; whose contributions throughout the whole process, from inception of the initial idea to final publication, have been invaluable.

In this regard, the editors would also like to acknowledge the authors of chapters included in this book that served as referees for articles written by other authors. Thanks goes to all those who provided constructive and comprehensive reviews.

In closing, we would like to our families and loved ones for their patience, love, and support throughout this project. May they be blessed with eternal happiness.

Editors,
Evon M. O. Abu-Taieh, PhD
Asim Abdel Rahman El Sheikh, PhD
Abid Al Ajeeli, PhD

Chapter I

Methodologies and Approaches in Discrete Event Simulation

Evon M. O. Abu-Taieh,

The Arab Academy for Banking and Financial Sciences, Jordan

Asim Abdel Rahman El Sheikh,

The Arab Academy for Banking and Financial Sciences, Jordan

Abstract

This chapter aims to give a comprehensive explanatory platform of simulation background. As this chapter comprises of four sections, it reviews simulation definitions, forms of models, the need for simulation, simulation approaches and modeling notations. Simulation definition is essential in order to set research boundaries. Moreover, the chapter discusses forms of models: scale model of the real system, or discrete and continuous models. Subsequently, the chapter states documentation of several reasons by different authors pertaining to the question of "why simulate?," followed by a thorough discussion of modeling approaches in respect to general considerations. Simulation modeling approaches are discussed with special emphasis on the discrete events type only: process-interaction, event scheduling, and activity scanning, yet, a slight comparison is made between the different approaches. Furthermore, the chapter discusses modeling notations activity cycle diagram (ACD) with different versions of the ACD. Furthermore, the chapter discusses petri nets, which handle concurrent discrete events dynamic systems simulation. In addition, Monte Carlo simulation is discussed due to its important applications. Finally, the fourth section of this chapter reviews Web-based simulation, along with all three different types of object-oriented simulation and modeling.

Introduction

The aim of this chapter is to serve as a comprehensive explanatory platform of simulation. It is imperative to define simulation in order to set boundaries for the research. As such, the chapter is comprised of four main sections: an overview of simulation modeling approaches, modeling notations, Petri Nets, Monte Carlo simulation, and Web-based simulation and object oriented simulation.

The chapter starts with a review of simulation definitions, forms of models, the need for simulation, simulation approaches and modeling notations. Next, the chapter discusses forms of models: scale models of the real system, or discrete and continuous models. Furthermore, the question *"why simulate?"* is addressed, through discussing a number of reasons by different authors in this section. Modeling approaches are discussed next with general considerations. Simulation modeling approaches are analyzed, emphasizing especially the discrete types, namely process-interaction, event scheduling, and activity scanning, with a minuscule comparison between the different approaches. Moreover, the chapter discusses the different modeling notations of the activity cycle diagram (ACD).

Bearing in mind that Petri Nets handles concurrent discrete events dynamic systems simulation, which is outside the scope of this book, nonetheless, various simulation packages, namely Optsim (Artifex), have used Petri Nets. Therefore, it has been reasoned to be imperative to further discuss Petri Nets in the second section of this chapter, particularly that the idea of Petri Nets was developed to answer the question of concurrency, which naturally arises constantly when discussing simulation. The Petri Nets will be discussed by interpreting the formal definition of Petri Nets, describing the classical Petri Nets and the different classifications of Petri Nets. However, the section does not include the parallel discrete event simulation languages (PDES); although some papers, such as Low et al. (1999) are available for concerned to read.

Furthermore, the third section tackles an interrelated topic—Monte Carlo simulation—noting that Monte Carlo is particularly important, since four simulation packages have used it, namely: *Crystal Ball, BuildSim,* and *Decision Script* and *Pro.*

Finally, the fourth section of this chapter reviews Web-based simulation, along with all three different types of simulation and modeling, in addition to the different programming languages and environments, through which simulation can be done using object-oriented perspective. Then, the idea of object-oriented simulation is discussed in comparison to algorithms perspective. As such, modeling principles and aims from an object-oriented perspective is introduced as a reminder to the reader. Considering that Web-based simulation and object oriented is to give an overview of the object-oriented simulation perspective, consequently discussing object-oriented perspective entails discussing the Web-based technology, particularly, since the effect of object-oriented is visibly seen on the Web-based technology.

Simulation Definitions

In their book, Paul & Balmer have quoted Pidd, defining simulation as:

Analyst builds a model of the system of interest, writes computer program which embody the model and uses a computer to initiate the system's behavior when subject to a variety of operating policies. Thus the most desirable policy may be selected (Paul & Balmer, 1998).

Notice how the definition is really a description of the process of building a simulation project. Nylor defined simulation as:

(computer) simulation is a numerical technique for conducting experiments on a digital computer, which involves certain types of mathematical and logical models that describes the behavior of a business or economic system (or some component thereof) over extended periods of real time (Paul & Balmer, 1998).

Nylor describes computer simulation a "technique." Mize & Cox defined it as:

The process of conducting experiments on a model of a system in lieu of either (i) direct experimentation with the system itself, or (ii) direct analytical solution of some problem associated with the system (Paul & Balmer, 1998).

On the other hand Mize & Cox define it as the "process." Shannon defines simulation as:

Simulation is the process of designing a model of a real system and conducting experiments with this model for the purpose of either understanding the behavior of the system or of evaluating various strategies for the operation of the system (Paul & Balmer, 1998, p. 1).

A more comprehensive definition is "Simulation is the use of a model to represent over time essential characteristics of a system under study" (El Sheikh, 1987).

Forms of Models

There are three forms of models known and listed by Paul & Balmer (1998, p. 3):

1.   A scale model of the real system, for example, a model aircraft in a wind tunnel or a model railway.

2.   A physical model in different physical system to the real one, for example, colored water in tubes has been used to simulate the flow of coal in a mine. More common in the use of electrical circuits – analogue computers are based on this idea.

3.   A set of mathematical equations and logical relationship.

Within the umbrella of the third model type, and when the equations can not be solved analytically or numerically, Paul & Balmer say the term "simulation" falls (Paul & Balmer, 1998, p. 3). Also in the same context, Paul & Balmer describe digital simulation generally as a procedure that involves:

calculation of state changes in each part of the system separately. These calculations will be repeated sequentially, reflecting the dynamics of the system through time. Each set of calculations depends in some way upon the results of previous calculations. Such procedure is sometimes referred to as "Digital Simulation" (Paul & Balmer, 1998, p. 3).

Also, known by the simulation industry that there are two distinct types of models: discrete and continuous. In "continuous systems the change through time are predominantly smooth, and conveniently described by sets of difference equations" (Paul & Balmer, 1998, p. 3); while discrete system changes at specific points in time and a model of such a system is concerned only with these events (Paul & Balmer, 1998, p. 3).

Why Simulate?

Simulation allows experimentation, although computer simulation requires long programs of some complexity and is time consuming. Yet what are the other options? The answer is direct experimentation or a mathematical model. Direct experimentation is costly and time consuming, yet computer simulation can be replicated taking into account the safety and legality issues. On the other hand, one can use mathematical models yet mathematical models cannot cope with dynamic effects. Also, computer simulation can sample from nonstandard probability distribution. One can summarize the advantages of computer simulation, best described, by Banks (1999, 2000) and recited again in AutoMod (www.automod.com:

Simulation, first, allows the user to experiment with different scenarios and, therefore, helps the modeler to build and make the correct choices without worrying about the cost of experimentations.

The second reason for using simulation is the time control. The modeler or researcher can expand and compress time as s/he pleases, just like pressing a fast forward button on life. The third reason is like the rewind button on a VCR: seeing a scene over and over will definitely shed light on the answer of the question, "why did this happen?"

The fourth reason is "exploring the possibilities." Considering that the package user would be able to witness the consequences of his/her actions on a computer monitor and, as such, avoid jeopardizing the cost of using the real system; therefore, the user will be able to take risks when trying new things and diving in the decision pool with no fears hanging over her/his neck.

As in chess, the winner is the one who can visualize more moves and scenarios before the opponent does. In business the same idea holds. Making decisions on impulse can be very dangerous, yet if the idea is envisaged on a computer monitor then no harm is really done, and the problem is diagnosed before it even happens. Diagnosing problems is the fifth reason why people need to simulate.

Likewise, the sixth reason tackles the same aspect of identifying constraints and predicting obstacles that may arise, and is considered as one major factor why businesses buy simulation software. The seventh reason addresses the fact that many times decisions are made based on "someone's thought" rather than what is really happening.

When studying some simulation packages, that is, *AutoMod*, the model can be viewed in 3-D. This animation allows the user "to detect design flaws within systems that appear credible when seen on paper or in a 2-D CAD drawing" (www.automod.com). The ninth incentive for simulation is to "visualize the plan."

It is much easier and more cost effective to make a decision based on predictable and distinguished facts. Yet, it is a known fact that such luxury is scarce in the business world. Moreover, building consensus, as Banks (1999) explains it, is the tenth incentive for simulation.

The American saying "wish for the best and plan for the worst" holds true, as such, that in the business world many would suggest the "what if" scenario. Nevertheless, before trying out the "what if" scenario many would rather have the safety net beneath them. Therefore, simulation is used for "preparing for change", which is suggested by Banks (2000) as the eleventh incentive on the list of simulation incentives.

In addition, the thirteenth reason is evidently trying different scenarios on a simulated environment; proving to be less expensive, as well as less disturbing, than trying the idea in real life. Therefore, simulation software does save money and effort, which denotes a wise investment.

Moreover, training any team using a simulated environment is less expensive than real life. As such, two examples come to mind—pilots and airplanes, as well as stockbrokers. The number of lives that can be saved, as well as the cost effectiveness of training a pilot on a simulated plane would be a positive feature; the worst that can happen is losing the pilot and the plane. The second (and more civil) example is training stock brokers to trade on-line: needless to say, such idea is less stressful for management and more cost effective. The idea of training the team escaped neither Drappa & Ludewig (2000) nor Banks (1999, 2000) (www.automod.com); furthermore, it is described by Banks (2000) as follows:

Training the team – *Simulation models can provide excellent training when designed for that purpose. Used in this manner, the team provides decision inputs to the simulation model as it progresses. The team, and individual members of the team, can learn from their mistakes, and learn to operate better. This is much less expensive and less disruptive than on-the-job learning.*

In any field listing the requirements can be of tremendous effort, for the simple reason that there are so many of them. As such, the fourteenth reason crystallizes in avoiding overlooked requirements and imagining the whole scene, or the trouble of having to carry a notepad to write on it when remembering a forgotten requirement. Banks recited "specifying requirement" as one of the reasons why we need simulation.

While these recited advantages are of great significance, yet many disadvantages still show their effect, which are also summarized best by Banks (2000) and again recited by AutoMod (www.automod.com. The first hardship faced in the simulation industry is:

Model building requires special training – Model building is an art that is learned over time and through experience. Furthermore, if two models of the same system are constructed by two competent individuals, they may have similarities, but it is highly unlikely that they will be identical (www.automod.com).

Since no two modelers can not agree on a model, which can be due to the human nature, the second hardship, a natural consequence of the first, is the difficulty of interpreting the results of the simulation. Another mishap accounted for in simulation and analysis would be the fact that both are time consuming and costly, as Banks (2000) states.

It is worth stressing the simple fact that simulation is not the cure of all diseases nor is the solution for all problems, hence, certain types of problems can be solved using mathematical models and equations, as stated by Banks (1999):

Simulation may be used inappropriately – Simulation is used in some cases when an analytical solution is possible, or even preferable. This is particularly true in the case of small queuing systems and some probabilistic inventory systems, for which closed form models (equations) are available.

Modeling Approaches

Pidd suggested, in his book (1998), three aspects need to be considered when planning a computer simulation project:

- Time-flow handling
- Behavior of the system
- Change handling

The flow of time in a simulation can be handled in two manners: the first is to move forward in equal time intervals. Such an approach is called *time-slice*. The second approach is *next-event*; which increments time in variable amounts or moves the time from state to state. On one hand, there is less information to keep in the time-slice approach; on the other hand, the next-event approach avoids the extra checking and is more general.

The behavior of the system can be deterministic or stochastic: deterministic system, of which its behavior would be entirely predictable, whereas, stochastic system, of which its behavior cannot be predicted but some statement can be made about how likely certain events are to occur.

The change in the system can be discrete or continuous. Variables in the model can be thought of as changing values in four ways (Pidd, 1998):

1.   Continuously at any point of time: thus, change smoothly and values of variables at any point of time.

2.   Continuously changing but only at discrete time events: values change smoothly but values accessible at predetermined time.

3.   Discretely changing at any point of time: state changes are easily identified but occur at any time.

4.   Discretely changing at any point of time: state changes can only occur at specified point of time.

Others define 3 and 4 as discrete event simulation as follows: "a discrete-event simulation is one in which the state of a model changes at only a discrete, but possibly random, set of simulated time points" (Schriber & Brunner, 2002). Mixed or hybrid systems with both discrete and continuous change do exist; in fact simulation packages try to include both: "Four packages out of 56 simulation packages claimed to have hybrid system" (Abu-Taieh, 2005).

General Considerations

To start the simulation model, the following two aspects are recommended by Pidd (1998) to decide the level of details and accuracy needed:

1.   Nature of the system being simulated.

2.   Nature of study being carried.

  a.   What are the objectives of the study?

  b.   What is the point of simulation?

  c.   What are the expected results?

Simulation Modeling Approaches

There are four significant simulation approaches or methods used by the simulation community:

- Process interaction approach.
- Event scheduling approach.
- Activity scanning approach.
- Three-phase approach.

There are other simulation methods, such as *transactional-flow approach,* that are known among the simulation packages and used by simulation packages like *ProModel, Arena, Extend,* and *Witness* (GoldSim Web). Another, known method used specially with continuous models is *stock and flow* method. A brief description is given in following paragraphs, yet an explicit description is better read in detail in Pidd (1998).

Process Interaction Approach

The simulation structure that has the greatest intuitive appeal is the *process interaction method.* In this method, the computer program emulates the flow of an object (for example, a load) through the system. The load moves as far as possible in the system until it is delayed, enters an activity, or exits from the system. When the load's movement is halted, the clock advances to the time of the next movement of any load.

This flow, or movement, describes in sequence all of the states that the object can attain in the system as seen in . In a model of a self-service laundry, for example, a customer may enter the system, wait for a washing machine to become available, wash his or her clothes in the washing machine, wait for a basket to become available, unload the washing machine, transport the clothes in the basket to a dryer, wait for a dryer to become available, unload the clothes into a dryer, dry the clothes, unload the dryer, and then leave the laundry. Each state and event is simulated.

Process interaction approach is used by many commercial packages, among them *Automod.*

Transaction Flow Approach

Transaction flow approach was first introduced by GPSS in 1962 as stated by Henriksen (1997). *Transaction flow* is a simpler version of *process interaction approach,* as the following clearly states: "Gordon's transaction flow world-view was a cleverly disguised form of process interaction that put the process interaction approach within the grasp of ordinary users" (Schriber et al., 2003).

In *transaction flow approach* models consist of entities (units of traffic), resources (elements that service entities), and control elements (elements that determine the states of the entities and resources) described by Henriksen (1997), Schriber and Brunner (1997) and GoldSim Web. Discrete simulators, which are generally designed for simulating detailed processes, such as call centers, factory operations, and shipping facilities, rely on such approach. Many discrete simulators use this approach, such as "*ProModel, Arena, Extend, and Witness*" (www.goldsim.com). Some scholars like Schriber and Brunner (1997) describe

Figure 1. A process interaction executive (Pidd, 1998)

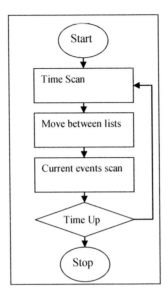

such approach as "transaction flow world view" and they add it "often provides the basis for discrete-event simulation."

In the transaction flow world view, a system is visualized as consisting of discrete units of traffic that move ("flow") from point to point in the system while competing with each other for the use of scarce resources. The units of traffic are sometimes called "transactions," giving rise to the phrase "transaction flow" (Schriber & Brunner, 1997).

The same scholars go on to describe best fitted applications to such approach: "manufacturing, material handling, transportation, health care, civil, natural resource, communication, defense, and information processing systems, and queuing systems in general" (Schriber & Brunner, 1997).

Another approach that is worth mentioning is the *stock and flow* approach, which is used in dynamic systems. This approach was "developed by Professor Jay W. Forrester at MIT in the early 1960s" (GoldSim Web). Stock and flow is based on system dynamics that are built using three principal element types (stocks, flows, and converters). System dynamics software, such as; "*Stella, iThink, Vensim*, and *Powersim*" (www.goldsim.com).

Event Scheduling Approach

The basic concept of the *event scheduling method* is to advance time to the moment when something happens next (that is, when one event ends, time is advanced to the time of the next scheduled event). An event usually releases a resource. The event then reallocates available objects or entities by scheduling activities, in which they can now participate. For example, in the self-service laundry, if a customer's washing is finished and there is a basket available, the basket could be allocated immediately to the customer, who would then begin unloading the washing machine. Time is advanced to the next scheduled event (usually the end of an activity) and activities are examined to see whether any can now start as a consequence, as can be seen in . *Event scheduling approach* has one advantage and one disadvantage as Schriber et al. (2003) states: "The advantage was that it required no specialized language or operating system support. Event-based simulations could be implemented in procedural languages of even modest capabilities." While the disadvantage "of the event-based approach was that describing a system as a collection of events obscured any sense of process flow" (Schriber et al., 2003). As such, "In complex systems, the number of events grew to a point that following the behavior of an element flowing through the system became very difficult" (Schriber et al., 2003).

Many simulation packages adopted the *event based approach*, of which are *Supply Chain Builder, factory explorer, GoldSim, and ShowFlow*, as stated by (Abu-Taieh, 2005). Furthermore, it is worth mentioning that in a later section all approaches will be compared.

Figure 2. An event scheduling executive (Pidd, 1998)

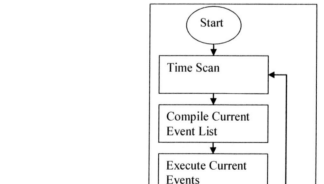

Activity Scanning Approach

The third simulation modeling structure is *activity scanning. Activity scanning* is also known as the *two phase approach*. *Activity scanning* produces a simulation program composed of independent modules waiting to be executed. In the first phase, a fixed amount of time is advanced, or scanned. In phase two, the system is updated (if an event occurs), as seen in Figure 3. Activity scanning is similar to rule-based programming (if the specified condition is met, then a rule is executed).

Many simulation packages adopted the *activity scanning* approach, one of which is *First-STEP*.

Three-Phase Approach

The next simulation modeling structure is known as the three-phase method. Described by Keith Douglas Tocher in 1963 in his book, *The Art of Simulation* (Odhabi et al 1998), then discussed in detail by Pidd and others in a series of research papers. The *three-phase* approach has, as the name suggested, A phase, B phase, and C phase, as seen in Figure 4; all phases will be overviewed next.

In the A phase, time is advanced until there is a state change in the system or until something happens next. The system is examined to determine all of the events that take place at this

Figure 3. An activity scanning executive (Pidd, 1998)

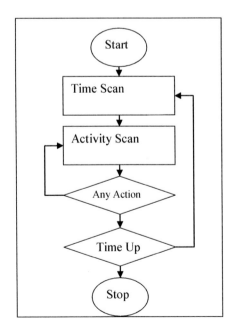

time (that is, all the activity completions that occur). The A phase is defined formally by Pidd and Cassel (1998) as "the executive finds the next event, and advances the clock to the time in which this event is due."

The B phase is the release of those resources scheduled to end their activities at this time. The B phase defined formally by Pidd and Cassel (1998) as "executes all B activities (the direct consequence of the scheduled events) which are due at the time."

The C phase is to start activities, given a global picture of resource availability. The C phase is defined formally by Pidd and Cassel (1998) as "the executive tries to execute all of the C activities (any actions whose start depends on resources and entities whose states may have changed in the B phase)."

Possible modeling inaccuracies may occur with the activity scanning and Three-Phase modeling methods, because discrete time slices must be specified. If the time interval is too wide, detail is lost. This type of simulation will become less popular as computing power continues to increase and computing costs continue to decrease (Pidd, 1998).

Figure 4. A three-phase executive (Pidd, 1998)

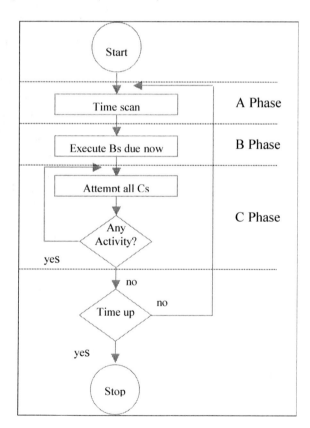

Comparison Between Simulation Approaches

In comparing the four approaches:

The event scheduling method is simpler and only has two phases so there is no Cs and Bs, this allows the program to run faster since there is no scanning for the conditional events. Those aforementioned advantages reiterate the disadvantages of the method since there are only two phases then all events are mixed (no Bs and Cs) then the method is not parsimony, which means it is very hard to enhance (Pidd, 1998).

The *activity scanning* approach is also simpler than the *three phase* method since it has no calendar, and it supports the parsimonious modeling. However this approach is much slower than *three phase* since all activities are treated as conditional. Moreover, the executive has two phases. Usually this approach is confused with the *three phase* method (Pidd, 1998).

The *process interaction* "shares two common advantages. First; they avoid programs that are slow to run. Second, they avoid the need to think through all possible logical consequences of an event" (Pidd, 1998). Yet, as Pidd (1998) claims this approach suffers from DEADLOCK problem, but this approach is very attractive for novice modelers. Although Schriber et al. (2003) say that, "process interaction was understood only by an elite group of individuals and was beyond the reach of ordinary programmers." In fact, Schriber et al. (2003) add: "Multi- threaded applications were talked about in computer science classes, but rarely used in the broader community." This indicates that the implementation of *process interaction* was very difficult to implement. The obvious contradiction, in the aforementioned quote, is due to the mix up between the *process interaction* approach and the *transaction flow approach*.

To see the complete idea of the origins of transaction flow, best stated by Schriber et al. (2003):

This was the primordial soup out of which the Gordon Simulator arose. Gordon's transaction flow world-view was a cleverly disguised form of process interaction that put the process interaction approach within the grasp of ordinary users. Gordon did one of the great packaging jobs of all time. He devised a set of building blocks that could be put together to build a flowchart that graphically depicted the operation of a system. Under this modeling paradigm, the flow of elements through a system was readily visible, because that was the focus of the whole approach.

The *three-phase* approach allows to "simulate parallelism, whilst avoiding deadlock" (Pidd & Cassel, 1998). Yet, *three-phase* has to scan through the schedule for bound activities, and then scans through all conditional activities, which consequently slows it down. Yet, many forgo the time spent in return for solving the deadlock problem. In fact, *three-phase* is used in distributed systems when referring to operating systems, databases, and so forth, under different names among them *three-phase commit* (see Tanenbaum & Steen, 2002).

The next section discusses the different modeling notations: activity cycle diagram (ACD) with different versions of the ACD, extended activity cycle diagrams (X-ACD), and hierarchy activity cycle diagrams (H-ACD).

Modeling Notations

Notations used to model in simulation are many. Yet we discuss here the activity cycle diagram (ACD), and two enhancements that were based on ACD: the extended ACD known as X-ACD and hierarchy ACD known as H-ACD. Modeling notations are discussed here to introduce the modeling approach known as *four phase*.

Activity Cycle Diagrams (ACD)

Activity cycle diagram (ACD) is a representation of the states that an *entity* goes through. "Activity cycle diagrams are one way of modeling the interactions of the entities and are particularly useful for systems with a strong queuing structure" (Pidd, 1998, p. 46).

Figure 5. Symbols for activity cycle diagrams (Pidd, 1998)

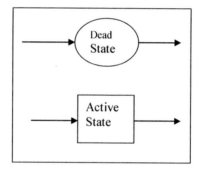

Figure 6. Generic activity cycle diagram

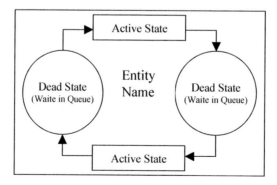

An entity is any component of the model which can be imagined to retain its identity through time. Entities are either idle, in notional or real queues, or active, engaged with other entities in time consuming activities (Paul, 1993).

The symbols used in the ACD are shown in Figure 5.

Activity cycle diagram makes use of two symbols shown in Figure 5. *Active state* usually involves the co-operation of different entities. On the other hand, a *dead state* or *passive state* involves no cooperation between different entities, generally a state where the entity waits for something to happen (Pidd, 1998). The diagram shows the life history of each entity and displays graphically their interactions. *Figure 6* shows a generic ACD.

Figure 7. Life cycles of each entity in PUB example

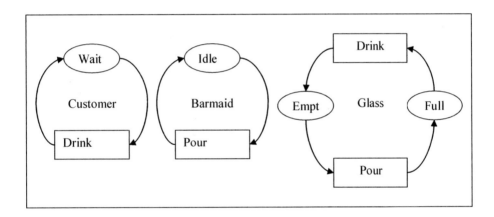

Figure 8. Life cycles of each entity in PUB example completed using ACD

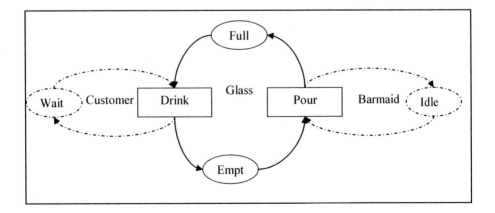

The best example used by Pidd (1998) and Paul (1993) is the Pub example. In the Pub there are three entities: Customer, Barmaid, and Glass. The customer has two states: drink (active state), and wait (passive state). The Barmaid has two states: pour (active states) and idle (passive state). On the other hand the entity Glass has two passive states, Full and Empty, and two active states, Drink and Pour. , taken from Paul (1993) below shows the life cycle of each entity using ACD.

Figure 8 shows the whole picture and how the entities work together in the PUB.

It is recommended that an entity alternates between the two states; in other words, an entity must go through a dead state after an active state. ACDs have the advantage of parsimony in that they use only two symbols, which describe the life cycle of the system's objects or entities.

The activity cycle diagrams (ACD) is known for some good qualities. First, ACD is very simple, with only two symbols for the modeler to remember, which makes ACD very easy to use. Second, ACD is parsimonious as described by Pidd (1998). Third, ACD is very "useful for understanding and communication" (Elsheikh, 1987). Fourth, ACD has two extra desirable aspects "comprehension, communication and generality" (Doukidis, 1985; Elsheikh, 1987).

The Four-Phase Method (FPM)

The four-phase method (FPM) modeling approach was first described by Odhabi & Paul in 1995. This "approach has many advantages in meeting the needs of the analysis and model

Figure 9. An illustration of the four-phase method (Odhabi et al., 1998)

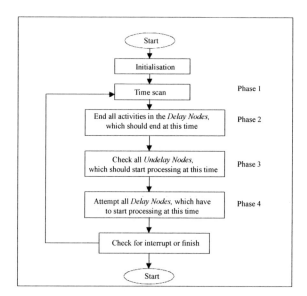

building phases when using an Object-Oriented approach" (Odhabi et al., 1997). FMP is as simple as *three-phase* yet has the power of the process interaction approach. As the following quote states:

The FPM has the simplicity of the Three-Phase method and the power of the process interaction approach in modeling complex system behavior. The FPM appears to be a simple, high-level system and enhances the conceptualization of simulation problems and the understanding of system behavior. At the same time, it reduces the time to produce working, valid simulation models by supporting the task of the modeler (Odhabi et al., 1997).

Before explaining the FPM, one must review hierarchical activity cycle diagrams (H-ACD) and extended activity cycle diagrams (X-ACD), both best described by Odhabi et al. (1997).

As such, an overview of FPM can be attained from the following quote, which best describes the approach; in addition, a diagram that gives an overview of the FPM is illustrated in :

Phase 1: Check the finish times of all the Delay Nodes currently in progress. Find the earliest of these, and advance the clock to this time (Odhabi et al., 1998).

On one note *time scan* phase or the A phase in the three-phase approach would be consequently brought to mind. On another note, the following quote by Odhabi et al. (1998) explains phases 2 and 3.

Phase 2: For the Delay Nodes, finish all the processing scheduled to be completed at this time, and move the relevant entities into the internal queue.

Phase 3: Check all of the UnDelay Nodes identifying all those which should start processing at this time. Perform the relevant processes (with duration time zero). Repeat the check until there are no UnDelay Nodes with processes to start at this time.

Note that Phases 2 & 3 are very much like the move between lists phase in the process-based approach. The final phase, 4, is explained by Odhabi et al. (1998) as follows:

Phase 4: Start the processing of any relevant Delay Nodes. Calculate when the Delay Node will finish its processing, and record this time. When all relevant Delay Nodes have been processed, check for an interrupt or any specified finishing conditions. If the model is not due to terminate, return to phase 1 and begin the execution cycle again.

The Petri Nets will be discussed next by interpreting the formal definition of Petri Nets, describing the classical Petri Nets and the different classifications of Petri Nets.

Petri Nets

This section gives an overview and formal definition of Petri Nets and the many classifications of Petri Nets. Petri Nets handles concurrent discrete events dynamic systems simulation, which is outside the scope of this chapter. Nevertheless, some simulation packages, namely *Optsim (Artifex)*, use Petri Nets. Therefore, it is imperative to discuss Petri Nets in this section, especially that the idea of Petri Nets was developed to answer the question of concurrency, which naturally arises constantly when discussing simulation. The Petri Nets will be discussed by interpreting the formal definition of Petri Nets, describing the classical Petri Nets and the different classifications of Petri Nets. However, the section does not include the parallel discrete event simulation languages (PDES), although some papers, such as Low et al. (1999) are available for concerned to read.

Petri Nets has been under development since the "beginning of the 1960's" (History, 2004), where Carl Adam Petri defined the language "in his Ph.D. thesis (Kommunikation mit Automaten)" (History, 2004). "It was the first time a general theory for discrete parallel systems was formulated" (History, 2004).

Definition of Petri Nets

In order to understand Petri Nets, a comprehensive definition must be initially realized. Following are two formal definitions of Petri Nets; one that calls it *technique,* while the other definition calls it *language.* The first definition is:

A formal, graphical, executable technique for the specification and analysis of concurrent, discrete-event dynamic systems; a technique undergoing standardization (PetriNets, 2004).

The second definition is: "Petri Nets is a formal and graphical appealing language, which is appropriate for modeling systems with concurrency"; while "[t]he language is a generalization of automata theory such that the concept of concurrently occurring events can be expressed" (History, 2004).

Back to the first quoted definition and to shed more light on the expressions used. First, the word *formal* means that:

The technique is mathematically defined. Many static and dynamic properties of a Petri Net (and hence a system specified using the technique) may be mathematically proven (PetriNets, 2004).

Using the word *graphical* means that:

The technique belongs to a branch of mathematics called graph theory. A Petri Net may be represented graphically as well as mathematically. The ability to visualize structure and behavior of a Petri Net promotes understanding of the modeled system (PetriNets, 2004).

Whereas the word *executable* means:

A Petri Net may be executed and the dynamic behavior observed graphically. Petri Net practitioners regard this as a key strength of the Petri Net technique, both as a rich feedback mechanism during model construction and as an aid in communicating the behavior of the model to other practitioners and lay-persons (PetriNets, 2004).

The word "*specification*: System requirements expressed and verified (by formal analysis) using the technique constitutes a formal system specification" (PetriNets, 2004).

When using the word *analysis* in the definition, it means:

Analysis: A specification in the form of a Petri net model may be formally analyzed, to verify that static and dynamic system requirements are met. Methods available are based on Occurrence graphs (state spaces), Invariants and Timed Petri Nets. The inclusion of timing enables performance analysis. Modeling is an iterative process. At each iteration, analysis may uncover errors in the model or shortcomings in the specification. In response, the Petri Net is modified and re-analyzed. Eventually a mathematically correct and consistent model and specification is achieved (PetriNets, 2004).

The word *concurrent*, when used in the context of Petri Nets, means:

The representation of multiple independent dynamic entities within a system is supported naturally by the technique, making it highly suitable for capturing systems which exhibit concurrency. (PetriNets, 2004).

Petri Nets is used in *discrete event dynamic system*, and in order to put a formal definition for such term, the following quote explains it best, since the technique as mentioned above is undergoing standardization: *"Discrete event dynamic system*: a system which may change state over time, based on current state and state-transition rules, and where each state is separated from its neighbor by a step rather than a continuum of intermediate infinitesimal states"(PetriNets, 2004).

Overview of Classical Petri Nets

"The Classical Petri Nets consist of four types of modeling elements, namely: (1) places, (2) transitions, (3) arcs, and (4) tokens" (Sawhney et al., 1999). Each of the modeling elements is symbolized as it is shown in *Figure 10*, taken from the same source.

Figure 10. Modeling elements of classical Petri Nets (Sawhney et al., 1999)

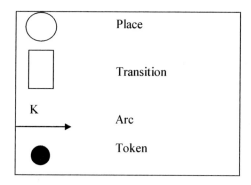

Petri Nets have two types of properties. The first type "is similar to the flowchart of a computer program and is called the static property" (Sawhney et al., 1999). The second type "resembles the execution of the computer program and is called the dynamic property (Shyam Kishore Bajpai, 1982; Sawhney et al., 1999). "Places, transitions, and arcs together are used to develop a static picture of a process while tokens provide the dynamic simulation capabilities to Petri Nets. They are initialized at a given place, which may contain zero or more tokens (Sawhney et al., 1998, 1999).

Some enhancements were added to the Classical Petri Nets so as to allow the CPN to model more complex systems as Sawhney et al. (1999) claims. *Figure 11* illustrates the enhanced Petri Nets modeling symbols.

Hierarchical transition "represents not a single task, but a group of recurrent work tasks at a lower level of the process" (Sawhney et al., 1999). The same source says, "A lower level sub-model, models these work tasks and constitute a module that can be repeatedly invoked by the higher level model using hierarchical transitions" (Sawhney et al., 1999).

When "resources are shared by activities at different levels of a construction process. The use of *Fusion Places* enhances the resource modeling features of Petri Nets and permits the modeling of such situations" (Sawhney et al., 1999).

As for the use of the token (Sawhney et al., 1999) best explain it in the following quote:

Tokens are normally used to model resources in a construction process. Resources with different attributes must thus be modeled by different tokens. In enhanced Petri Nets, the modeler has the possibility to define more than one type of token for a Petri Net. This is achieved by assigning a color or type to the token. These types of tokens are called colored or typed tokens, while the resulting Petri Nets are called Colored Petri Nets (Jensen, 1992; Sawhney et al., 1999).

Finally, the "*Probabilistic Arcs* provide a way to model a situation in which one activity/ transition is more likely to occur than another" (Sawhney et al., 1999).

Figure 11. Enhanced Petri Nets modeling symbols (Sawhney et al., 1999)

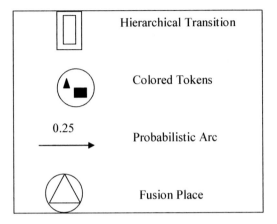

Classification of Petri Nets

Petri Nets were classified in different ways. The following classifications were recited in Class Web, among others:

1.   Petri Net systems of *level (1)* are characterized by "Boolean tokens," that is, places are marked by at most 1 unstructured token.

2.   Petri Net systems of *level (2)* are characterized by "integer tokens," that is, places are marked by several unstructured tokens - they represent counters.

3.   Petri Net systems of *level (3)* are characterized by high-level tokens, that is, places are marked by structured tokens where information is attached to them.

There are many classifications, sub-classifications, and extensions for Petri Net Systems that were not included in this research since they are outside the scope of this work. But since the question of concurrency always arises, when discussing simulation, then it was inevitable to give an overview about Petri Nets; particularly, that some packages like *Optsim (Artifex)* have used Petri Nets. As such, the section first discussed the origins of Petri Nets, providing explicate history, and the many classifications of the subject at hand.

Monte Carlo Simulation

Monte Carlo method is formally defined by the following quote as:

Numerical methods that are known as Monte Carlo methods can be loosely described as statistical simulation methods, where statistical simulation is defined in quite general terms to be any method that utilizes sequences of random numbers to perform the simulation (Introduction to Monte Carlo methods, 1995).

Noting that three words stand out in the definition: loosely, random, and statistical, where statistical simulation is defined as "method that utilizes sequences of random numbers to perform the simulation" (*Introduction to Monte Carlo methods*, 1995), therefore, the word random appears in the definition.

Monte Carlo' was coined by Metropolis (inspired by Ulam's interest in poker) during the Manhattan Project of World War II, because of the similarity of statistical simulation to games of chance (Introduction to Monte Carlo methods, 1995).

In 1987 Metropolis wrote: "Known to the old guards as statistical sampling: in it new, surroundings and owing to its nature there was no denying its new name of the Monte Carlo Method" (Metropolis, 1987).

Application

The applications of Monte Carlo are diverse, as such, following is a suggested list of those applications by *Introduction to Monte Carlo methods* (1995):

- Nuclear reactor simulation
- Quantum chromodynamics
- Radiation cancer therapy
- Traffic flow
- Stellar evolution
- Econometrics
- Dow Jones forecasting
- Oil well exploration
- VSLI design

Many simulation packages use the Monte Carlo method like *Gauss, Crystal Ball, vanguards, Decision Pro*, and *Decision Script*.

Finally, the last section of this chapter reviews Web-based simulation, along with all three different types of simulation and modeling. In addition to the different programming languages and environments, through which simulation can be done, using object oriented perspective.

Web-Based Simulation and Object-Oriented Simulation

The most common two ways to approach a model are from an algorithmic perspective and object-oriented perspective (Booch et al., 1999). The aim of the section is to give an overview of the object-oriented simulation perspective. Consequently, when discussing the object-oriented perspective, then the Web-based technology must be therefore discussed, particularly since the effect of object-oriented is visibly seen on the Web-based technology. In addition, this section introduces some environments and languages for Web-based simulation.

This section is laid out in three parts: first, Web-based simulation is discussed with all three different types of simulation and modeling. Second, the different programming languages and environments, through which simulation can be done using object-oriented perspective is discussed. Then, the idea of object-oriented simulation is discussed in comparison to algorithms perspective. In the third section the principles and aims of modeling from an object-oriented perspective are also introduced as a reminder to the reader.

Web-Based Simulation

There are certain types of applications that can utilize Internet computing. The most likely, as Kuljis and Paul (2000) stated:

- Applications that deal with huge quantities of data like, for example, meteorological modeling.
- Applications that enable users on multiple sites to collaborate in model design.
- Applications that require direct input from a customer like, for example, manufacturing that offers customized products.
- Applications in education and training, which increasingly have to cater for distance learning students.

Like most application types of software, computer simulation took advantage of the Internet technology.

This section describes the three known Web-based simulation approaches: remote simulation & animation, local simulation & animation, and remote simulation & local visualization. Notwithstanding the fact that Web-based simulation cannot be discussed in isolation of discussing the object-oriented approach and object-oriented programming languages for Web-based simulation.

Simulation and Animation Basic Approaches

There are three Web-based simulation and animation approaches suggested by Lorenz et al. (1997). In addition, Whitman et al. (1998) suggested three approaches, noting that both have the same idea. Following, an overview is given about each approach.

Remote Simulation and Animation Approach

The first approach suggested by Lorenz et al. (1997) is remote simulation & animation, but named as *server hosted simulation* by Whitman et al. (1998). In this approach, as can be seen in *Figure 12*:

The user specifies values of parameters for a simulation model in an HTML data entry form, submits the form to a server and starts the simulation by pushing the START-button on the form. A Common Gateway Interface (CGI) is used to transfer the data to the server. A CGI script starts the simulation after the data have been received. When the simulation has finished, the CGI-script prints the results (including the URL of files that have been created to show results) in a new HTML page and transmits it back to the client. This technique is called dynamic document generation (Lorenz et al., 1997).

Although Lorenz et al. (1997) state that this approach is suitable for the existing simulation & animation software, yet it has two disadvantages. The first disadvantage is the fact that this approach is not well qualified for "observing the dynamic process." The second

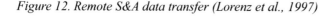

Figure 12. Remote S&A data transfer (Lorenz et al., 1997)

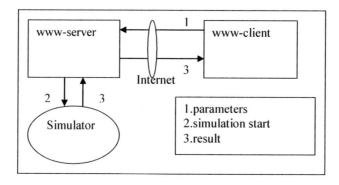

disadvantage is the fact that this approach does not allow the user to interrupt the simulation while running. However, Whitman et al. (1998) recapitulates that the advantage of this type is using a familiar tool and its syntax; the use of existing models, and the execution of the simulation engine on a more powerful machine than most client machines, thus enabling the reuse of existing models.

Local Simulation and Animation Approach

The second approach is also described by Lorenz et al. (1997), and is named by Whitman et al. (1998) as *client executed simulation*. In this approach the user loads the Java Applet on his/her computer and the work is done on the client's machine, as can be seen in *Figure 13*. The current simulation & animation software is not suited for such technology as Lorenz et al. (1997) explain. As such, Whitman et al. (1998) state that the disadvantage "is the lack of maturity of the language and few existing models;" however, it is worth mentioning that the two papers were written in 1997 and 1998 respectively, and some tools were developed since then.

Remote Simulation and Local Visualization Approach

The third approach is also described by Lorenz et al. (1997) and recited from another paper. The following quote describes how this approach works:

The simulation runs remotely on a simulation server. The results are transferred to the client and visualized locally. The user begins by loading some applets. After these applets have started, a connection to the Java server is built and simulation data are transmitted to the Web browser. The data can change continuously, delayed only by the executing simulation model and transmission time on the Internet. The user can interact with the model by us-

Figure 13. Client site simulation with loaded applets (Lorenz et al., 1997)

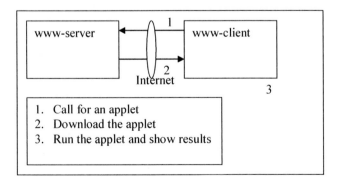

Figure 14. Remote simulation and local visualization (Lorenz et al., 1997)

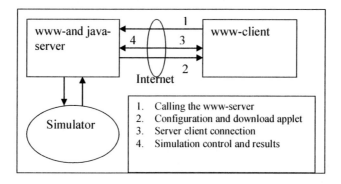

ing buttons on the HTML page or by clicking into a graphical representation of the model (Lorenz et al., 1997).

Moreover, *Figure 14* provides an elaboration on how the approach works. The same approach was named by Whitman et al. (1998) the hybrid client/server simulation. In addition, Whitman et al. (1998) assert that this approach "attempts to combine the advantages" of the previously mentioned two approaches.

Environment and Languages for Web-Based Simulation

There are several Java-based discrete simulation environments, asserts Kuljis & Paul (2000). In order to look at and define a few like simjava, DEVSJAVA, JSIM, JavaSim (J-Sim Web), JavaGPSS, Silk, WSE, and SRML, an overview is given here:

The first in this list of definitions is simjava. First the following quote by Kuljis and Paul (2000) is given:

Simjava is a process based discrete event simulation package for building working models of complex systems. It includes facilities for representing simulation objects as animated icons on screen.

It is worth stressing that *simjava*, based on this definition, is a *simulation package*. Hence, a more technical detailed definition by Page et al. (1997) is given next:

A simjava simulation is a collection of entities each running in its own thread. These entities are connected together by ports and can communicate with each other by sending and receiving event objects. A central system class controls all the threads, advances the simulation time, and delivers the events. The progress of the simulation is recorded through trace messages produced by the entities and saved in a file (Page et al., 1997; Howell & McNab, 2004).

The second on the list is DEVSJAVA environment, which was built using Java and is based on discrete event system specification (Kuljis & Paul, 2000). "A user of DEVSJAVA is able to experiment with any DEVS model from any machine at any time and to interactively and visually control simulation execution" (Kuljis & Paul, 2000).

The third on the list is JSIM, which is described by Kuljis and Paul (2000) as: "Java based simulation and animation environment supporting web based simulation as well component-based technology." The component-based technology that JSIM utilizes in this case is Java Beans as Kuljis and Paul (2000) claim in their paper. The idea is to build up the environment from reusable software components that "can be dynamically assembled using visual development tools "(Kuljis & Paul, 2000).

JavaSim (later the name was changed to J-Sim) is "a set of Java packages for building discrete event process-based simulation" (Kuljis & Paul, 2000). JavaSim was renamed because the word Java is a trademark owned by SUN Microsystems. J-Sim is an implementation of a simulation tool kit named C++SIM developed in the University of Newcastle as Kuljis & Paul (2000) explain in their paper. Back to the technical specification of J-Sim, the official Web site of J-Sim describes it as "component-based, compositional simulation environment" (www.j-sim.org). Yet, the Web site adds: "unlike the other component-based software packages/standards, components in J-Sim are autonomous" (www.j-sim.org).

The fifth on the list is JavaGPSS compiler. "The *JavaGPSS* compiler is a simulation tool which was designed for the Internet" (Kuljis & Paul, 2000). JavaGPSS was built so that GPSS can be run on the Internet, claim Kuljis & Paul in their paper.

Silk is a "general-purpose simulation language based around a <u>process-interaction</u> approach and implemented in Java" (Kuljis & Paul, 2000). The purpose of Silk is "to encourage better discrete-event simulation through better programming by better programmers" (Kilgore, 2003). "The *Silk* language is an opportunity to make simulation more accessible without sacrificing power and flexibility" (Healy & Kilgore, 1997). *Figure 15* shows graphical modeling using Silk JavaBeans.

The *WSE* (Web-enabled simulation environment) is the seventh item on the list. WSE "combines web technology with the use of Java and CORBA" (Kuljis & Paul, 2000). Also

Iazeolla and D.Ambrogio (1998) claim that the WSE environment provides transparent access to simulation models and tools (location transparency, distribution transparency, and platform independence); dynamic acquisition, instantiation and/or modification of simulation models; global availability (Internet based interaction); and plug-and-use architecture to easily embed simulation models and tools (Kuljis & Paul, 2000).

SRML (simulation reference markup language) and the simulation reference simulator (*SR* Simulator) are both developed by Boeing and used in many projects (Reichenthal, 2002). Like HTML, *SRML* represents simulation models and as a Web browser represents universal

Figure 15. Graphical modeling using Silk-based JavaBeans (Healy & Kilgore, 1997)

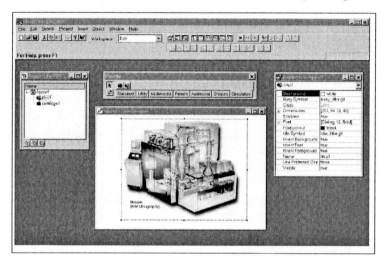

client application so does *SR*. SRML binds the declarative with procedures and like HTML contains both declarative and procedural definitions. "*SRML* is an XML-based language that provides generic simulation markup for adding behavior to arbitrary XML documents" (Reichenthal, 2002).

Why Java has Potential with Web-Based Simulation

Java has the following features listed by Ferscha and Richter (1997) and recited by Lou et al. (2000) that facilitate Web-based simulation:

- Transparency of network heterogeneity
- Transparency of platform and operating-system heterogeneity
- Transparency of user-interface heterogeneity

In addition, Lou et al. (2000) recited the following advantages of using Java for Web-based simulations

- Java has built-in support for producing sophisticated animations.
- Java has built-in threads making it easier to implement the process/resource interaction worldview.
- Models implemented as Java applets can be made widely accessible through Web browsers.

- Java's universal portability eliminates the need to port to a different platform, to re-compile or to relink.

Kuljis and Paul (2000) recited more reasons why java has the potential to dominate Web-based simulation based on Kilgore et al. (1998):

- Java will become a common foundation for all simulation tools because it is the only object-oriented programming environment that effectively supports standardized components.
- Java will foster execution speed breakthroughs through convenient and robust support for distributed processing of simulation experiments on multiple processors.
- Java will improve the quality of simulation models as the development of application-specific software components redirects the emphasis of simulation software firms toward modeling and away from modeling development environments.
- Java will expose the benefit of computer simulation to a larger audience of problem-solvers, decision-makers and trainers since models can be distributed and executed over the Internet using standard browser software on any operating system and hardware platform.
- Java will accelerate simulation education because students already familiar with object-oriented design, Java syntax and Java development environments will no longer require instruction in specific simulation tools.

In conclusion, *Table 1* (Kuljis & Paul, 2000) summarizes the difference between Web simulation and classical simulation. As can be seen in , the Web technology rids the simulation from two main hindering qualities: common standards and platform independence.

Table 1. Comparison between Web features and simulation (Kuljis & Paul, 2000)

Web Features	Web-Based Simulation	Classical Simulation
Common Standards	Yes	No
Platform independence	Yes	No
Interoperability	Generally not supported	Not supported
Ease of navigation	Varies	Varies
Ease of use	Difficult to use	Difficult to use
No specialist knowledge required	Specialist knowledge required	Specialist knowledge required
Not affected by change	Affected by change	Affected by change
Unstructured	Structured	Structured

Nevertheless, some problems, which hindered many software industries and were solved using Internet technology, are still lingering in the simulation arena: problems like need of specialist knowledge, interoperability, ease of navigation, effect of change and the inherited problem of structure.

Object-Oriented Simulation

"In software, there are several ways to approach a model. The two most common ways are from algorithm perspective and from an object-oriented perspective." (Booch et al., 1999). Using the algorithm approach, the "developer focuses on issues of control and the decomposition of longer algorithm into smaller ones" (Booch et al., 1999). Yet such an approach creates a weak system that is very hard to maintain (Booch et al., 1999). In addition:

Structured approaches were the often missing horizontal consistency between the data and behavior part within the overall system model, and the vertical mismatch of concepts between the real world domain and the model as well as between the model and the implementation (Engels & Groenewegen, 2000).

On the other hand, the object-oriented approach has proved to be mainstream because it can build all sorts of systems in different problem domains (Booch et al., 1999). Still, a number of problems stems from such approach, for example; "what's the structure of a good object-oriented architecture? What artifacts should the project create? Who should create them? How should they be measured?" (Booch et al., 1999). In fact, the purpose of the unified modeling language (UML) is "[v]isualizing, specifying, constructing, and documenting object-oriented systems" (Booch et al., 1999).

Object-oriented simulation was defined by (Joines & Roberts, 1998) as follows: "An object-oriented simulation (OOS) consists of a set of objects that interact with each other over time" (Joines & Roberts, 1998).

The success of object-oriented modeling approaches was hindered in the beginning of the 1990s, due to the fact that more than fifty object-oriented modeling approaches have claimed to be definitely the right one. This so-called *method war* came to an end by an industrial initiative, which pushed the development of the meanwhile standardized object-oriented modeling language UML (unified modeling language) (Engels & Groenewegen, 2000).

Through the eyes of UML there are four objectives to achieve and four principles for modeling while upholding the essence of object-oriented method: encapsulation, inheritance, and polymorphism.

Principles of Modeling

Although the principles of modeling stay almost the same whether using object-oriented simulation or any other way, yet the principles are better articulated by Booch et al. (1999, pp. 7-9) as follows:

1.   The choice of what models to create has a profound influence on how a problem is attacked and how a solution is shaped.
2.   Every model may be expressed at a different level of precision.
3.   The best models are connected to reality.
4.   No single model is sufficient. Very nontrivial system is best approached through a small set of nearly independent model.

The Aims of a Model

Although discussed previously, again, the aims of the model are better stated by Booch et al. (1999) as follows:

1.   Models help us to visualize a system as it is or as we want it to be.
2.   Models permit us to specify the structure or behavior of system.
3.   Models give us a template that guides us in constructing a system.
4.   Models document the decision we have made.

The aim of this section was to show that the advantages of object-oriented simulation, when compared to the algorithmic perspective of simulation, outweigh the disadvantages. Moreover, there are many object-oriented programming languages that have been developed especially for object-oriented simulation. In addition, there are Web-based simulation approaches that have already used the simulation object-oriented technology. In fact, as *Table 1* suggests, Web-based simulation encourages common standards and platform independence.

Summary

The goal of this chapter has been to give an overview of the simulation science, with emphasis on discrete rather than continuous event simulation. As such, first, the chapter discussed simulation definitions, approaches, modeling notations, and the need for simulation. Furthermore, the chapter has provided an overview of Petri Nets, particularly, since the question of concurrency always arises upon discussing simulation, although there are many classifications, sub-classifications, and extensions for Petri Net Systems that were not included in this chapter since they are outside the scope of this work. Nevertheless, some packages like *Optsim (Artifex)* have used Petri Nets, as such; the section has discussed the origins of Petri Nets, providing explicit history, and the many classifications of the subject at hand.

Furthermore, the chapter has discussed the Monte Carlo simulation. Monte Carlo simulation was discussed in this chapter due to its importance and uses. Then the chapter has showed that the advantages of object-oriented simulation outweigh the disadvantages, in the case of being compared to the algorithmic perspective of simulation. Likewise, the chapter elaborates on the object-oriented programming languages that have been developed especially for object-oriented simulation. In addition, there are Web-based simulation approaches that already used the simulation object-oriented technology.

References

Abu-Taieh, E. (2005). *Computer simulation using Excel without programming.* Unpublished PhD Thesis.

Banks J. (1999, Dec 5-8). Introduction to simulation. In P.A. Farrington, H.B. Nembhard, D.T. Sturrock, & G.W. Evans (Eds.), *Proceedings of the 1999 Winter Simulation Conference,* Phoenix, Arizona, U.S. (pp. 7-13). New York: ACM Press.

Banks J. (2000, Dec 10-13). Introduction to simulation. In J.A. Joines, R.R. Barton, K. Kang, & P.A. Fishwick (Eds.), *Proceedings of the 2000 Winter Simulation Conference,* Orlando, Florida (pp. 510-517). San Diego, CA: Society for Computer Simulation International.

Booch, G., Rumbaugh, J., & Jacobson, I. (1999). *The unified modeling language user guide.* Reading, MA: Addison-Wesley.

Brooks Software. (n.d.). *Simulation and modeling software.* Retrieved September 1, 2003, from www.automod.com

Drappa, A., & Ludewig, J. (2000, Dec 10-13). Simulation in software engineering training. In J.A. Joines, R.R. Barton, K. Kang, & P.A. Fishwick (Eds.), *Proceedings of the 2000 Winter Simulation Conference,* Orlando, Florida (pp. 199-208). New York: ACM Press.

El Sheikh, A. (1987). *Simulation modeling using a relational database package.* Unpublished PhD Thesis. The London School of Economics.

Engels, G., & Groenewegen, L. (2000). Object-oriented modeling: A roadmap. In *Proceedings of the Conference on The Future of Software Engineering* (pp. 103-116). Special Volume published in conjunction with ICSE 2000. New York: ACM Press.

Ferscha, A., & Richter, M. (1997, Dec 7-10). Java based conservative distributed simulation. In S. Andradóttir, K.J. Healy, D.H. Withers, & B.L. Nelson (Eds.), *Proceedings of the 1997 Winter Simulation Conference,* Atlanta, Georgia (pp. 381-388). New York: ACM Press.

Goldsim. (n.d.). *Monte Carlo simulation software for decision and risk analysis.* Retrieved September 1, 2003, from www.goldsim.com

Healy, K.J., & Kilgore, R.A. (1997, Dec 7-10). Silk: A Java-based process simulation language. In S. Andradottir, K.J. Healy, D. Withers, & B. Nelson (Eds.), *Proceedings of the 1997 Winter Simulation Conference,* Atlanta, Georgia (pp. 475-482). New York: ACM Press.

Henriksen, J. (1997, Dec 7-10). An introduction to SLX™. In S. Andradóttir, K.J. Healy, D.H. Withers, & B.L. Nelson (Eds.), *Proceedings of the 1997 Winter Simulation Conference,* Atlanta, Georgia (pp. 559-566). New York: ACM Press.

Introduction to Monte Carlo methods. (1995). Retrieved April 1, 2004, from www.phy.ornl. gov/csep/CSEP/BMAP.html

Joines, J., & Roberts, S. (1998, Dec 13-16). Fundamentals of object-oriented simulation. In D.J. Medeiros, E.F. Watson, J.S. Carson, & M.S. Manivannan (Eds.), *Proceedings of the 1998 Winter Simulation Conference,* Washington, DC (pp. 141-149). Los Alamitos, CA: IEEE Computer Society Press.

J-Sim. (n.d.). Retrieved April 10, 2004, from www.j-sim.org

Kilgore, R., Healy, K., & Kleindorfer, G. (1998, Dec 13-16). The future of Java-based simulation. In D.J. Medeiros, E.F. Watson, J.S. Carson, & M.S. Manivannan (Eds.), *Proceedings of the 1998 Winter Simulation Conference,* Washington, DC (pp. 1707-1712). Los Alamitos, CA: IEEE Computer Society Press.

Kilgore, R. (2003, Dec 7-10). Object-oriented simulation with Sml and Silk in .Net and Java. In S. Chick, P.J. Sánchez, D. Ferrin, & D.J. Morrice (Eds.), *Proceedings of the 2003 Winter Simulation Conference,* New Orleans, LA (pp. 218-224). Winter Simulation Conference

Kuljis, J., & Paul, R. (2000, Dec 10-13). A review of Web based simulation: Whither we wander? In J.A. Joines, R.R. Barton, K. Kang, & P.A. Fishwick (Eds.), *Proceedings of the 2000 Winter Simulation Conference,* Orlando, FL (pp. 1872-1881). San Diego, CA: Society for Computer Simulation International.

Lorenz, P., Dorwarth, H., Ritter, K., & Schriber T. (1997, Dec 7-10). Towards a Web based simulation environment. In S. Andradóttir, K.J. Healy, D.H. Withers, & B.L. Nelson (Eds.), *Proceedings of the 1997 Winter Simulation Conference,* Atlanta, GA (pp. 1338-1344). New York: ACM Press.

Luo, Y., Chen, C., Yücesan, E., & Lee, I. (2000, Dec 10-13). Distributed Web-based simulation optimization. In J.A. Joines, R.R. Barton, K. Kang, & P.A. Fishwick (Eds.), *Proceedings of the 2000 Winter Simulation Conference,* Orlando, FL (pp. 1785-1793). San Diego, CA: Society for Computer Simulation International.

Low, Y., Lim, C., Cai, W., Huang, S., Hsu, W., Jain, S., & Turner, S. (1999). Survey of languages and runtime libraries for parallel discrete-event simulation. *Simulation, 72*(3), 170-186.

Metropolis, N. (1987). The beginning of Monte Carlo method. *Los Alamos Science* (Special Issue, Stanislaw Ulam 1909-1984), (15), 125-130

Metropolis, N. (1987). *The beginning of Monte Carlo method.* Retrieved April 18, 2004, from http://jackman.stanford.edu/mcmc/metropolis1.pdf

Odhabi, H., Paul, R., & Macredie R. (1997, Dec 7-10). The four phase method for modelling complex systems. In S. Andradóttir, K.J. Healy, D.H. Withers, & B.L. Nelson (Eds.), *Proceedings of the 1997 Winter Simulation Conference,* Atlanta, GA (pp. 510-517). New York: ACM Press.

Odhabi, H., Paul, R., & Macredie R. (1998, Dec 13-16). Making simulation more accessible in manufacturing systems through a 'four phase' approach. In D.J. Medeiros, E.F. Watson, J.S. Carson, & M.S. Manivannan (Eds.), *Proceedings of the 1998 Winter Simulation Conference,* Washington, DC (pp. 1069-1075). Los Alamitos, CA: IEEE Computer Society Press.

Page, E., Moose, R., & Gri, S. (1997, Dec 7-10). Web-based simulation in Simjava using remote method invocation. In S. Andradóttir, K.J. Healy, D.H. Withers, & B.L. Nelson (Eds.), *Proceedings of the 1997 Winter Simulation Conference,* Atlanta, GA (pp. 468- 474). New York: ACM Press.

Paul, R.J., & Balmer, D.W. (1998). *Simulation modelling.* Lund, Sweden: Chartwell-Bratt Student Text Series.

Paul, J.R. (1993, Dec 12-15). Activity cycle diagrams and the three-phase method. In *Proceedings of the 25th Conference on Winter Simulation,* Los Angeles, CA (pp. 123-131). New York: ACM Press.

Petri Nets. (n.d.). Retrieved April 18, 2004, from http://www.petrinets.info/graphical.php

Petri Nets World. (n.d.). Retrieved April 18, 2004, from http://www.daimi.au.dk/PetriNets/faq/

Pidd, M., & Cassel, R. (1998, Dec 13-16). Three phase simulation in Java. In D.J. Medeiros, E.F. Watson, J.S. Carson, & M.S. Manivannan (Eds.), *Proceedings of the 1998 Winter Simulation Conference,* Washington, DC (pp. 267-371). Los Alamitos, CA: IEEE Computer Society Press.

Pidd, M. (1998). *Computer simulation in management science* (4th ed.). Chicester, UK: John Wiley & Sons.

Reichenthal, S. (2002, Dec 8-11). Re-introducing Web-based simulation. In E. Yücesan, C.-H. Chen, J.L. Snowdon, & J.M. Charnes (Eds.), *Proceedings of the 2002 Winter Simulation Conference,* San Diego, CA (pp. 847-852). Winter Simulation Conference.

Sawhney , A., Mund, A., & Marble, J. (1999, Dec 5-8). Simulation of the structural steel erection process. In P.A. Farrington, H.B. Nembhard, D.T. Sturrock, & G.W. Evans (Eds.), *Proceedings of the 1999 Winter Simulation Conference,* Phoenix, AZ (pp. 942-947). New York: ACM Press.

Schriber, T., & Brunner, D. (1997, Dec 7-10). Inside discrete-event simulation software: How it works and why it matters. In S. Andradóttir, K.J. Healy, D.H. Withers, & B.L. Nelson (Eds.), *Proceedings of the 1997 Winter Simulation Conference,* Atlanta, GA (pp. 14-22). Winter Simulation Conference.

Schriber, T., & Brunner, D. (2002, Dec 8-11). Inside discrete-event simulation software: How it works and why it matters. In E. Yücesan, C.-H. Chen, J.L. Snowdon, & J.M. Charnes (Eds.), *Proceedings of the 2002 Winter Simulation Conference,* San Diego, CA (pp. 97-107). Winter Simulation Conference.

Schriber, T.J., Ståhl, I., Banks, J., Law, A.M., Seila, A.F., & Born, R.G. (2003, Dec 7-10). Simulation Text-Books, old and new (Panel). In S. Chick, P. J. Sánchez, D. Ferrin, & D.J. Morrice (Eds.), *Proceedings of the 2003 Winter Simulation Conference,* New Orleans, LA (pp. 238-245). Winter Simulation Conference.

Tanenbaum, A.S., & Steen, M.V. (2002). *Distributed systems principles and paradigms.* Upper Saddler River, NJ: Prentice Hall.

Whitman, L., Huff, B., & Palaniswamy, S. (1998, Dec 13-16). Commercial simulation over the Web. In D.J. Medeiros, E.F. Watson, J.S. Carson, & M.S. Manivannan (Eds.), *Proceedings of the 1998 Winter Simulation Conference,* Washington, DC (pp. 355-339). Los Alamitos, CA: IEEE Computer Society Press.

Chapter II

Distributed Simulation in Industry

Roberto Revetria, Università degli Studi di Genova, Italy

Roberto Mosca, Università degli Studi di Genova, Italy

Abstract

This chapter introduces the basic concepts of distributed simulation applied to real life industrial cases with particular reference to IEEE 1516 High Level Architecture(HLA): one of de facto standards for distributed simulation. Starting from a concise introduction HLA, the chapter proposes a simple hands-on with a complete implementation example of HLA. Successfully achieved the ability to create small federations, the reader is guided by two real life application of distributed simulation: the first is related to a supply chain modeling for the aerospace industry while the second one is focused on logistic platforms modeling. The authors have edited this chapter keeping in mind usual difficulties that can be encountered in real life projects: for such purpose a reference implementation and full code examples are provided in order to ensure a smooth but effective learning curve. The reader will also find suggestions for proper management of HLA-based simulation projects.

Introduction

Originally, modeling was divided into discrete simulations and continuous simulations. Then, some other simulation models appeared, such as Monte Carlo simulations (time-independent) and continuous/discrete mixed simulations. Also the architecture of the simulations has been developing during these years, especially parallel and distributed simulation, in order to speed up simulation time (the first) and to improve the interoperability among different systems (the latter). One of the latest steps was to introduce the concept of human interactivity in simulation ("human-in-the-loop"), from which sprung real-time and scaled real-time simulations. Finally, with the development of HLA, dissimilar simulation components could interoperate in the same framework.

HLA: Introduction

In 2000, HLA became an IEEE standard for distributed simulation. It consists of several federates (members of the simulation), that make up a federation (distributed simulation), work together and use a common runtime infrastructure (RTI).

The RTI interface specification, together with the HLA object model template (OMT) and the HLA rules, are the key defining elements of the whole architecture.

Figure 1. Continuous and discrete simulations

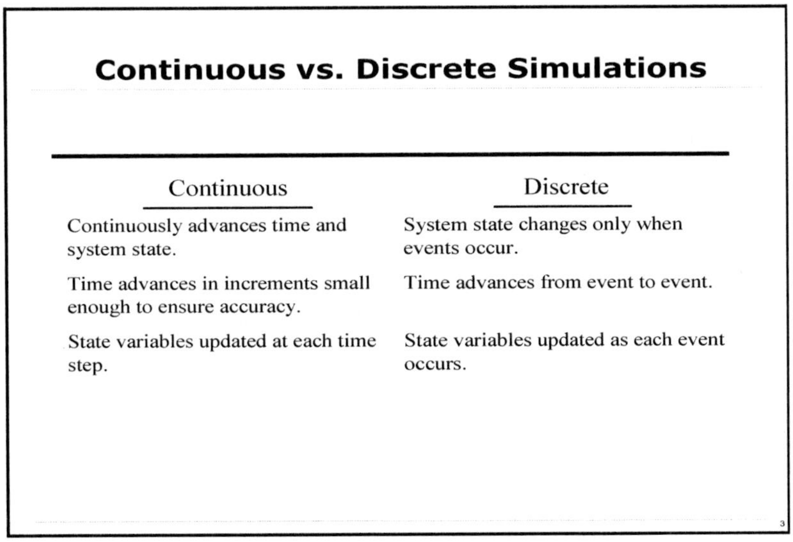

Continuous vs. Discrete Simulations

Continuous	Discrete
Continuously advances time and system state.	System state changes only when events occur.
Time advances in increments small enough to ensure accuracy.	Time advances from event to event.
State variables updated at each time step.	State variables updated as each event occurs.

HLA Object Models

Whereas the interface specification is the core of the transmission system that connects different software systems, regardless of platform and language, the object model template is the language spoken over that line.

HLA has an object-oriented world view, that is not to be confused with OOP (object-oriented programming) because it doesn't specify the methods of objects, since in the common case this is not an info to be transferred between federates. This view does only define how a federate must communicate with other federates, while it doesn't consider the internal representation of each federate. So, a simulation object model (SOM) is built, which defines what kind of data federates have to exchange with each other. Furthermore, a meta-object model, the federation object model (FOM), collects all the classes defined by each participant to the federation in order to give a description of all shared information.

The object models, then, describe the objects chosen to represent real world, their attributes and interactions and their level of detail.

Both the SOM and the FOM are based on the OMT, that is, on a series of tables that describe every aspect of each object. The OMT consists of the following fourteen components:

- Object model identification table
- Object class structure table

Figure 2. A distributed simulation under HLA

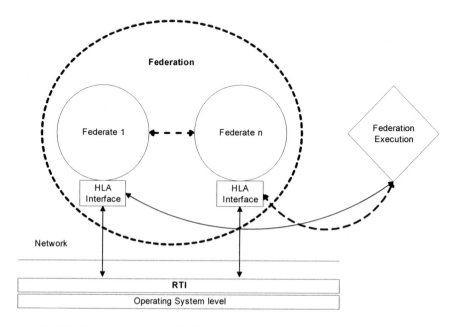

- Interaction class structure table
- Attribute table
- Parameter table
- Dimension table
- Time representation table
- User-supplied tag table
- Synchronization table
- Transportation type table
- Switches table
- Data type tables
- Notes table
- FOM/SOM lexicon

HLA does not mandate the use of any particular FOM, however, several "reference FOMs" have been developed to promote a-priori interoperability. That is, in order to communicate, a set of federates must agree on a common FOM (among other things), and reference FOMs provide ready-made FOMs that are supported by a wide variety of tools and federates. Reference FOMs can be used as-is, or can be extended to add new simulation concepts that are specific to a particular federation or simulation domain.

The RPR FOM (real-time platform-level reference FOM) is a reference FOM that defines HLA classes, attributes and parameters that are appropriate for real-time, platform-level simulations.

HLA Interface Specification

The interface specification document defines how HLA compliant simulators interact with the runtime infrastructure (RTI). The RTI provides a programming library and an application programming interface (API) compliant to the interface specification.

A federate communicates with the RTI using its RTI ambassador and vice versa. The interface specification has six categorizes of services, of which time and data distribution management are the most significant:

- Time management permits to effectively synchronize the simulation clocks of a wide range of types of simulations.
- Data distribution management allows an efficient transmission of data among federates and results in a considerable reduction of the amount of data transferred.

Federation Management

These kind of services are used to start, join, resign and close a federation execution.

Declaration Management

With these services, federates specify which data types they would like to send or receive. Federate declares data it will send, or publish, to the federation:

- Data sent at the end of each time-step (*Publish Object Class Attributes*).
- Data sent at arbitrary times (events) (*Publish Interaction Class*).

Federate declares data that it is interested in receiving, or subscribing to, from other federates:

- Data received regularly at the end of other federates' time-step (*Subscribe Object Class Attributes*).
- Data received at arbitrary times (events) (*Subscribe Interaction Class*).

Figure 3. Federation management life cycle

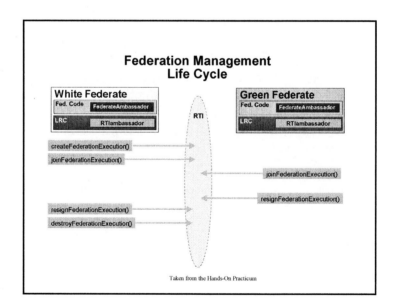

Figure 4. Declaration management services

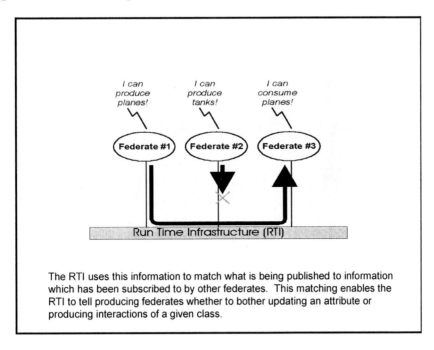

The RTI uses this information to match what is being published to information which has been subscribed to by other federates. This matching enables the RTI to tell producing federates whether to bother updating an attribute or producing interactions of a given class.

If no other federate is interested in the data from objects of a particular class, for example, simulated by federate "A," then federate "A" could save simulation overhead if it stopped sending the data at all. There is a mechanism in the HLA for handling this situation. A federate may express an interest to the RTI in controlling the sending of unnecessary messages by using the RTI service "enable attribute relevance advisory switch" for updates, or the RTI service "enable interaction relevance advisory switch" for interactions. When any other federate subscribes to the data published by a federate (and who has issued the above enable commands to the RTI), the RTI will call the federate's "turn updates on for object instance" or "turn interactions on call-back routines."

Also, there is no need to register an object of a certain class if no federate exists that has subscribed to it. The RTI will call the federate routines "start/stop registration for object class" to control the object registration calls if the federate has previously called the RTI service "enable object class relevance advisory switch." Calling the RTI service may stop this advisory: "disable object class relevance advisory switch."

Object Management

RTIambassador service and *FederateAmbassador* callback methods support object management. Object management includes:

- instance *registration* and instance *updates* on the object production side and

- instance *discovery* and *reflection* on the object consumer side.

- methods associated with sending and receiving interactions, controlling instance updates based on consumer demand, and other miscellaneous support functions.

Registering/Discovering/Deleting Object Instances

To create an object, the federate must first have published that object class using the declaration management services of the RTI. The publishing federate then registers the object using the declaration management services. This results in a *discover* callback by the RTI to any subscribing federate.

Figure 3 illustrates the interactions required to register and to discover object instances. The *RTI_RTIambassador()'s* method *registerObjectInstance()* informs the *Local RTI Component* that a new object instance has come into existence. The method requires the object class of the new object instance and an optional object name. The method returns an *RTI_RTIObjectInstanceHandle* that the LRC uses to identify the particular object instance.

The *RTI_RTIambassador's* method *deleteObjectInstance()* is called to remove a registered object. The *FederateAmbassador's* callback *removeObjectInstance()* informs federates that a previously discovered object no longer exists. The *RTI_RTIambassador's* method *localDeleteObjectInstance()* effectively "undiscovers" an object instance. This method

Figure 5. Object management methodology

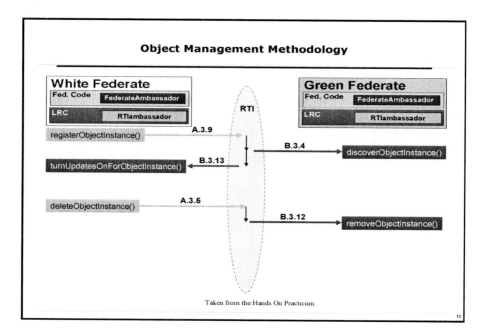

does not ensure the object will be permanently undiscovered. This service is intended to be used when a federate discovers an object as an instance of an object class but would like to subscribe to object classes that extend the discovered class and then rediscover the instance based on the new subscriptions. The object instance will be rediscovered upon the next *updateAttributeValues()* invocation that meets the receiving federate's subscriptions.

Updating and Reflecting Object Attributes

To update one or more attributes associated with a registered object instance, a federate must prepare an *RTIAttributeHandleValueMap*. An *Attribute_ HandleValueMap* (AHVM) identifies a set of attributes and their values.

Attribute updates are provided for an object instance via the *RTI_RTIambassador's* method *updateAttributeValues()*. The method requires an *RTI_ ObjectInstanceHandle,* which the LRC uses to identify an object instance, an AHVM, and a descriptive character string (RTI_User Supplied Tag).

Reflection is the counterpart to attribute updates. The *FederateAmbassador's* callback method *discoverObjectInstance()* informs the federate that a new object instance has come into existence. The method provides an object handle that will be used to identify the object for subsequent updates, etc. The method also identifies the object class of the new object instance. It is extremely important to note that the *RTI_ObjectInstanceHandle* is a local

Figure 6. Object management updates

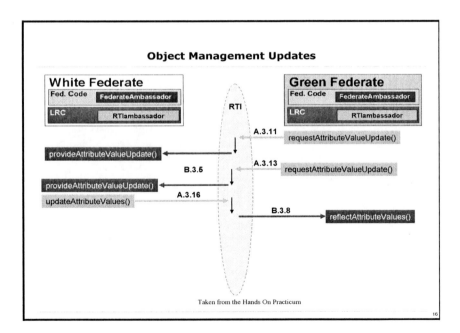

(numeric) representation maintained by the LRC. The same object instance is typically known by different handle values in each federate.

Sending/Receiving Interactions

Interactions are constructed in a similar fashion to the way attribute updates are constructed. Recall that objects persist, interactions do not. Each interaction is constructed, sent, and forgotten. Interaction recipients receive, decode, and apply the interaction. Recall that, as for objects, to send an interaction, the federate must have published that interaction class (per declaration management) and to receive an interaction, a federate must have subscribed to that interaction class (per declaration management).

Change Transport and Ordering Mechanisms

Object attribute updates and interactions are conveyed between federates using one of two data transportation schemes—"reliable" and "best effort." For objects, the transportation scheme is specified at the level of individual attributes. For interactions, the transportation scheme is specified at the interaction level (i.e., not the parameter level). By default, the transportation scheme is specified per object/attribute name and per interaction name in the *FOM document data* (FDD) file

As regards ordering, messages containing events or updates may be delivered in time stamp order (TSO) or in receive order (RO).

In the first case, they have an associated "time-stamp" and must be delivered to the receiving federate in the correct order with respect to this time-stamp.

In the second case, other types of messages, such as informational messages, may not have an associated time stamp, and are delivered upon arrival, without regard to the time when the message was sent.

Ownership Management

The attribute ownership services of the RTI are intended to facilitate the coordination of distributed, object state computation. They do this by providing the simulations a means to negotiate the transfer of attribute ownership. When a simulation owns an object attribute, it has the responsibly to model that attribute, provide new values for that attribute to the RTI when they change in a significant way, and respond to requests from the RTI for values of that attribute. Within the RTI, ownership is represented with tokens. An object gets a token for each of its attributes when that object is created. Conceptually, the attribute ownership services deal with the manipulation of these tokens. When a simulation successfully takes ownership of an attribute it is said to hold the ownership token for that attribute. When the RTI needs the current value of an attribute, it makes a request to the simulation that holds the token for that attribute. How these tokens are managed is the essence of the ownership management design. The ownership management services are partitioned into two symmetric groups based when how the ownership transfer is initiated. A simulation can request:

- ownership acquisition (pull) or
- ownership divestiture (push).

If ownership acquisition is desired, the RTI locates the specified ownership token and invokes a service in the holder of the token requesting its release. If the simulation is willing to release the token, the RTI informs the original requester that it now owns the attribute and the token is moved to the acquiring simulation. If a simulation wishes to divest ownership, the RTI invokes a service in all simulations that could possibly model that attribute. Zero or more simulations will respond with a desire to take ownership of the specified attribute. The RTI will arbitrarily select a recipient, notify the divesting simulation that it need no longer model the attribute, and manipulate the ownership token accordingly.

Time Management

Types of Simulation Implementations

While the observance of a distinction between continuous and discrete simulations has long been a practice in the simulation community at large, HLA literature often uses a different terminology and refers to *time-stepped* and *event-driven* simulation.

Time-stepped means that time advances in equal increments. This would apply to the simulation of an electronic digital circuit, for example, in which the outputs of its electronic components change only synchronously with the pulses of a system clock. It also applies to many implementations of continuous models in which the step size of the numerical integration algorithm is constant (fixed-step integration), that is, the increase of a capital investment due to the annual interest rate. Some integration algorithms use a variable step-size (to control errors) but still communicate with other simulation modules at fixed communication intervals comprising several steps. These would still qualify as time-stepped.

Time-stepped federates will usually use the time advance request (TAR) to advance their logical time to the next time step.

If, however, a variable step-size simulation communicates with other modules after each time step, then the end of each step could be regarded as an event, in which case the process would be classified as event-driven.

Event-driven federates will frequently use the next event request (NER) to advance to the time stamp of their next local event.

The Role of Time

There are three different types of time in distributed simulations:

- Physical time is the time of the real system that has been modeled, for example, an airplane taking off at 7 a.m. and landing at 1 p.m. on April 5, 2004.

- Simulation (or logical) time is the time of the simulation. One unit of logical time may represent one second or one hour of physical time.

- Wall clock time is the time of the execution of the simulation. Referring to the airplane example, it could be from 2 p.m. to 3 p.m. on May 12, 2004, supposing the physical system over the period of 6 hours takes one hour to execute.

Real-Time vs. Non Real-Time Simulations

A simulation can also be characterized as either a real-time or a non real-time simulation. Many simulations execute as rapidly as possible, as determined by the time taken to complete all the necessary calculations. In some cases, the simulated behavior might be generated in a period of time that is much greater than the duration of the actual behavior. Simulations of elementary particle motions or high-speed electrical phenomena might fall in this category. Other simulations might execute much faster than real-time, such as in the simulation of an astronomical system, or in human population studies. Most simulations like these would be referred to as non real-time. There are, however, a number of situations that might be qualified as real-time.

One type of real-time simulation occurs when human players provide inputs to an executing simulation and so influence the subsequent course of the simulation. Such simulations do not necessarily guarantee that the timing of the results of the simulation will accord precisely with the timing of the behavior of the system being simulated. Many simulated training exercises, computer war games, and management games fall into this category.

There is a special situation in which the time scale of the simulation bears a fixed relationship to the time scale of the real system (say 100 times faster or 1,000 times slower). This can be called scaled real-time, but the most restrictive definition of real-time simulation requires that the outputs of the simulation occur with exactly the same timing as the corresponding outputs of the real system.

Each of these approaches to simulation requires special handling of the advancement of time and, in order to be able to provide for them, we need to understand how the time management features of the HLA would work in each case.

Schemes for Time Management in HLA

First of all, it is important to recognize that time management in the HLA is not about constraining the federates in a federation to execute in real-time. HLA does not support the concept of wall clock time for a federation as a whole. Each federate has a logical time and the purpose of time management is to coordinate the advance of the logical times of all federates.

The HLA allows flexibility in the way a particular federate handles time. It supports a number of time-management schemes, including the following:

1.   *No time management*, in which each federate advances time at its own pace.
2.   *Conservative synchronization,* in which federates advance time only when it can be guaranteed they will receive no past events.
3.   *Optimistic synchronization,* in which a federate is free to advance its logical time, but has to be prepared to roll-back its logical time if an event is received in its past.
4.   *Activity scan,* in which federates proceed through periods during which they exchange messages at the same time until they agree to advance their logical times together.

In all cases the logical time belonging to a federate conforms to the following:

1.   It has an initial value
2.   Its value is not tied to any system of units. Unit logical time could represent a microsecond, a second, an hour, a decade or whatever the federation convention requires.
3.   It is well-ordered. The RTI can determine which of two times is the greater.
4.   It is always greater than the initial time.
5.   It is effectively discrete. It has an epsilon value representing the smallest possible difference between two logical times.
6.   It can assume a value of positive infinity, which is greater than any other value.

Each federate determines its own degree of involvement in the time management process. It can choose not to participate in time management (it is said to be neither time-regulating nor time-constrained). This is the default state of a federate when it joins a federation. It can choose to be time-regulating, in which case it is capable of generating time-stamp ordered (TSO) events. It can choose to be time-constrained, in which case it is capable of receiving TSO events. It can also choose to be both time-regulating and time-constrained, in which case it can both generate and receive TSO events. The important point about TSO events (introduced in the object management paragraph) is that it is possible to determine the order in which they were sent, and they are delivered to the receiving federates in this order. Receive ordered (RO) events, on the other hand, are delivered in the order in which they are received and, lacking time stamps, it is impossible to determine the order in which they were sent. Since a federate that is not time-constrained cannot receive TSO events, an event that is sent with a time stamp (sent as a TSO event) will be received as a RO event, without the time stamp, if the receiving federate is not time-constrained. RO events bear no relationship to logical time.

To summarize, for a time-constrained federate operating with conservative synchronization:

1.   TSO events are delivered to the federate in time-stamped order, irrespective of the order in which the originating events are sent.
2.   No event will be delivered to the federate with a time stamp less than the current logical time of the federate.

To guarantee the above conditions without the possibility of a deadlock in which no federate is able to advance its clock, it is necessary for each time-regulating federate to declare a "lookahead," which is a time period beyond its current logical time for which it is forbidden to send events. In other words, it is a notice period after which events can be sent and it gives the receiving, time-constrained federates the leeway to advance their own logical time free of concern that in doing so they will be open to the danger of receiving events with a time stamp less than their current logical time, that is, in their past.

Looking at this from the perspective of a time-constrained federate, capable of receiving events from a number of other federates, there is a time, beyond its current logical time, before which it knows that no federate will send it an event. It is able, therefore, to advance its time to this new time without fear of subsequently receiving an event in its past. This time is known as the *greatest available logical time (GALT)* of the federate.

Note that a federate is allowed to change its lookahead during the federation execution. It must request such a change and if it is reducing its lookahead by an amount *delta*, then it must wait until it has advanced its own logical time by at least *delta* to avoid violating the time-management guarantees.

Real-Time Execution

Many HLA federations execute at a rate that is approximately equal to real-time. In other words, events appear to occur at a normal rate, but there is no attempt to ensure precise agreement in time with real events. In the HLA literature, this is what "real-time simulation" means. That is, intervals between events are perceived as normal by an observer. Federations do not naturally execute in this way. Unless the real-time property is designed into a federation, it will simply execute as fast as the computers are able. To achieve the appearance of real-time, it may be necessary to make a federate go into a waiting state so as to slow the simulation down.

Imagine a federation that models traffic flowing through an intersection. Vehicles, traffic signals, and pedestrians might all be represented by different federates. Vehicles move through the intersection when the light is green and their intended path (straight on, left or right turn) is not blocked by other vehicles or pedestrians. Similarly pedestrians cross the road when the signal says WALK and the path is clear. Traffic signals change state, either according to a strict time schedule, or controlled by the vehicles passing a sensor in a particular location. If all of these operations were programmed without regard to time-management, it is likely that the whole simulation would run very much faster than real-time. Constraints must be introduced which make the traffic lights wait an appropriate time before changing from red to green, or which ensure that vehicles and pedestrians don't appear to move at supersonic speed through the intersection.

Consider a vehicle that is stopped at the traffic signal intending to move straight ahead when the signal changes. Assume there are no obstacles to forward progress when the light changes. The logical time of the vehicle federate, T, coincides with the time at which the light change occurs. In this particular run, this occurs at a wall clock time of Tc. The vehicle federate does a calculation that determines it will take, say, 5 seconds to clear the intersection, but the calculation is completed in 0.1 second at a wall clock time of $Tc+0.1$

sec. The federate advances its logical time to $T+5$ seconds and must wait for another 4.9 seconds for wall clock time to catch up. After the waiting period is complete, the federate issues its updated position. In other words, the federate waits until wall clock time catches up before it issues (and receives) updates and then it immediately proceeds to a new set of calculations of the next set of updates.

The appearance of running in real-time can be maintained as long as the computers can process the necessary information faster than real-time and then wait for real-time to catch up.

Time management is effected by the sending and receiving of events (an event is characterized by its ability to have an associated time stamp). A federate sends an event when it calls one of the following:

1.   *update attribute values*
2.   *send interaction*
3.   *delete object instance*

A federate receives an event when the RTI calls one of the following on it:

1.   *reflect attribute values*
2.   *reflect interaction*
3.   *remove object instance*

Figure 7. Synchronization of logical time to real-time

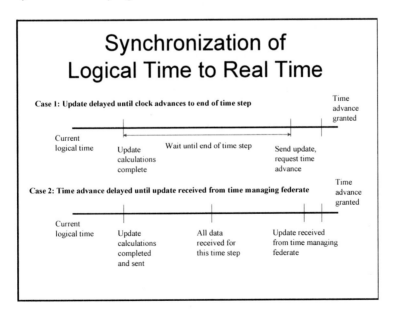

A federate that is time-regulating should make regular requests to the RTI to allow it to advance its clock because this affects the rest of the federation. It is these requests that co-ordinate the advance of time in the federation, not the sending of time-stamped events. The reception of events is controlled by the advance of time, but events do not advance time.

Conservatively synchronized federates, such as we are discussing here, will call one of two RTI services to advance time, either *time advance request* or *next message request()*. The corresponding callback service in either case is *time advance grant*. Conservative federates are constrained to not advance their local time until permitted by the *time advance grant*. This advance in time is limited by the GALT. Optimistic federates are not constrained by the GALT and can advance local time into the future, beyond the GALT. Therefore, these federates require additional time-management services.

Synchronization of Federate Timing

One of the features that makes the HLA so flexible is that, as mentioned earlier, it does not dictate a timing regime to any federate. Each federate is autonomous with respect to its determination of a timing regime. This has some consequences that, for those who may be familiar with different simulation environments, may be unexpected. As an example, let us examine a real case regarding a simulation dealing with population growth. The simulation was realized by California State University at Chico.

In accordance with normal HLA procedures, the country federates of the simulation join the federation one at a time. The first federate to join creates the federation and starts to execute. Federates are made both time-regulating and time-constrained. Each country federate calculates its own population increase during each time step and publishes this information to all other country federates. It also receives the population data from all the other federates.

In the original version of the simulator the second federate, C2, to join the federation was, in general, a little out of step with the original federate, C1, so that its time stamps differ by a fixed amount from the time stamps of C1. This problem was fixed in the modified version. Moreover, the populations of C1 and C2 were reported as 194.774 and 167.1 respectively at time 670 and 671, respectively. The model for population growth used in these two countries was identical with identical data, so why the difference? The reason is that C2 has taken one large step from time=0 to time=671 and has calculated the population to be 100 + 67.1*0.01*100 = 167.1. On the other hand, C1 has been advancing in steps of 10 and increasing its population by 1% in each step, which leads to the higher figure of 194.774.

Hence we need a federation in which all the federates start executing and advancing their logical times at the same instant in time in order to avoid unexpected results.

The federation, then, needs some way of recognizing that the last federate has joined. Two methods have been selected:

Figure 8. Time-regulating and time-constrained federates

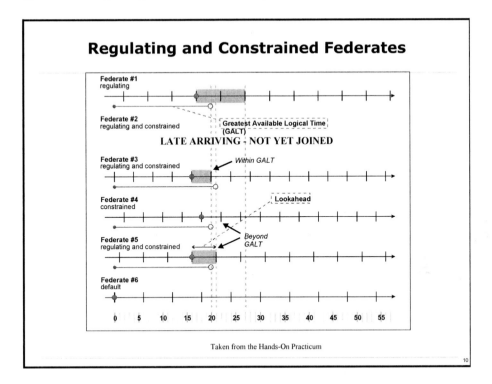

1.   A synchronizing federate would join first and would then count the countries as they joined until some agreed number of countries had joined. It could then generate an interaction that would effectively start all federates advancing their times at once. Note that we are still not synchronizing the federation with wall clock time, but merely ensuring that the logical times of the federates all advance together.

2.   A modified federate can be created, to be used specifically as the last federate to join, so that when this modified federate joined the federation, it generates the required interaction to start time advancing.

The problem with the late-arriving federate is that, if it requires to be both time-regulating and time-constrained, it must assume a time that guarantees that it cannot generate a TSO message earlier than the GALT of the remaining federates. This is illustrated in *Figures 8* and *9*, in which Federate 2 is a late-arriving federate and is assigned an initial time of 20.

It is important to recognize that three things must have happened for an event to be delivered TSO:

Figure 9. Late arriving federate

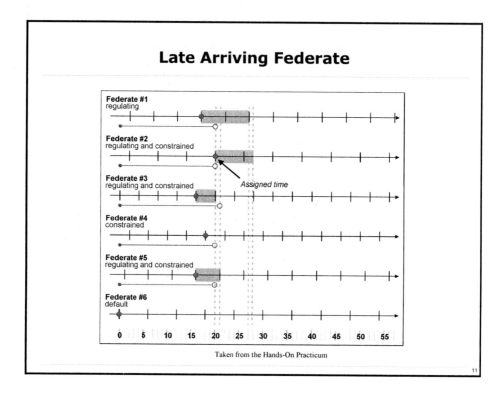

1.   The sender must be time-regulating
2.   The receiver must be time-constrained
3.   The event itself must be designated TSO.

Review of Time Management Services

Events are sent and received by the local RTI component (LRC) of the appropriate federates. Each LRC contains two queues, a FIFO receive queue and a time-stamp queue. Events that meet the above TSO criteria are placed in the time-stamp queue, which is ordered based on the time value of its contents. The receive queue contains RO events (events that do not meet all the criteria) entered in the order in which they arrive. Since there are no time constraints on the delivery of RO events, the receive queue is drained any time the federate provides sufficient time to libRTI.

In the example illustrated by *Figure 9*, Federate #3 generates a TSO event. Federate #6 sees this as a RO event. The event does not arrive as a TSO event because Federate #6 is unconstrained and therefore unable to receive events in time-stamped order. The same event is received by Federate #2, which is time-constrained, as a TSO event.

The *RTIambassador* and *FederateAmbassador* functions associated with time regulation and time constraint are:

enableTimeRegulation()	*enableTimeConstrained()*
timeRegulationEnabled()	*timeConstrainedEnabled()*
disableTimeRegulation()	*disableTimeConstrained()*

By default, federates are not time-regulating. A federate uses the member function *enableTimeRegulation()* to request that it be made time-regulating. The LRC uses the callback function *timeRegulationEnabled()* to inform the federate that the request has been granted and informs the federate of its (possibly new) logical time to which it must advance to ensure that the GALT of existing federates will be respected. The time-regulating status can be cancelled dynamically by calling *disableTimeRegulation()*, which takes effect immediately.

By default, federates are not constrained. A federate uses the member function *enableTimeConstrained()* to request that it be made time-constrained. The LRC uses the callback function *timeConstrainedEnabled()* to inform the federate that the request has been granted. The time-constrained status can be canceled dynamically by calling *disableTimeConstrained()*,

Several additional time-management functions, listed below, are available to allow federates to make inquiries about time status and to modify lookahead (their purpose is self-explanatory):

queryLogicalTime()

queryLookahead()

modifyLookahead()

queryGALT()

queryLITS().

Although the *tick()* method is not part of the HLA specification, it is an important method for RTI 1.3-NG, which is not fully multithreaded and which uses *tick()* to provide an opportunity for the LRC to do its work. Failure to "tick the LRC" can lead to federation-wide problems. A late-arriving federate can be prevented from joining a federation because information that needs to be passed to the LRC of existing federates is blocked by the failure of the federate to tick the LRC. This could cause the entire federation to stall.

Data Distribution Management

The DDM services are used to avoid the transmission and the reception of unnecessary data. This is provided by multi-dimensional routing spaces, that is, the producers of data declare an update region associated with a specific attribute update or interaction, so that receiving federates know immediately if the region they are interested in can match producers' area.

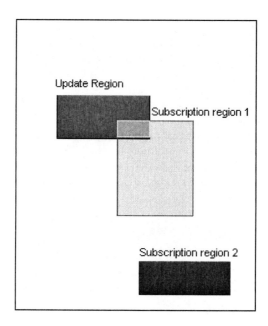

The HLA Rules Revisited

The HLA rules are divided into federation rules and federates rules. A definition and a description of each rule are presented later on.

Rules 1 through 5 deal with federations and 6 through 10 with federates.

- Rules 1 and 6 specify that the federation and all the federates must be documented, and how this should be done (FOM and SOM).
- Rule 2 requires that the objects defined in the FOM must be represented in the federates and not in the RTI, and Rule 3 specifies that data exchange between federates (i.e., FOM data) occurs via the RTI. Rule 4 introduces the HLA interface specification and its role in defining the interaction between the federates and the RTI. Taken together, Rules 2, 3 and 4 define the specific roles played by the federates and by the RTI in an HLA federation.

Figure 10. Federation rules, reviewed

Federation Rules

1. Federations shall have a FOM, documented in accordance with the OMT.
2. All representation of objects in the FOM shall be in the federates, not in the RTI.
3. During a federation execution, all exchange of FOM data among federates shall occur via the RTI.
4. During a federation execution, federates shall interact with the RTI in accordance with the HLA interface specification.
5. During a federation execution, an attribute of an instance of an object shall be owned by only one federate at any given time.

19

- Rule 5 establishes the important requirement that only one federate can own an attribute of an object instance at any one time, but it does not rule out the possibility that the ownership of such an attribute might be passed from one federate to another during a federation execution.

- Rules 7, 8 and 9 deal with the control and transfer of relevant object attributes. Rule 7 establishes that a federate can update the object attributes it owns and reflect (receive values of) objects in which it is interested, and that are not owned by it. It also allows federates to send and receive interactions as specified in the SOM. Rule 8 allows federates to exchange ownership of attributes dynamically (during a federation execution) and Rule 9 gives a federate control of the conditions under which it is required to provide updates of the attributes it owns.

- Rule 10 deals with time management. Note that it requires that federates be free to manage their own local time *"in a way which will allow them to coordinate data exchange with other members of a federation."* This is rather non-specific but Rule 10 needs to be interpreted in conjunction with Rule 4, which states that *"federates shall interact with the RTI in accordance with the HLA interface specification."* In other words, most of the restrictions that apply to the management of time are found in the RTI interface specification.

Figure 11. Federate rules, reviewed

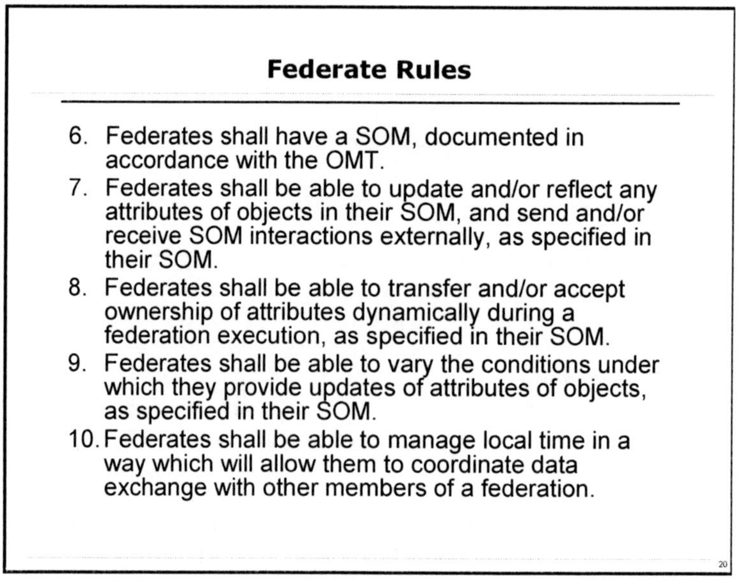

Federate Rules

---

6. Federates shall have a SOM, documented in accordance with the OMT.
7. Federates shall be able to update and/or reflect any attributes of objects in their SOM, and send and/or receive SOM interactions externally, as specified in their SOM.
8. Federates shall be able to transfer and/or accept ownership of attributes dynamically during a federation execution, as specified in their SOM.
9. Federates shall be able to vary the conditions under which they provide updates of attributes of objects, as specified in their SOM.
10. Federates shall be able to manage local time in a way which will allow them to coordinate data exchange with other members of a federation.

20

Navi: The Simplest Case Study

This simple implementation of a set of HLA federates can be used to demonstrate the basic concept of HLA and is more realistic than a classic HelloWorld application.

Proposed application must simulate a set of vessels able to interact together in a collaborative environment. Each ship must be controlled in bearing and speed by a GUI where all the participating federates will be displayed.

Consequently federates are be identical and based on a simple three Java class schema. For the implementation of such exercise we will need a GUI, a simple ship motion model and a FederateAmbassador implementation for holding all necessary call backs from the RTIExecution.

In order to keep the complexity of the example reasonable, a Java-based implementation has been chosen, proposed example is made upon RTI1.3v7 (formerly available free of charge and with restrictions from http://hla.dmso.mil) and implemented through the Java Binding 1r2v4,5,6,7. GUI will be designed as a Java awt applet for running with Java version 1.1.7 and above.

GUI main panel will hold three areas:

1. **Top panel:** Bearing and speed controls, user will control his ship by changing the values and updating it inside the model.

2. **Center panel:** Visualization, the user will see his ship as a red square and the other federates' ships as blue squares.

3. **Bottom panel:** Messages, coordinates positions updating and federates' messages will be displayed here.

The last panel is also used both by RTIAmbassador and FederateAmbassador to describe the federation activities such as joining, resigning and update attributes.

The RTI callback are visualized in this area as well as the interaction message and the value of the federates' attributes.

In such a way the operator is informed also of the background events and can better understand the HLA transactions.

Every federate is a simulation tool able to model the behavior of a ship: it must create, initialize and register an object Vessel and it must be responsible for its position updating during the entire simulation exercise. Through the FederateAmbassador, it must discover all the remote instances of the similar class Vessel and will be responsible for keeping the local copy updated by reflecting its position attributes. Moreover, the federate must investigate constantly about collisions among itself and any of the other Vessels, launching a collision interaction every time another essel is overlapping its position. Time advance policy is time-stepped with an interval ΔT in which a single ship is supposed not to change its position much than its average size, in formula:

$$\Delta\tau \le \frac{1}{\omega_i}\min(\lambda_i)$$

Where:

$\Delta\tau$ is the time step duration,

ω_i is the speed of the HLA i-th object

λ_i is the average HLA i-th object size.

The FOM must define an object class, ",Vessel," and an interaction class called "Message:" according to the OMT, these class must be described in the ".FED" file that will be placed in the configuration folder of the RTIExecution directory.

The HLA object class "Vessel" has basically five attributes and two degree of freedom:

1. Name, unique it identifies the ship;

2. Position X, for simplicity is given in pixel related to the background image on the visualization panel;

3. Position Y, similar to the X position;

Figure 12. Object model for object class vessel

```
(class Vessel
    (attribute Name reliable timestamp)
    (attribute PositionX reliable timestamp)
    (attribute PositionY reliable timestamp)
    (attribute Bearing reliable timestamp)
    (attribute Speed reliable timestamp)
)
```

4.   Bearing, it is the orientation of the vector speed, is graphically displayed in the visu-
      alization panel;
5.   Speed, for simplicity is given in pixel/update and is displayed in the visualization
      panel at length of the vector bearing;

For each attribute of the object class "Vessel" the FOM specifies reliable (TCP/IP) as transportation protocol and time stamp as delivery option.

Similar to the definition of the object class "Vessel," the interaction class "Communication"
with only one parameter:

1.   Message, the text of the broadcasted information.

The specification of the transportation protocol and the delivery option is given at interaction
class level, since publication and subscription will be performed one-for-all.

The FOM is completed by adding the MOM and the "Navi.fed" file is automatically generated by using one of the commercially available packages for object model development
(i.e., Aegis OMDT).

The SOM of the Navi federation is very simple since there is only one object class and only
one interaction class.

Figure 13. Object model for interaction class communication

```
(class Communication reliable timestamp
    (parameter Message)
)
```

Figure 14. Discovered instances enumeration

```
Enumeration ships = Vessel.sRemoteVessels.elements();
while (ships.hasMoreElements()){
  Vessel remoteShip = (Vessel)ships.nextElement();
  ta.append(remoteShip.toString()+"\n");
}
```

The DM of the federation requires that each federate must publish and subscribe all attributes of the object class Vessel and the interaction class Communication; each update in any attribute of the simulated objects will be reflected by all the federates in their remote instance object lists.

Basically each simulated ship is a Java instance of the HLA object class "Vessel," the motion law is given by the following differential equations integrated at finite difference:

Figure 15. Java implementation of the update/reflect attribute value

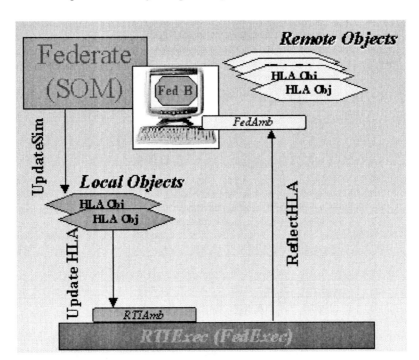

$$\frac{\partial x}{\partial t} = v \cdot \cos(\theta)$$

$$\frac{\partial x}{\partial t} = v \cdot \operatorname{sen}(\theta)$$

where:

θ: is the direction of the speed vector [°]

v: is the module of the speed vector [pixel/update]

Implemented federation time management is strictly conservative (each federate is both time constrained and time regulating) and the advance time request is time-stepped.

At the beginning of the simulation the first federate tries to start the fedex.exe process on its computer, asks for federation creation and tries to join the federation itself.

Figure 16. Creating a federation execution and managing joining

```
fedamb = new NaveJavaFedamb();
try {
  rtiamb = new RTIambassador();
  try {
  rtiamb.createFederationExecution("Nave", "Nave.fed");
    ta.append("***FED_HW: created federation execution\n");
  }
  catch (FederationExecutionAlreadyExists ex) {
    ta.append("***FED_HW: federation execution already exists \n");
  }
  int tries = MAX_NUM_TRIES;
  while (tries > 0) {
    try {
      rtiamb.joinFederationExecution(fedname,"Nave",fedamb);
      break;
    }
    catch (FederationExecutionDoesNotExist ex) {
      ta.append("***FED_HW: federation does not exist \n");
      tries--;
      rtiamb.tick(0.4, 0.5);
    }
  }
catch (RTIexception ex) {
ta.append("***FED_HW: caught exception " + ex +"\n");
  return;
}
```

Figure 17. Federate initialization

```
try {
  Vessel.Initialize(rtiamb);
  ta.append("****FED_HW: established RTI<->SIM handle mapping \n");
  Vessel.PublishAndSubscribe();
  ta.append("****FED_HW: initialized DM disposition \n");
  rtiamb.enableTimeConstrained();
  while (!fedamb.mConstraintEnabled) {
    rtiamb.tick(0.01, 0.5);
  }
  ta.append("****FED_HW: time constraint enabled @ " +
  fedamb.mCurrentTime + "\n");
  rtiamb.enableTimeRegulation( EncodingHelpers.encodeDouble( fedamb.mCurrentTime ),
      rtiamb.queryLookahead() );
  while (!fedamb.mRegulationEnabled) {
    rtiamb.tick(0.01, 0.5);
  }
  ta.append("****FED_HW: time regulation enabled @ "+fedamb.mCurrentTime + "\n");
  rtiamb.enableAsynchronousDelivery();
  ta.append("****FED_HW: asynchronous delivery enabled \n");
```

Other federates now discovered that the federation execution already exists and try to join it.

Since each federate must know each single object class instance, and each interaction class available in the federation through their RTI mapping handles, an static initialization procedure is used.

Figure 18. Time advancing mechanism

```
myShip = new Vessel(fedname, positionX, positionY, bearing, speed);
ta.append("****FED_HW: registered new Vessel instance \n");
for (int ticknum = 1; ticknum <= total_ticks; ticknum++) {

  ta.append("****FED_HW: Event Loop Iteration #" + ticknum +"\n");
  myNave.Update(fedamb.mCurrentTime + 1.0);
  ta.append(myShip.toString()+"\n");
  fedamb.mTimeAdvanceGrant = false;
  rtiamb.timeAdvanceRequest( EncodingHelpers.encodeDouble (fedamb.mCurrentTime + 10.0 ) );
  while (!fedamb.mTimeAdvanceGrant) {
    rtiamb.tick(0.01, 0.5);
  }
  ta.append("****FED_HW: time grant to" +fedamb.mCurrentTime + "\n");
}
```

Figure 19. Remote object instance discovering

```
public void discoverObjectInstance(int oid, int oclass, String theObjectName)throws ObjectClassNotKnown {
  if (oclass != Vascello.hVessel){
    throw new ObjectClassNotKnown("unknown object class handle");
  }
    Vessel.sRemoteVessels.put(new Integer(oid), new Vessel(oid));
    System.out.println("****FED_HW: discovered instance " + oid);
}
```

Figure 20. GUI updating process

```
g.setColor(Color.red);
myShip.toPaint(g);
g.setColor(Color.blue);
Enumeration ships = Vascello.sRemoteVessels.elements();
while (ships.hasMoreElements()){
  Vessel remoteVessel = (Vessel)ships.nextElement();
  remoteVessel.toPaint(g);
}
```

Each federate creates, now, its HLA object Vessel and registers as well as publishes and subscribes all attributes, after successfully achieved this point it starts to update its attribute values.

As soon as a new HLA object Vessel instance is discovered, subscribing federates start to create a new Java object Vessel with empty attributes, awaiting the first updating in order to complete the knowledge about the new entity.

Graphical representation of the simulation simply repaints every simulated step the federate Java object Vessel itself and all the discovered instances in the discovered Java object Vessel list.

At the end of the simulation each federate resigns from the federation and tries to destroy the federation execution if empty.

Figure 21. Navi Federate's GUI

Installing, Configuring and Testing RTI1.3v7 Java Binding

The DMSO RTI1.3v7 installers are available for several platforms, these installations notes are specific for Win32 OS, for other installation please refer to the appropriate documentation.

After successfully installing the RTI packager, setting the following environment variables will complete the installation procedure:

- RTI_HOME should be set to the directory in which the RTI was installed. Typically, this will be c:\Program Files\DMSO\RTI{version}.

- RTI_CONFIG should be set to the location of the directory containing the RTI's RID and FED configuration files. Typically, this will be %RTI_HOME%\config.

- RTI_MESSAGE_VERSION should be set to an integer value. This variable can be used to run multiple independent instances of the RTI on a network. A typical installation should use the same RTI_MESSAGE_VERSION for all machines on the network.

- RTI_SAVE_PATH should be set to the location of the directory where the RTI save/restore procedure will take place.

- The PATH environment variable should include the %RTI_HOME%\bin\WIN32 directory (i.e., c:\winnt\system32;%RTI_HOME%\bin\WIN32). In the English version OS, the "Program Files" directory contains a blank space that can cause problems in configuration, in this way the PATH variable should be set within ' " ' (i.e., "c:\winnt\system32;c:\Program Files\Adobe\bin;%RTI_HOME%\bin\WIN32").

The Java binding is a thin layer of C++ code that exposes the native C++ API of the RTI to Java applications. The Java binding should run on any configuration that supports the DMSO RTI1.3v7. The Java1r2v4,5,6,7 binding is distributed as a self-extracting executable. In addition to the normal RTI environment variables, it is necessary to set the following system or user environment variable:

- Set JAVA_BINDING_HOME variable whose value is the path of the directory where the Java binding is installed (including the Java directory itself.)

- Add the %JAVA_BINDING_HOME%\javalib directory to CLASSPATH environment variable.

- Add the %JAVA_BINDING_HOME%\lib\v7\WIN32 directory to PATH environment variable.

- Add "." (the current directory) to CLASSPATH environment variable.

Every application built on top of a Java binding can take advantage from implementation of a Federate Ambassador that simplifies the creation process of a valid federate. Applications should instantiate a concrete subclass of the FederateAmbassador interface. The NullFederateAmbassador class provides no-op implementations for all FederateAmbassador interface methods. Applications utilizing only a small number of FederateAmbassador callbacks may wish to derive their federate ambassadors from this class rather than from the FederateAmbassador interface.

The Java binding holds a reference to the Java FederateAmbassador while the application is joined to a federation, so it is not necessary for the application to maintain a reference in order to avoid garbage collection. The application should not hold on to references to Java objects corresponding to C++ objects that are passed as arguments to FederateAmbassador callbacks (namely, AttributeHandleSet, ReceivedInteraction, and ReflectedAttributes instances.) The corresponding C++ objects will be deleted after the return of the callback, invalidating the references. All other arguments are proper Java objects and will persist beyond the callback as long as a reference is outstanding. In particular, the byte-arrays returned by the getValue methods of ReceivedInteraction and ReflectedAttributes may be referenced after the callbacks in question return.

The Java binding represents time values as opaque sequences of bytes. The default fedtime library provided with the RTI uses C++ doubles to represent time values. Java applications should convert the byte arrays to and from Java doubles using, for example, the DataInput-Stream.readDouble and DataInputStream.writeDouble methods.

The DMSO RTI1.3v7 MOM specifies that MOM attribute and parameter values are encoded as null-terminated ASCII strings. Java does not use null-terminated strings, so is necessary to add a null character ('\0') to any strings that are transmitted as outgoing MOM attribute or parameter values.

Reuse and Extending the Federation Navi

This simple model can be adapted easily to simulate a traditional battleship game by means of a simple extension that introduces new subclasses and implementing new behavior. The inheritance concept in HLA, in fact, is the key for the extension of the model, in fact ships, torpedoes, submarines and mines behave basically in the same way: every object must be able to determine its position according its motion law and be informed of the other's position according to its sensor system.

Inheriting from Java object Vessel, the new model introduces other Java objects:

- Battle Ships have the general Vessel behavior: it can move, live and combat in a 2-D environment.

- Submarines have the ability to move in a 3-D environment and in such way they offer a more complex behavior;

- Torpedoes are a particular kind of Java object Vessel that have a fixed direction and a higher speed as well as a activation time.

- Depth charges are allowed to move only in the Z axis (first approximation) and have their own activation rules.

- Bombs (Naval Gun Projectiles) have a parabolic motion from the firing ship to the target.

Battle Ships may launch an attack just firing torpedoes, shooting Bombs against other Ships and launching Depth Charges to submarines.

Submarines are allowed to attack Ships firing torpedoes and have no defense against Depth Charges except escape from their active range.

Figure 22. FOM extension for battle field federation

```
(class Vessel
    (attribute Name reliable timestamp)
    (attribute PositionX reliable timestamp)
    (attribute PositionY reliable timestamp)
    (attribute Bearing reliable timestamp)
    (attribute Speed reliable timestamp)
    (class BattleShip)
    (class 3DObject
     (attribute PositionZ reliable timestamp)
      (class Submarine)
     (class Torpedo
        (attribute activationTime reliable timestamp)
        )
     (class DepthCharge
        (attribute firingDepth reliable timestamp)
        )
     (class Bomb)
     )
 )
```

In case of collision, explosion an interaction, derived from mother interaction Message, is broadcast informing each federate that something is appended to the battlefield. If damage suffered from an object exceeds a fixed limit the object is removed from the federation.

Distributed Simulation in Supply Chain Management

Introduction

The recent developments in manufacturing have boosted the practice of outsourcing, in which suppliers are continuously specializing and improving in order to meet the market changes. This point is not only a way to reduce production costs but is also a common practice to increase the flexibility to the market: new ideas and proof of concept can be obtained from the "external world" and transformed into real product. The "externalize & specialize" approach has turned in to several spin-off projects in which companies can gain business performance from a previously critical division. A production line that is underused can be turned into an interesting business unit by transforming it into a separate company and opening it to the open market (i.e., PUMA Experience for Gas Turbine Power Plants Maintenance). This issue requires a efficient level of control that is becoming extremely complex for highly distributed manufacturing systems, in which simulation is largely used due to the extremely non linear nature of the problems. Simulation, here, is often used to improve process. Until now all the members of a supply chain had to simulate separately: their processes and information are not shared among the supply chain partners. But, for taking advantage from the proposed methodology, a full-scale simulator model may be used in order to build a model that resembles reality more effectively. The high level infrastructure (HLA) is a standard framework that supports simulations composed of different simulation models. From now on the different models—parts of the total model—are called federates. In order to design a simulation composed of different federates on different computers it is necessary to connect them together and establish a communication protocol, this is done by an HLA that clearly separate simulation from communication process. The application of the HLA has many advantages since it offers interoperability, encourages reusability and makes it possible to use confident information in models without the necessity of being visible to other partners in the supply chain. In this way other partners don't have access to confident information and can't use it in a strategic way. The other partners will only see the results of simulation runs/steps and not the data behind the results. Because every partner now is more willing to use confidential information, the total quality of a model of a supply chain increases and hence the benefits for the partners. Furthermore, because each partner builds its own module, it is much easier to keep modules up-to-date both in term of data (i.e., directly from ERP). The University of Genoa was particularly involved in the Web Integrated Logistic Designer Project (WILD I & WILD II). These projects involve the development of a federation composed by simulators and dynamic programming systems (i.e., Nash Equilibrium Negotiation). The WILD Federation reproduces the supply chain and supports online the distribution among suppliers, main contractors, and outsourcers and was successfully tested for an supply chain in the aerospace industry.

To make the HLA accessible also for Commercial Off-of-the-Shelves Simulators (COTS) the HLA Operative Relay Using Sockets (HORUS™) has been developed. The HORUS™ manages the communication between the different simulation models and the HLA. To take advantage of interoperability and reusability, existing federates, implemented in various languages, can be integrated.

Achieving HLA Migration into the Civil Domain

Two important problems must be solved in order to move HLA into a civil (industrial) application:

1. Provide a physical interconnected network by using a TCP/IP and UDP connection that must be utilized at a reasonable cost.
2. Integrate a wide set of commercially available simulation tools since they are commonly used in the industry.

The first problem was solved in the falls of 2000 by implementing a demonstrator federation that could use the Internet as a wide area network (WAN) for the spanning exercise. This sub-project was also designed to provide a practical demonstration of the possibility of using commercially available simulation tools (Arena™ and Simul8™) as federates.

In this sub-project, named SHOW (Spanning an HLA federation Over a Wan), two teams worked in the Savona Logistics Lab (Italy) and at the Riga Technical University (Latvia). Initially, a proof of concept test was implemented in October 2000 on a previously implemented federation (NAVI Federation) that was set up to run over an Internet connection between Savona and Riga: performance statistics were acquired along with specifications for the consequent upcoming exercise. The problem that emerged in this experience referred mainly to the time advance synchronization affected by network latencies. A synchronization procedure was then designed and implemented for use in the SHOW project. A specific approach had to be developed to connect non-HLA compliant software into an HLA federation. In particular, two problems were solved:

1.   To implement a mechanism by which a simulator can communicate with the external world;

2.   To implement a procedure for controlling a simulator by means of the RTI driven message.

Although the simulations chosen were not HLA compliant, they were able to communicate with external software utilizing a standard extension: VBA™ for Simul8™ and VB™ for Arena™. In this scheme an internal event could be used to write information in an Excel™ file and to transfer it later to the RTI by means of an implemented Federate Middleware. Since a VB™ porting of the RTI API was not available, a Java™ binding was used to create

Figure 23. FOM architectural view

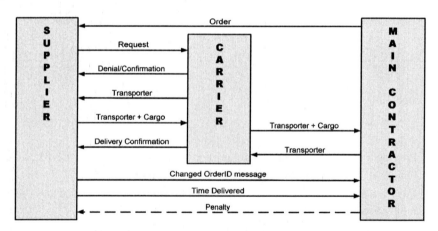

software that could communicate with the RTI execution on one side and with the federation on the other.

Communication with the federation was carried out using the RTI services while several methods for communicating between the simulators and the middleware were also considered. The authors propose a socket-based approach in which persistent (PCP/IT) connections were established between the middleware socket ports and the Excel™ VBA™ Module. Visual Basic for Application™ offers an ActiveX™ Winsock connection that could be used for such a purpose. A small supply chain was designed in the implemented federation in which the manufacturing process takes place by assembling parts utilizing specific machinery of an external supplier (subcontractor). The main contractor places orders with the subcontractor according to his production plan and then receives finished products as soon they are shipped from it. This implementation involved a federation object model as shown in *Figure 23*. A double subcontractor implementation (one in Arena™ and one in Simul8™) was used during tests in order to demonstrate the interoperability among different tools in the same federation.

The test was successfully conducted on December 12, 2000, in the presence of the delegation of project partners.

WILD project partners also tested several approaches to solve the outlined problem. These included file tracing for middleware-simulator communication and message-based RTI proxy implementation but all were inadequate due to the extreme severity of the data exchange and the performance level obtained. Up to the end of the WILD project no other integration methodologies were proven to be effective in integrating non-HLA simulation tools other than the socket-based one. In this way the authors began to design and implement a general-purpose integrator tool based on the socket bus to be applied to various applications without re-coding. Such a tool was named HORUS (HLA Operative Relay Using Sockets) and is based on a Java™ implementation of the middleware design to operate with a configuration file. A middleware implementation is generated at run-time inside HORUS that deals with both the time advancing mechanism and the synchronization issues. A user-friendly GUI is used to generate the configuration file in an intuitive manner and then to support federation execution and simulation management. Today, HORUS is one of the outstanding WILD project solutions for implementation of a complete civil domain HLA federation.

Continuous Integration as a Critical Issue

A distributed simulation system often requires distributed development in which subjects act as simulation designers of single federates. This requires exceptional coordination in order to produce measurable results. While in modeling and simulation projects developers must deal with a lack of information and the difficulties in creating a credible model, in HLA this subject can became even more critical. It is not always possible to test a single procedure due to the fact that more than one federate must be implemented and that dummy federate needs may also arise. In common simulation system design practice, developers are used to working top down by continuously refining high-level specifications into directly operative requirements. In HLA, this process must be carried out bottom up by ensuring that the simulation building blocks are consistent before being utilized in a simulation exercise. Intercommunication and data exchange processes must be tested separately prior to the final integration since errors cannot be traced back easily in a distributed environment and no debugger can follow an error across a distributed simulation exercise.

Processes have to be kept simple as possible and "fancy" solutions must be carefully reviewed: HLA development is not just a matter of interaction and class diagrams where "boxes" are talking with other "boxes." Data exchange must be accurately planned and efficiently verified before carrying out the final integration. The simulation accuracy level must be assured by unifying the entire simulation exercise with respect to the same requirements. In an HLA federation, it is not possible to require a hi-fidelity level for a parameter whose calculation relies on a rough-cut simulated input. In addition, the time advance mechanism must be properly evaluated since it may reduce computational speed and the usability of the results. In the proposed WILD example, the authors experimented with an incremental integration in which each single component was carefully tested prior to its integration. This was accomplished by running a test federation in which missing federates were substituted by a "dummy" one as required.

Preliminary releases of the WILD federation were used to test proof of concepts and data exchange, while the simulation algorithms were kept as simple as possible. In this way, on April 3, 2001, a preliminary federation implementation was available. An HLA implemented federation was carried out between Savona (University of Genoa) and Bari (Polytechnic). The WILD federation featuring the Piaggio P180 Supply Chain was implemented in a set of GG simulators designed and implemented by the DIP (University of Genoa) in Java™ and successfully integrated with the Arena™-based simulator of a supplier running in Bari. The Internet WAN used demonstrated that such an exercise could be carried out and that network latencies may subsequently affect the usability of the simulation results. The reliability rate calculated in a WAN federation run was a fraction of the one calculated with the simulation running on a LAN. In this way subsequent efforts were made to reduce network traffic. Final integration on the preliminary release was obtained on May 28, 2001, with nine simulators running at the following locations: MISS Genoa - Bari Polytechnic, L'Aquila University, Milan Polytechnic, Salerno University, Naples University, and Florence University.

Interoperability was carefully tested since three simulators were GG Java™ based, four simulators were based on Arena™ and two were built using Simple++™. Some of the locations were also running multiple federations (i.e., MISS - Genoa) interconnected in a LAN and using an RTI provider machine as an HLA gateway.

The implemented federation was also used to test the robustness of the HLA over non-dedicated networks. Experimental results indicated serious criticality in this application since the federation success rate dropped moving from LAN to WAN (Internet). This reliability issue must be taken into consideration with regard to future federation design since it may reduce the usability of the federation results.

Furthermore, the use of an HLA middleware for a non-HLA compliant simulation must be properly evaluated since is not always easy to convert pre-existing simulators into an HLA federation.

Another important issue is related to data warehousing: WILD RR_HLA federates were equipped with a JDBC-ODBC connection to obtain information and the startup configuration directly from a relational database. The implemented approach demonstrated how it is possible to connect federates to the ERP system by means of an interconnection database. In fact, this small DBMS is responsible for providing the updates from the real system and for maintaining all the scenario and hypothesis values in order to conduct a proper what-if analysis. In this way, ERP is unaffected by the simulation results until the final decision is made at the end of the experimental campaign.

Data are left untouched on the ERP while the decision-maker is free to define a virtual scenario for the experimental campaign. Rollbacks and mission rehearsals are always possible by reproducing the past situation.

Most of the critical issues of HLA development are based on the lack of competence in the field of distributed computing. In the industry as well as in the academia there is growing interest in simulation but, in the near future, its use will still be limited to implementation of commercial tools rather than the direct application of modeling and simulation techniques into a software implementation. Young researchers learn how to simulate simple tasks using a commercial suite without any knowledge of how a simulation exercise is effectively implemented in a computer program. Today, this type of use is adequate, but exercise problems and other difficulties will arise when these people start to consider a distributed simulation. In commercial simulation packages, objects are created on request by a few lines of code and flooded along the remaining simulation river into the same computer process. In HLA this is not possible: researchers must understand the distinction among local and

Figure 24. Integration architecture

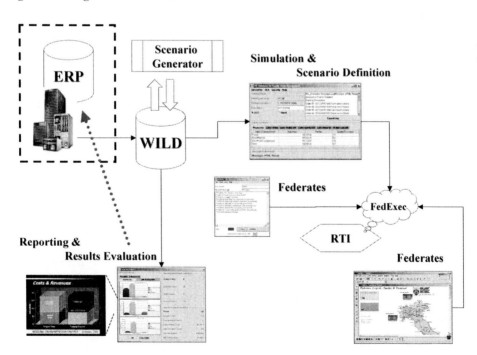

remote object instances. In addition, commercial tools do not allow any interaction in their internal event queues, while in HLA it is common to mix internal and external events. The best way to design and implement an effective HLA federate is to use a general-purpose language (i.e., C++) and, eventually, a simulation library designed to interact with an external event source.

A specific HLA interface could be also used if appropriate, however HLA is a multi-layer architecture in which federates talk by means of predefined interfaces (RTI API). Therefore, it's not a good idea to introduce additional layers between the RTI execution and the federates' software implementation.

Distributed Modeling Issues for an Innovative Supply Chain Optimizer

We can extend the test case outlined in the previous section by considering the integration of an S&ME (small & medium enterprise) in the supply chain.

Due to the specialization of the various subjects in the supply chain it is even more likely that the S&ME must carry out specialized operations (i.e., thermal treatment, recoating, special parts production, etc.) that are usually critical but not very expensive if compared to others in the production budget. In this case, the penalties usually applicable for each day of delay cannot exceed a set fraction of the value of the supply. In the worst case of delay S will pay a heavy penalty, from its point of view, and M will incur the same delay by paying a much higher penalty to its customer. We can express this in a formula as follows:

$$B = pM\delta t - \pi S\delta t \tag{1}$$

B	The Total Balance Due by M	€/unit
p	% Penalty Paid by M for Delay	-/day
π	% Penalty Paid by S for Delay	-/day
M	Value of M's Finished Product	€/unit
$S \ll M$	Values of S's Part	€/unit
δt	Due Date Delay	days

In this way it is easy to note that the only effect of applying penalties in the outsourcing contract will be a slight reduction in the balance to pay without considering the cost due to loss of image.

The only way to reduce the costs coming from unexpected delays in the distributed production process is to reduce the effect of a delay on the final assembly process.

Based on this approach the key aspect of the methodology is the value of receiving information in advance about a possible delay from the subcontractor to the main contractor.

In a real world application, the main contractor has several production processes running at the same time with different time advance mechanisms. If the assembly line is waiting for a part that is delayed, the production manager can change the work schedule by modifying the job priority and use the idle resource to increase other product time advance mechanisms. In this way, the final cost of the delay will be compensated by the increase resource utilization.

The possibility of rearranging the production schedule is strictly related to the possibility of evaluating the impact of the subcontractor delay on the main contractor production process. A simulation or a meta-model can be used to carry out such an evaluation. In the first case, the computational effort might be too high for an online, real-time application and thus the second approach must be used. The proposed methodology is outlined in the following process.

The subcontractor is evaluating a situation in which it will be not able to meet its due dates and decides to inform the main contractor, proposing a new due date and a discounted price. The main contractor will evaluate the proposal and decide whether to accept it or to refuse it. The simple negotiation process shown in *Figure 25* also identifies two critical processes: the subcontractor delay estimate and the main contractor delay evaluation. The output of the first process is an estimate of the delay in supplying parts from the subcontractor: it is measured in terms of days or unit of time (UT). Instead, the output of the second process is an estimate of the cost resulting from the Subcontractor's failure to meet the due date. It's

Figure 25. Basic negotiation process

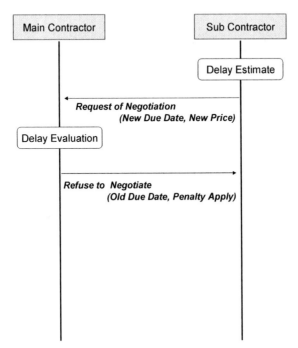

possible to identify several conditions and regimens in which such a process will take place. In particular, we can identify a static-deterministic scenario and a dynamic-stochastic one.

Static-Deterministic Case

In the first case we considered an assembly process described by a PERT graph where the various parts are required during the activities. The simplest delay estimation process can be summarized as follows:

Delay Estimate

- Step 1: The d_j - duration of j-th activity - are used to estimate the early start and finish dates as well as the late start and finish dates.
- Step 2: The forward and backward rules can be used to identify the *Critical Path*.
- Step 3: If the subcontractor has an expected delay of h days in the k-th activity with $k \in$ *Critical Path*, the subcontractor will have a supply delay of h days.

Delay Evaluation

- Step 1: The delay by the main contractor can be evaluated by using the same process applying the first part of the formula (1) to the estimate delay of h days:

$$C_h = pMh \tag{2}$$

In this case the main contractor will estimate the delay just by calculating the difference between the new due date (D_{new}) and the old due date (D_{old}).

$$h = D_{new} - D_{old} \tag{3}$$

- Step 2: The cost identified at Step 1 must be corrected by subtracting the difference between the original price (P_{old}) and the new discounted price (P_{new}).

$$G_h = P_{old} - P_{new} \tag{4}$$

The final difference between the cost of the delay (2) and the gain of the delay (4) will identify the cost of the opportunity to accept the negotiations.

- Step 3: The cost for the opportunity to refuse the negotiation could be calculated simply by applying the entire formula (1) to the estimated delay of h days.

$$B_h = pMh - \pi Sh \tag{5}$$

- Step 4: Choose the negotiation opportunity with a lower cost, so:

$$\min\left\{(C_h - G_h), B_h\right\} \tag{6}$$

In this example, the main contractor can also systematically refuse each negotiation by applying the penalty each time. In this case, the evaluation of the delay is completed in one step only.

Advanced Negotiation Process

In the real world the negotiation process is much more complicated and delay estimation and evaluation must investigate the impact of the stochastic and properly evaluate the delay-related costs.

A technique to make an early estimate of the probability to meet the due date can be introduced to extend the simple example shown in the previous section.

Static-Stochastic Case

In this case, introducing the concept of expected duration of the elementary j-th activity can extend the delay estimate.

Now, each d_j has to be calculated by evaluating three different duration estimates:

a	Optimistic Duration Estimate
b	Pessimistic Duration Estimate
M_0	Modal Duration Estimate

The beta distribution can be used to calculate the following parameter:

$$d_j = \mu_{(j)} = \frac{a + 4M_0 + b}{6}$$ Average duration of j-th activity

$$\sigma^2_{(j)} = \frac{(b-a)^2}{36}$$ Estimation Variance of duration of j-th activity

In this way we can define the following summary (7) as the probability factor that the j-th activity will be delayed:

$$Z_{(j)} = \frac{-(D_{\{D_{LF}^{(j)}\}} - D_{\{D_{EF}^{(j)}\}})}{\sqrt{\sigma^2_{\{D_{LF}^{(j)}\}} - \sigma^2_{\{D_{EF}^{(j)}\}}}}$$

(7)

where:

$Z_{(j)}$	Distributed as NID(0,σ)
$D_{\{D_{LF}^{(j)}\}}$	Late Finish Date for j-th activity
$D_{\{D_{EF}^{(j)}\}}$	Early Finish Date for j-th activity
$\sigma^2_{\{D_{LF}^{(j)}\}}$	Variance of Late Finish Date for j-th activity (: of σ on Backward Rule)
$\sigma^2_{\{D_{EF}^{(j)}\}}$	Variance of Early Finish Date for j-th activity (of σ on Forward Rule)

and by using the standard normal distribution table we can estimate the probability of a delay for the j-th activity.

Such a procedure can be summarized in the following steps.

Probability of Delay and Delay Estimation

- Step 1: The subcontractor calculates the critical path of its production process and evaluates the average durations as well as their variances.
- Step 2: For the desired activity the probability factor is calculated along with the subsequent probability.
- Step 3: Iteratively a new due date is proposed and the probability of delay is recalculated: if this value is lower than a predefined value (i.e., 2%), the new due date is set for the request to negotiate with the main contractor.

The main contractor can use the same approach to evaluate the cost of the delay simply by calculating the expected delay with regard to the completion date. The procedure for such an evaluation is shown in *Figure 26* and summarized in the following steps.

Figure 26. Due date estimation (dynamic)

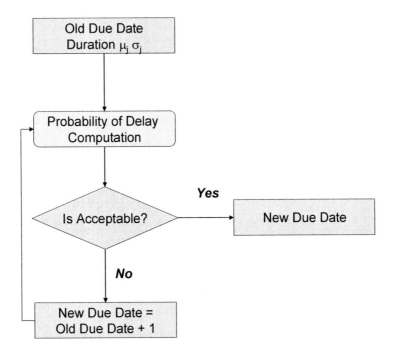

Delay Evaluation in a Static-Stochastic Case

*   Step 1: Recalculate the PERT of the final assembly process using the proposed due date: if the parts or the service are involved with an activity on the critical path it will not affect the final due date, so accept the proposal from the subcontractor.

*   Step 2: If the activity is on the critical path, use formula (7) to calculate the risk to fail the final due date. If the probability is too high, iterate the process until the new due date is determined.

*   Step 3: If the new due date still meets the customer needs, accept the negotiation or, in the opposite case, refuse it.

For a dynamic scenario the methodology must be extended to properly evaluate the economic impact of the expected delay according to the other entities in the supply chain. The resulting problem can be successfully addressed by using the nested simulation paradigm but, because the computational effort will be much too high in several cases, it cannot be used. In the dynamic scenario, the negotiation process is carried out iteratively until an agreement is reached or the negotiation is broken for each simulation event in which the subcontractor is requesting to specify

Figure 27. Extending the negotiation model

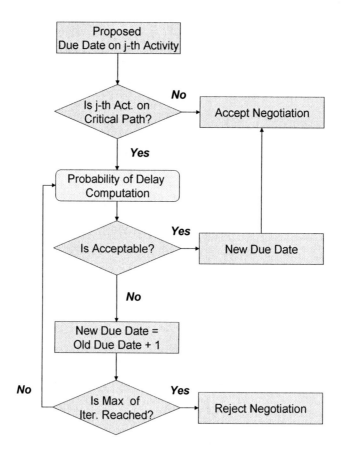

a new due date. In several cases the main contractor could also request a negotiation. This is the case in which the final assembly is late and the main contractor project manager requests a delay in the supply to reduce inventory costs.

The previously described conditions occur in particular when subcontractors have different relationships with other customers to whom they guarantee a higher priority. The SM&E main contractor does not have enough power to compete with a large outsourcer and so, sometimes, it must accept an unexpected delay in its supply. The general approach is proposed in *Figure 27* in which the main contractor has to use nested simulation inside the delay evaluation.

In order to take advantage of the practical features of the methodology without using a nested simulation, a meta-modeling approach is presented. The basic concept involves the possibility of substituting the nested simulator with a numerical model that can calculate the expected cost for a predefined delay. In this way we can identify the cost level of each offer presented by those involved in the supply chain and make the correct trade-off.

The cost resulting from rescheduling an activity in the final assembly can be calculated in terms of:

- Customer satisfaction costs
- Inventory costs
- WIP costs
- Inefficiency costs

The customer satisfaction costs are proportional to the perceived value of the product and to the amplitude of the delay. For a delay equal to or greater than the acceptable maximum, such a cost may be equal to or exceed the market value of the product (i.e., refuse to accept the product, damage claim, etc.)

$$K_h = \rho V \frac{h}{h_{max}}$$

(8)

Figure 28. Full negotiation process

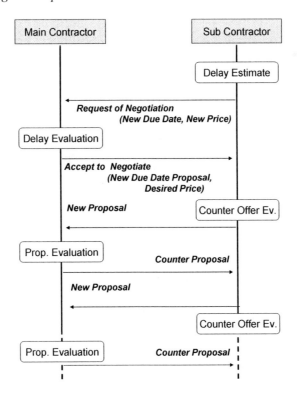

where:

ρ	Coefficient of Perceived Value
V	Market Value of the Product
h	Number of Units of Time of Delay
h_{max}	Maximum Delay Acceptable by Customer

The coefficient of perceived value is a subjective evaluation of the customer's expectation with regard to the product and may vary from less than one (no expectation) to several units (great expectation).

The inventory costs and the WIP costs are obtained by noticing that a delay in an assembly phase can postpone the utilization of the parts used in the next steps and decrease the inventory turnover ratio as well as "anticipate" some machining on other parts. In a meta-model such costs are then proportional to the assembly logic inventory level and to the amplitude of the delay. The assembly logic inventory parameter can be calculated as the difference between the budget cost of the work performed (BCWP) and the budget cost of the work scheduled (BCWS).

As known from the literature, the difference between BCWP and BCWS is a measure of the progress of material purchasing and work performed.

$$I_h = (BCWP - BCWS)ih \tag{9}$$

I_h	Inventory and WIP Costs for a Delay of h Uni Time
i	Coefficient of Inv. and WIP Costs [0.2 – 0.8]
h	Expected Delay

The inefficiency costs are obtained by noticing that the workforce was designed for a predefined work schedule and now, due to the delay in the supply, some work activities cannot take place while the resources have already been hired or acquired. Such costs are simply proportional to the net number of resources multiplied by their current retribution and the amplitude of the delay.

$$M_h = h\left(\sum wR + \sum mCo\right) \tag{10}$$

M_h	Inefficiency Cost
w	Number of Workers
R	Retribution per Worker per Unit of Time
m	Number of Machines
Co	Cost for each Machine per Unit of Time

By using the formulas (5) (9) (8) and (10) we can write the following expression that identifies the trade-off between the cost and the benefits of the negotiation:

$$T_h = h\left(\rho V \frac{1}{h_{max}} + (BCWP - BCWS)i + + \sum wR + \sum mCo - \pi S\right)$$

(11)

Modelling a Distributed Logistic Platform Using HLA

The proposed example is related to the modeling of a complete supply chain that produces business planes: the main assembly line is located in Genoa, Italy while the major supplier

Table 1.

Objects	Attribute	Supplier	Customer	Logistic
HLA Order	Name	Publish	Subscribe	-
	ID	Publish	Subscribe	-
	Quantity	Publish	Subscribe	-
	Date	Publish	Subscribe	-
	From	Publish	Subscribe	-
	To	Publish	Subscribe	-
HLAPF	Name	Subscribe	Publish	-
	ID	Subscribe	Publish	-
	Quantity	Subscribe	Publish	-
	From	Subscribe	Publish	-
	To	Subscribe	Publish	-
Pick_Up_ Location	Request_for_ Transportation	Publish	-	Subscribe
	Parking_Lot	Publish Subscribe		Publish Subscribe
	Last_Pick_ Up_List	Publish	-	Subscribe
Delivery_ Location	Items_ Delivered	-	Subscribe	Publish
	Parking_lot		Publish Subscribe	Publish Subscribe
	Items_ Collected	-	Publish	Subscribe
HLA Invoice	Deliveries_to_ be_Paid	Subscribe	-	Publish

and outsourcer are spread all around the Italian peninsula. The name for the project was fixed in WILD—Web Integrated Logistic Designer—and was a research project founded by the Italian Ministry of Research (MIUR), and the aim was to demonstrate the practical use of the proposed methodology in support the management of a complex supply chain.

Since the WILD project was designed based on a two step approach, the first phase has regarded the demonstration of a high level architecture federation able to simulate the various actors in a supply chain. On such part, the real-life example was modeled by a coordination of seven universities, each of them involved in the modeling and integration of a supplier into the federation. As a second step, WILD Federation was extended in order to include the transportation federate, the logistic federate and the negotiation process. Particularly this last topic was introduced in order to support the development of the new supply chain managing approach presented in the previous paragraph. The initial OMT was extended in order to provide a model for the integration of the logistic federate (carrier and terminal), and its general architecture is presented in the following table.

The various simulator were build by using several simulation tools (i.e., Arena™, Simple++™) and general purpose language (i.e., Java™, C/C++) in order to demonstrate the practical interoperability of the proposed HLA framework. The integration of the COTS simulator into HLA federation was obtained by implementing HORUS (HLA Operative Relay Using Sockets) middleware that relies all the events from the simulation and from the federation to the COTS simulator by using TCP/IP sockets.

Such an approach has demonstrated to be very effective in reducing the implementation efforts since it separates all the HLA programming, usually done in C/C++ or in Java, from the modeling phase. In this way people can concentrate only to define the simulation logic rather than hardcode the HLA software. The implementation of the outlined federation using HORUS and HORUS-based approach has required also the integration of the COTS package with a customized interface for the HORUS-COTS information exchange, this last part was implemented in MS-Visual Basic for Application™ (VBA).

Distributed HLA Models for Logistic Vectors

The logistic federate implements all the aspects and the logic of two logistic objects: the terminal and the carrier. The purpose of the terminal was, in this exercise, to coordinate the shipping process among federates and to assure that the logistics procedures will affect the shipment procedures as it will happen in reality. The basic logic was designed as a set of modules in which each component can be connected to the others without the need to redesign the entire federation.

Federate Carrier

The federate carrier is a basic logistic module that is modeling the process of transport products along a network between a source and a destination. The network design was fixed using an oriented graph. In such structure each node is a physical location while the arcs that connect the nodes are the feasible routes among the nodes. Each module can receive

Table 2.

Attribute	#1	#2	#3	#4	#5
Request_for_Transp.	FROM	TO	CARRIER	KIND	ITEMS
Last_Pick_ Up_List	FROM	TO	CARRIER	KIND	ITEMS
Park_Lot	FROM	TO	CARRIER	VEHICLE ID	CAPACITY
Items_Delivered	FROM	TO	CARRIER	KIND	ITEMS
Items_Collected	FROM	TO	CARRIER	KIND	ITEMS

a list of items to be shipped and it decides the routing among the various possibilities in the graph according to its internal logic. The HLA object designed for its implementation is the Pick_Up_Location. Each federate that has to ship a parcel must publish the attribute Request_for_Transportation with a list of items to be shipped coded as follows:

As soon as the federate logistic will reflect the update it will start to process the request by sending to the location identified by element #1 in the attribute an update of the Attribute Park_Lot with a list of vehicles for the pick-up procedure. As soon as the loading operations are finished, the supplier will update the attribute Last_Pick_Up_List with the list of items loaded on the vehicle. By using HLA the carrier will obtain a local copy of the items to be shipped and the routing procedure will identify the better route to get to destination. The routing is obtained by matching the elements #2 #4 #5 with the internal routing list and by process the shipping in the simulator itself. Each request for transportation is coded with the ID of the carrier (see #3 element on the attribute) so each "listening" carrier federate can identify its own simply by filtering the reflect Request_for_Transportation attribute. The modularity of the approach can guarantee also multi-carrier-based shipping: at the end of its transportation process each carrier simply updates the #3 element in the attribute with the value of the #2 element if it has reach its final destination, or the ID of a new carrier for the next leg of its journey. In this way on the next simulated step the next carrier federate will take care of it.

On its final destination the federate carrier will update its attribute Items_Delivered and the Parking_Lot attribute with the list of the vehicles in delivering, as soon as the unloading operation is over the customer federate will update its attribute Items_Collected releasing at the same the delivering vehicles.

Federate Terminal

The concept of federate terminal is an extension of the basic module logistic since the co-ordination of the loading/unloading operation has to be properly evaluated. It is a common practice, especially in the maritime sector, that when a vehicle is approaching the terminal the resource for the loading/unloading operation must be assigned in order to reduce the

Figure 29. HLA integration for federates carrier and terminal

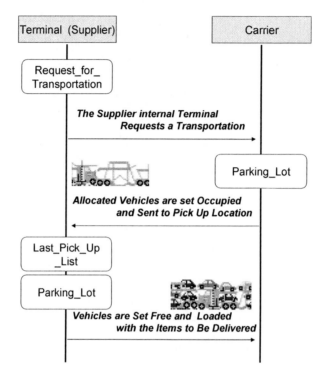

handling time to its minimum. Such event cannot be integrated in the simulator simply by means of a stochastic delay; this is because there is a kind of border event that depends from the availability of two classes of resources that belongs to different simulators: the handling resource on one hand (terminal) and the vehicle approaching (carrier) on the other. In order to reproduce such an event, the terminal resource must reserve its resource by receiving the vehicle object in its internal event queue and process it according to its internal priority rules.

For this purpose the carrier, in receiving the load for example, will send its vehicle to the supplier's event queue until the supplier will send them back loaded. In this way if the loading resources are occupied the carrier will have its vehicle idle on supplier location until the starting of the load event: by applying such procedures the real-life terminal logic will be respected. In the same way when the carrier notifies the customer that the shipment is ready to be delivered, the customer will receive on its internal event queue the list of the delivering vehicle that will be used until the customer will release it.

HLA Models for Implementing Distributed Negotiation in Supply Chain Management

The implementation of the negotiation procedure is a key aspect of the WILD II HLA Implementation since it will allow the entire federation to work as supply chain DSS by applying the rules outlined in the previous paragraphs. In such a way managers will be able to improve the performance of their production system by sharing common objectives and take the proper decisions.

In order to integrate the negotiation procedures in the WILD federation, the OMT has to be extended by introducing the object negotiation with the following attributes.

Step 1

Each time a federate needs to redefine a due date it has to register a new instance of the object negotiation and update the following attributes:

- ID_Facility: with its proper ID
- ID_Order: with the ID of the order to be negotiated
- New_Deadline: with the proposed deadline
- Agreement: fixed to 1
- Counter: fixed to 1
- Id_Turn: with the ID of the federate that will respond to the request.

Table 3.

Attribute	Description	UM
ID_Order	Is The ID of the Negotiating Order	-
ID_Facility	Is The ID of the Facility that Requests Negotiation	-
New_Deadline	N° Of Days of Delay Proposed by Initiator	days
PayOff	Is the Penalty to be Applied	€/day
Date_Sub	Is The Date Proposed by Subcontractor (Supplier)	Date
Date_Main	Is the Date Proposed by Main Contractor (Customer)	Date
Agreement	State of the Negotiation = 0: Sleeping Order = 1 Start Negotiation = 2 Negotiation On-Going = 3 Agreement Found = 4 Negotiation Broken	-
Counter	Counter of Negotiation Attempt	-
ID_Turn	Is the Next Federate that has to Reply to the Offer	-

Step 2

The responding federate (Id_Turn) will notify the initiating federate the penalty to be paid for the proposed delay by updating the following attributes:

- PayOff: with the penalty per day to be applied
- Counter: increment by 1
- ID_Turn: with the ID of the originating federate

Step 3 (Starting the Bidding Phase)

At this point a new due date will be proposed and the negotiation will enter in the bidding phase. The originating federate (main or sub) will propose a new due date by updating the attributes:

- Date_(Sub or Main): with the new due date
- Agreement: set to 2
- Counter: increment by 1
- Id_Turn: with the counterpart ID

Step 4: Evaluating the Proposal

The counterpart can have the opportunity to accept the proposal and stop the negotiation by updating the following attributes:

- Agreement: set to 3
- New_Deadline: set to the proposal accepted

In the case in which the counterpart is not satisfied with the proposal it can continue the negotiation by updating the following attributes:

- Date_(Sub or Main): with the new due date
- Counter: increment by 1
- Id_Turn: with the counterpart ID

When the maximum number of interactions is reached without an agreement the originating federate will have the opportunity to make a new proposal, if such last proposal will

Figure 30. HLA implemented negotiation approach

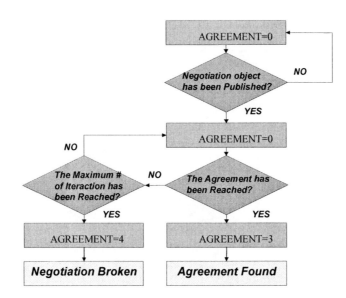

New Deadline

be refused the negotiation will end and the attribute agreement will be set to 4 (negotiation broken). The entire life cycle of the object negotiation is shown in *Figure 30*.

Experimental Results

The proposed architecture could be used to support decision-making. In fact, the methodology in the proposed application was used to investigate the efficiency of the logistics chain used to deliver the items from suppliers to main contractors.

The federation integrates a main contractor who is responsible for placing orders with a set of supplier federates and a carrier federate in charge of modeling the physical movement of the items on the network as well as the routing algorithm and the resource management.

The carrier must manage the following resources:

- *Drivers:* management policy must define the mode (i.e., sleepers, relay, meet-and-turn, etc.) and the crew (i.e., one or two drivers, etc.)
- *Infrastructure Resources:* the manager must decide on the number of facilities to be used (i.e., cranes, loading/unloading bays, refueling areas, maintenance workshops, etc.)

Figure 31. Modified WILD federation

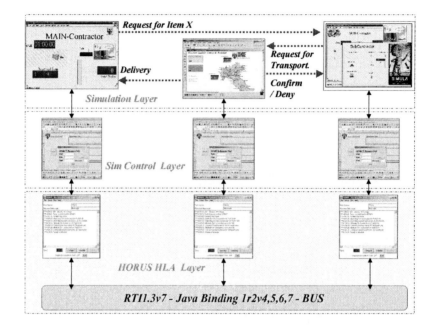

- *Ancillary Operators:* the decision maker must indicate the maintenance and administrative staff.

The experimental campaign was carried out on a WILD modified federation designed as follows (see *Figure 31*).

Simulation models are controlled by a VB external application by means of an RPC (remote procedure call) application that is connected via socket with the HORUS delegate federate. The HLA WILD modified federation is then only limited to the three HORUS modules, but all the controlled simulators are kept under an "HLA" conservative synchronized state.

The performance-indicators are set to determine the model performance levels and to make a comparison when adjustments are made to one or more federates. In this way we can define the influence of parameters and, by using design of experiments, perform a sensitivity analysis and identify the most effective parameters.

The following parameters were considered:

1.   Number of Transportation Requests not Fulfilled: This performance-indicator is one of those necessary to make a clear judgment about carrier performance levels.

2. By looking at this performance-indicator, it is also possible to extract information regarding the use of the transporters. When the percentage is greater then 100, the conclusion can be drawn that all transporters are in use either for transportation and/or maintenance. If this happens it might be useful to use more transporters in the carrier if this is possible.

3. Transportation costs: The costs may differ depending on the number of drivers used. Very high costs (higher than normal) could indicate an expensive carrier. To use this performance-indicator experiments must be performed to estimate the normal costs under current conditions. Then, when other parameters are used, say for another carrier, comparisons can be made.

4. Degree of utilization: The degree of utilization is calculated for the cranes. Certain conclusions can be reached by considering the degree of utilization. First, when this indicator, from one of the objects mentioned above, increases to a certain level (above a percentage of 80, for example), then it is recommended to increase the number of cranes. This reduces the work of the other crane(s) and the system can handle larger amounts of cargo. So, with an increase in cargo, some over-capacity is available to handle this increasing amount and the system won't be overloaded.

5. Time_Delivered and Penalties: The Time_Delivered is a message to the supplier. It contains information about what time the main contractor received the cargo. This indicator helps to set the penalties defined by the main contractor to the supplier. Based on expected time delivered and the time delivered we can calculate whether an order is too late. High costs may indicate a carrier with a poor performance level, a supplier production process that's not efficient enough or overly optimistic expected delivery times in the main contractor.

The carrier simulator requires a credible model for simulating the transportation process so as to design a point-to-point algorithm. The distances between the different destinations and terminals were generated using the Michelin route planner. The distances shown in Table 4 are used in the carrier to calculate the best way to combine orders with different destinations. This calculation is performed using an algorithm that utilizes the so-called saving matrix.

At the beginning, each delivery is assigned to a separate delivery trip and a distance matrix is calculated. The distance matrix provides the distance between the distribution center (DC) and the various destinations. Since a trip that is going from location A to B can be always be made by using an intermediate destination C, a simple formula could be written to estimate the advantage (Saving) of the new path versus the old one.

$$S(A, B) = D(DC, A) + D(DC, B) - D(A, B) \qquad (12)$$

A simple search algorithm is then used to find the best cluster among the different destinations to be served according to the capacity of the vehicle. This problem is known in the literature as the capacitated routing vehicle problem (CRVP).

The commercial speed is used to calculate the travel time for a route; in this case a mediated value of 80 km/h is considered. Normally, transporters drive around 100 km/h in

Table 4.

Amsterdam	-	1236	1730	73	1760	1370	910	1160	1220	2300	660
Genoa	1236	-	520	1223	1530	2100	780	630	1320	2180	980
Rome	1730	520	-	1490	2040	2610	1290	1140	1830	2690	1490
Rotterdam	73	1223	1490	-	1767	1602	987	1181	1261	2408	704
Madrid	1760	1530	2040	1767	-	3100	2310	2380	2930	650	2315
Oslo	1370	2100	2610	1602	3100	-	1350	1660	1490	3640	1602
Prague	910	780	1290	987	2310	1350	-	1140	1830	2910	987
Vienna	1160	630	1140	1181	2380	1660	1140	-	690	3049	627
Warsaw	1220	1320	1830	1261	2930	1490	1830	690	-	3530	560
Lisbon	2300	2180	2690	2408	650	3640	2910	3049	3530	-	3003
Berlin	660	980	1490	704	2315	1602	987	627	560	3003	-

West-European countries. The 80 km/h rate is based on traffic jams and border operations (customs). Just like the number of drivers, it is also possible to change the commercial speed of the transporters.

A transport cost calculation depends on the distance driven, the number of drivers used during the trip and the total amount of time spent on the trip. Loading and unloading operations are computed based on statistics or, in a more detailed way, by considering a set of sub-models. To do this a crane simulator was used, and the various times were computed as follows:

$$C = \sum_{j=1}^{N} \sum_{i=1}^{n} C_{ij} = \sum_{j=1}^{N} \sum_{i=1}^{n} \left(c_j k_{ij} \oint_j \frac{dl}{v(l)} \right) \tag{13}$$

where:

- C: is the total loading/unloading cost
- C_{ij}: is the cost of loading/unloading the i-th item on the j-th batch
- c_j is the cost factor for the j-th batch and depends on work organization (i.e., number of crane operators, ancillary activities, etc.)
- k_{ij}: is the unit cost for the labor related to the i-th item of the j-th batch.
- $v(l)$: is the speed function of the crane during loading/unloading operations.

Several driver configurations are possible: this makes it possible to compare the differences of using one, two or more drivers with regard to costs and delivery time. The cost calculation

Table 5.

	One driver	Two drivers
Food	30	60
Sleep	40	-
Loan [?]	480	960
Fixed costs transporter [16 ? /h]	384	384
Total costs [?]	934	1404
Costs per hour [?]	38.92	58.50

depends on the number of drivers. According to Italian law, if only one driver is used, he can drive only for eight hours and then must stop for eight hours. Owing to this condition, the average driving time in one day is about twelve hours. When two drivers are used they can drive together for sixteen hours a day. After this, they must stop to rest and the next day they can continue their journey. The results of the cost calculation for one and two drivers are presented in the following table. Hourly-based costs are related to twenty-four hours and include the transporter's fixed costs. In addition to its fixed costs the transporter also has some variable costs. These costs increase depending on distance traveled. The experiments were in Table 5.

The penalties the main contractor can impose on the supplier are based on the difference between the expected delivery time and the effective one for a set of orders. The main contractor does this normally to guarantee a proper delivery time and is based on a mutual agreement among the parties (contract). The expected delivery time is calculated by increasing the time an order is sent by seventy. This value is calculated based on the average time an order is processed by the supplier (50 hours), the average route_time (15 hours) and the five hours that can be lost when objects are sent from one module to another. The difference is that when a loading process is present, the transporter does not need to wait for the crane to return to its initial position when the last piece of cargo is loaded. It can leave when that piece is loaded. When unloading, a transporter does not have to wait for the crane to return to its initial position with the last piece of cargo. The time a transporter takes to drive to the crane from the parking lot is five minutes. In addition, it takes five minutes to leave the crane and to drive to the exit. The time for cargo to be positioned so that the crane can pick it up also takes five minutes. The same can be said for unloading cargo at a terminal. Only for the main contractor this time does not influence the performance of the transporters. When a transporter has traveled for a certain Route_Time, the Route_Time raises the maintenance counter. After a transporter returns from its last destination it looks if the maintenance counter exceeds 4,500 hours. If this limit is reached, maintenance is performed. Maintenance process time is calculated using a triangular distribution, with the following parameters: lower-bound, 4h; mode, 12h; upper bound, 36h. The production process for the ordered goods is simulated in the same way as the maintenance process. The production process takes a certain amount of time based on the triangular distribution (lb: 25h, md: 50h, ub: 75h).

A factorial design was used for the statistical analysis. Data were obtained directly by the implemented federation and the results are presented below. Due to the extremely long run-time for such experiments, a replicated half-fractional factorial 2^{2-1} design was preferred. The experiments performed focus on three factors. The factors, along with their levels, were:

Table 6.

Response: Costs						
ANOVA for Selected Factorial Model						
Analysis of variance table [Partial sum of squares]						
	Sum of			Mean	F	
Source	Squares	DF	Square	Value		Prob > F
Model	8.371E+015	3	2.790E+015	6590.74		< 0.0001	signif
A (transporters)	4.600E+014	1	4.600E+014	1086.48	< 0.0001
B (drivers)	7.601E+015	1	7.601E+015	17952.15	< 0.0001
C (orders)	3.106E+014	1	3.106E+014	733.59	< 0.0001
Pure Error	1.694E+012	4	4.234E+011		
Cor Total			8.373E+015	7	

The Model F-value of 6590.74 implies the model is significant. There is only
a 0.01% chance that a "Model F-Value" this large could occur due to noise.

Values of "Prob > F" less than 0.0500 indicate model terms are significant.
In this case A, B, C are significant model terms.
Values greater than 0.1000 indicate the model terms are not significant.
If there are many insignificant model terms (not counting those required to support hierarchy),
model reduction may improve your model.

Std. Dev.	6.507E+005	R-Squared	0.9998
Mean 3.976E+007		Adj R-Squared	0.9996
C.V. 1.64		Pred R-Squared	0.9992
PRESS 6.774E+012		Adeq Precision	166.947

The "Pred R-Squared" of 0.9992 is in reasonable agreement with the "Adj R-Squared" of 0.9996.

Final Equation in Terms of Coded Factors:

Costs =
+3.976E+007
+7.583E+006   * A
-3.082E+007   * B
-6.231E+006   * C

Final Equation in Terms of Actual Factors:

Costs =
+3.97614E+007
+7.58280E+006   * Number of Transporters
-3.08231E+007   * Number of Drivers
-6.23082E+006   * Number of Orders

- Number of transporters (10 or 15 per terminal);
- Number of drivers (1 or 2 per experiment);
- Number of orders (30 or 60 orders per 24 hours).

The objective functions measured were costs and penalties. The ANOVA results are presented and discussed in Tables 6 and 7.

Table 7

	Sum of		Mean		F		
Response: **Penalties** ANOVA for Selected Factorial Model Analysis of variance table [Partial sum of squares]							
Source	Squares	DF	Square		Value	Prob > F	
Model 4.321E+008		3	1.440E+008		36.43	0.0023	significant
A	2.159E+008	1	2.159E+008		54.59	0.0018	
B	1.565E+008	1	1.565E+008		39.57	0.0033	
C	5.984E+007	1	5.984E+007		15.13	0.0177	
Pure Error	1.582E+007	4	3.954E+006				
Cor Total			4.480E+008	7			

The Model F-value of 36.43 implies the model is significant. There is only
a 0.23% chance that a "Model F-Value" this large could occur due to noise.
Values of "Prob > F" less than 0.0500 indicate model terms are significant.
In this case A, B, C are significant model terms.
Values greater than 0.1000 indicate the model terms are not significant.
If there are many insignificant model terms (not counting those required to support hierarchy), model reduction may improve your model.

Std. Dev.	1988.47	R-Squared	0.9647
Mean	44753.38	Adj R-Squared	0.9382
C.V.	4.44	Pred R-Squared	0.8588
PRESS	6.326E+007	Adeq Precision	13.679

The "Pred R-Squared" of 0.8588 is in reasonable agreement with the "Adj R-Squared" of 0.9382.

Final Equation in Terms of Coded Factors:
Penalties   =
+44753.38
+5194.37   * A
-4422.38   * B
-2734.87   * C

Final Equation in Terms of Actual Factors:
Penalties   =
+44753.37500
+5194.37500   * Number of Transporters
-4422.37500   * Number of Drivers
-2734.87500   * Number of Orders

Figure 32. Total costs (1 driver)

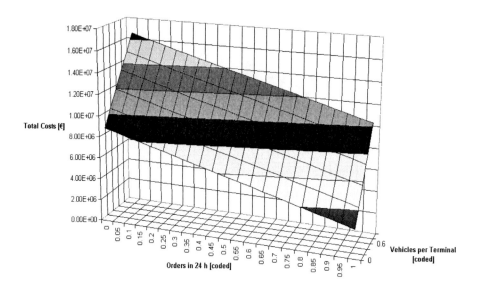

Figure 33. Total costs (2 drivers)

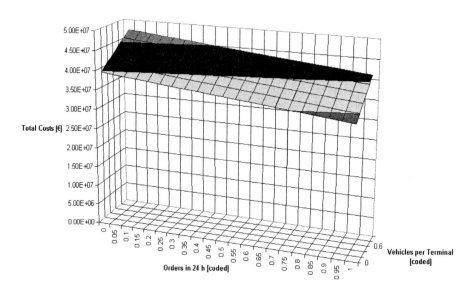

Figure 34. Total penalties paid (2 drivers)

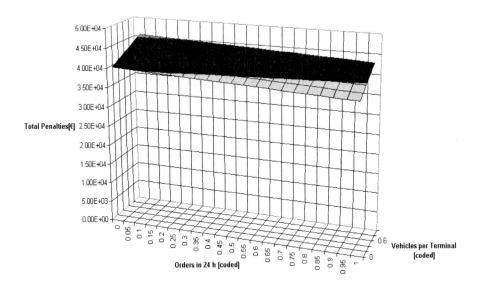

Figure 35. Total penalties paid (1 driver)

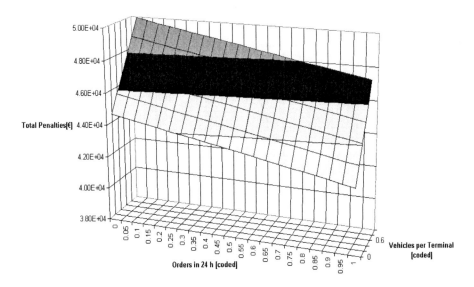

As indicated, two different regression meta-models were found to estimate the two objective functions. In this way managers can obtain important information from their decisions and the improvement process can be increased. Owing to the explicit nature of the regression equation, we can investigate the correct level of resources that must be used to reduce the overall costs and increase the performance levels, depending on the different scenarios.

The regression meta-models are shown in the following figures, respectively, for one driver for delivery (Figure 32) and for two drivers (Figure 33).

As shown in *Figures 34* and *35*, the behavior of the penalties response surface is more stable with two drivers than with just one driver. This is due to the fact that using a backup driver could prevent the delivery from being late. On the other hand, the costs increase when two drivers are used and the solution with just one driver is evidently more stable. As indicated in the previous example, managers can obtain significant advantages from the proposed methodology. That's because they can receive credible answers not only from their managed systems but also in relation to the other entities involved in a supply chain. Furthermore, the regression meta model can explicitly present the complex relationships among the parameters and help decision makers to improve their systems.

Acknowledgment

The author is very grateful to Dr. Blomjous P.E.J.N. and Dr. Van Houten S.P.A. for their contribution in the experimentation campaign and to Dr. Eng. Matteo Maria Cagetti for his extraordinary help in editing and organizing this chapter.

Source Code Availability

The example "Navi" is available online free of charge at http://gestionale.inge.unige.it/php/navi/, as well as with a short HLA online course. HORUS middleware is available from http://www.mindrevolver.com. WILD Project Final Reports are available at http://st.itim.unige.it/wild/index.html.

A full comprehensive course on HLA is available at http://www.ecst.csuchico.edu/~hla/courses.html.

References

Anonymous. (1999). *HLA outreach program*. Retrieved December 18, 1999, from http://www.ecst.csuchico.edu/~hla/courses.html

Beamon, B. (1999). Measuring supply chain performance. *International Journal of Operation and Production Management, 19*(3), 275-292.

Bruzzone, A., & Revetria, R. (2003, July 20-24). VV&A for innovative concept design in new vessels simulation and virtual prototypes. In *Proceedings of SCSC2003 Montreal*, Quebec.

Bruzzone, A.G., & Giribone, P. (1996, April 26). Decision-support systems and simulation for logistics: Moving forward for a distributed, real-time, interactive simulation environment. In *Proceedings of the Annual Simulation Symposium IEEE*, Boston (pp. 4-9).

Bruzzone, A.G., Mosca, R., & Revetria, R. (2001). *Gestione integrata di sistemi produttivi interagenti: Metodi quantitativi avanzati per la quick response.* Genova, Italy: DIP Press.

Bruzzone, A.G., Mosca, R., & Revetria, R. (2002). *Gestione della supply chain mediante federazione di simulatori interagenti: Compendium.* Genoa: DIP Press.

Bruzzone, A.G., Mosca, R., & Revetria R. (2002). Web integrated logistics designer and intelligent control for supply chain management. In *Proceedings of Summer Computer Simulation Conference 2002*, San Diego, CA.

Bruzzone, A.G., Mosca, R., & Revetria, R. (2002, Jan 27-29). Cooperation in maritime training process using virtual reality based and HLA compliant simulation. In *Proceedings of XVIII International Port Conference*, Alexandria, Egypt.

Bruzzone, A.G., Mosca, R., & Revetria, R. (2002). Supply chain management dynamic negotiation using Web integrated logistics designer (WILD II). In *Proceedings of MAS2002*, Bergeggi.

Bruzzone, A.G., & Revetria, R. (2003, July). Advances in supply chain management: An agent based approach for supporting distributed optimization. In *Proceedings of SCSC2003*, Montreal.

Department of Defense. (1998). *Defense Modeling and Simulation Office "HLA Rules"* (version 1.3). DoD Document.

Hamilton, Jr., J.A., Nash, D.A., & Pooch Udo, W. (1999). *Distributed simulation.* New York: CRC Press.

IEEE 1516-2000. (2000). *Standard for modeling and simulation (M&S) high level architecture (HLA) - Framework and rules 2000.* IEEE Standards.

Kindler, E. (2003). Nesting simulation of container terminal operating with its own simulation model. *JORBEL, 40*(3-4), 169-181.

Kuhl, F., Weatherly, R., Dahmann, J., Kuhl, F., & Jones, A. (1998). *HLA: Creating computer simulation systems: An introduction to the high level architecture.* New York: Prentice Hall.

Merkuriev, Y., Merkurieva, G., Bruzzone, A., Revetria, R., & Diglio, G. (2001). Advances In HLA based education for supply chain management. In *Proceedings of HMS2001*, Marseille, France.

Revetria, R., Blomjous, P.E.J.N., Van Houten, S.P.A. (2003, Oct 26-29). An HLA federation for evaluating multi-drop strategies in logistics. In *Proceedings of ESS2003*, Delft, NL.

Rumbaugh, B., Blaha, M., Premerlane, W., Eddy, F., & Lorensen, M. (1991). *Object-oriented modeling and design*. Englewood Cliffs, NJ: Prentice Hall.

Zeigler, B. (1990). *Object-oriented simulation with hierarchical modular models*. Boston: Academic Press.

Chapter III

Object-Oriented Modeling and Simulation of Optical Burst Switched Mesh Networks

Joel J. P. C. Rodrigues, University of Beira Interior, Portugal

Mário M. Freire, University of Beira Interior, Portugal

Abstract

This chapter presents an object-oriented approach for the development of an optical burst switching (OBS) simulator, called OBSim, built in Java. Optical burst switching (OBS) has been proposed to overcome the technical limitations of optical packet switching and optical circuit switching. Due to the high costs of an OBS network infrastructure and a significant number of unanswered questions regarding OBS technology, simulators are a good choice for simulation and estimation of the performance of this kind of networks. OBSim allows the simulation and evaluation of the performance of IP over OBS mesh networks. A detailed description of the design, implementation and validation of this simulation tool is presented.

Introduction

Optical burst switching (OBS) (Baldine et al., 2002; Qiao & Yoo, 1999; Turner, 1999; Wei & McFarland, 2000) is a technical compromise between circuit switching (wavelength routing) and optical packet switching (Murthy & Gurusamy, 2002). OBS has been proposed to overcome the technical limitations of optical packet switching, namely the lack of optical random access memory, and to the problems with synchronization since it does not require optical buffering or packet-level processing, and it is more efficient than circuit switching if the traffic volume does not require a full wavelength channel. In OBS networks, IP datagrams are assembled into very large-sized packets called data bursts. These bursts are transmitted after a burst header packet (set-up message), with a delay of some offset time. Each burst header packet contains routing and scheduling information and is processed at the electronic level before the arrival of the corresponding data burst. OBS has some special characteristics (Xu, 2002), such as: *i*) Granularity: the size of a transmission unit in OBS is between OCS and OPS; *ii*) Data and control separation: control information is transmitted in a separate channel; *iii*) Unidirectional reservation: resources are reserved using a unidirectional messaging system (assuming one-way reservation scheme); *iv*) Burst with variable size: the size of each burst may not be fixed (Qiao & Yoo, 2000); *v*) No buffering of data: once the data is sent, it must reach destination only with the delay inherent to the medium – the propagation delay of the signal in the optical fiber (assuming that no limited buffering is used; e.g., fiber delay lines). More details about OBS and a comparison with other optical switching paradigms may be found in Rodrigues et al. (2005).

OBS technology raises a number of significant questions related with the analysis of the performance of different resource reservation protocols. Several network parameters may be taken into account; namely, the network size, the network topology, the number of channels per link, the number of edge nodes per core, the edge to core node delay, the propagation delay between core nodes, the burst offset length, the processing time of setup messages and the optical switch configuration time. These OBS parameters may have a significant impact on the network performance. Therefore, these questions may be answered with a toll that simulates the behavior of an OBS network, given the inexistence of such networks in the real-world, although there are some testbeds, as reported in Baldine et al. (2005), Baldine et al. (2003), Baldine et al. (2002) and McAdams et al. (1994).

Previous works about optical networks simulators are based on packet traffic (e.g., IP networks), which is significantly different from the bursty traffic in an OBS network, since bursts are transmitted through the OBS network in a transparent way, in the sense that the network does not recognizes neither the end of burst nor its content. Therefore, new tools are needed in order to include the specific features of OBS traffic at the network layer. This chapter presents a proposal of an object-oriented approach for the development of an OBS simulator, called OBSim, built in Java. OBSim supports studies to evaluate the performance of the following resource reservation protocols: JIT (Wei & McFarland, 2000), JumpStart (Baldine et al., 2005; Baldine et al., 2002; Baldine et al., 2003; Zaim et al., 2003), JIT$^+$, (Teng & Rouskas, 2005), JET (Qiao & Yoo, 1999), and Horizon (Turner, 1999).

This simulation tool is designed to implement a model of OBS networks based on objects and it was programmed in an object-oriented programming (OOP) built model, with the following objectives: *i*) to compare the performance of different resource reservation protocols based on the burst loss probability; *ii*) to study the influence of different network profiles on the performance of OBS networks; *iii*) to evaluate the performance of OBS networks for different network topologies defined by the user; *iv*) to compare OBS performance with the performance of other technologies; and *v*) to test new OBS resource reservation protocols.

The remainder of this chapter is organized as follows. The second section analyses methodologies for performance evaluation in OBS networks. The third section presents an overview of the modeling and simulation techniques and the followed burst traffic model. The fourth section describes the design of OBSim simulator, including OBSim design considerations, its architecture, session traffic and scenario generation, and the user interface of OBSim. The fifth section discusses the validation of the simulator results and main conclusions of this chapter are presented in the sixth section.

Methodologies for Performance Evaluation in OBS Networks

Regarding network performance, research and development tools fall into three different categories: analytical tools, *in situ* measurements, and simulators (Bartford & Landweber, 2003; Jeruchim et al., 1992). All of these three techniques can be used during a network project. However, the evaluation of performance parameters may involve prototypes development and may need laboratorial experiments or *in situ* measurements. Thus, this technique is very expensive and slow, and furthermore it is not flexible. By these reasons, that technique is only used at the final stage of a project to test the robustness and the performance of the overall projected system. Analytical techniques are usually useful in the beginning of these kinds of projects, because these techniques are based on mathematic models that produce results that move far away from results obtained by measure because optical networks are very complex. However, they are very efficient from the computational point of view. Simulation techniques allow modeling in detail of the system to be analyzed, and they are more flexible than analytical tools in terms of different network topologies with changing numbers of nodes, links and interconnections. Therefore, optical network simulators are recognized as essential tools to evaluate the performance of these systems, mainly in the phase that there are several possibilities to consider or in phases that one intends to optimize and compare networks' performance. However, developing optical networks requires deep knowledge, not only from programming and simulation environment but also devices and sub-systems (e.g., edge and core nodes, links, reservation protocols…) that comprises the system to simulate. Furthermore, developing simulators requires long-time programming and the correction of errors. On the other hand, it is possible to buy simulators if they are commercially available, but this kind of simulator, with deep detail of the simulation parameters, if available, is usually very expensive.

Concerning analytical models for performance evaluation of OBS networks, there are several proposals in the literature. These techniques are efficient from a computational point of view, but the utilization of these techniques requires very restrictive assumptions that may make impossible a detailed description of the network (Freire, 2000). Teng (2004) and Teng and

Rouskas (2003, 2005) propose analytical models to evaluate the performance of JIT, JET and Horizon, but their proposals only evaluate a single OBS node. Another analytical model to evaluate the performance of OBS networks was proposed by Rosberg et al. (2003). This model only considers JIT and JET protocols and it presents a limitation concerning the study of low-connectivity topologies. These authors evaluate their proposal under an irregular topology, but it is not possible to apply the model to low-connectivity topologies because it is assumed that each blocking event occurs independently from link to link along any route. Another analytical models for JIT, JET, TAW, and differentiated intermediate node initiated signaling were presented in Jue & Vokkarane (2005). These models only evaluate the end-to-end delay of each OBS signaling technique.

Simulators may be developed using high-level language programming, such as PASCAL, C/C++, Java, or FORTRAN, or they may be developed using specific simulation language, such as MATLAB (MathWorks, 1994-2005a), Ptolemy II (Center for Hybrid and Embedded Software Systems (CHESS) University of California at Berkeley, 2005; Lee, 2005) (set of Java packages developed at the University of California at Berkley), SSFNet (Nicol et al., 2003; The SSFNet project) or SIMSCRPT II.5 (CACI Products Company, 2005), among others. Some of the simulation languages possess specific modules for communications. MATLAB is one these cases because it has a specific toolbox for communications. Furthermore, MATLAB is associated with another simulation tool, called Simulink (MathWorks, 1994-2005b). Simulink is a platform for multi-domain simulation and model-based design for dynamic systems. It provides an interactive graphical environment and a customizable set of block libraries, and can be extended for specialized applications. However, running simulations using MATLAB simulators is slower than running simulations using high-level language programming simulators. Even so, this problem could be minimized through the conversion to MEX files and through the use of a C/C++ compiler. MEX-files are dynamically linked subroutines that MATLAB can automatically load and execute. They provide a mechanism by which it is possible to call C and Fortran subroutines from MATLAB as if they were built-in functions. Ptolemy II is a software framework developed as part of the Ptolemy project. It is a Java-based component assembly framework with a graphical user interface called Vergil. Vergil itself is a component assembly defined in Ptolemy II. The Ptolemy project studies modeling, simulation, and design of concurrent, real-time, and embedded systems. The focus is on assembly of concurrent components. Scalable simulation framework network models (SSFNet) are composed by open-source Java models of protocols [IP, TCP, User Datagram Protocol (UDP), Border Gateway Protocol, version 4 (BGP4), Open Shortest Path First (OSPF), and others], network elements (hosts, routers, links, LANs), and various classes for realistic multi-protocol, multi-domain Internet modeling and simulation.

Nowadays, powerful simulators are available commercially, such as Optsim (RSoft Design Group), LinkSIM (RSoft Design Group), and OPNET Modeler (OPNET). However, these solutions did not include modules for OBS networks.

Previous works were focused on simulating traffic for several types of networks and they were primarily designed to simulate TCP/IP traffic. The main example is ns-2 (The network simulator ns-2, 2002), developed on C++ and based on a project started in 1989 (called REAL network Simulator, 2002), which has been widely used for network protocol performance studies (Bhide & Sivalingam, 2000). This network simulator is available for free from ISI (The network simulator ns-2, 2002) and it is documented in The *Network Simulator ns-2:*

Documentation. There have been also other developments in the area of simulation, such as Owns (Bhide & Sivalingam, 2000; Wen et al., 2001), being this simulator an extension to the ns-2. OWns is an optical wavelength division multiplexing network simulator and it did not simulate OBS networks. In Institute of Communication Networks and Computer Engineering (2004) it was found the IND Simulation Library, which is an object-oriented class library for event-driven simulation implemented in C++. These classes have been designed to support of communication networks performance evaluation. While developing OBSim, one had access to the simulator developed by Teng & Rouskas from North Carolina State University, U.S., and to the OBS-ns simulator released by DAWN Networking Research Lab from University of Maryland (DAWN Networking Research Lab, 2004). Both simulators were developed under C++ programming language. The first was not published and it does not have documentation support. However, there are published results obtained by this simulator in Teng (2004) and Teng & Rouskas (2003, 2005) and these results were used to validate the results of OBSim. On the other hand, OBS-ns presents several bugs and limitations to simulate OBS networks. To solve this problem, more recently, Optical Internet Research Center (OIRC) released the optical burst switching simulator (OOS) as a new version of OBS-ns by fixing its errors (Optical Internet Research Center, 2005). Another simulator, called NTCUns, belongs to SimReal Inc. (SimReal), a virtual company operated by the Network and System Laboratory (NSL), Department of Computer Science and Information Engineering, National Chiao Tung University (NCTU), Taiwan. NTCUns is a network simulator capable of simulating wired and wireless IP networks, optical networks, and GPRS cellular networks. Recently, this simulator incorporated several modules to construct the protocol stacks used in OBS networks. Therefore, OBSim is a tool that also gathers contributions from all these previous works in the area of network simulation.

Simulation techniques are a good resource when networks under study are: *i*) very complex to analyze using analytical tools or *ii*) very expensive to investigate by measure or by prototype developing, or by both reasons. In this chapter, it is described the design and development of a new simulator.

OBS Networking Modeling

Regarding network simulation, research and development tools fall into three different categories: analytical tools, *in situ* measurements, and simulators (Bartford & Landweber, 2003). OBSim is an *i*) event driven, *ii*) stochastic, and *iii*) symbolic simulator. Event driven simulators are a class of models in which data flows to the pace of events of some type; other simulators may be activity driven or time driven, namely when the simulator responds to some kind of user interaction (internal or external, user initiated or not) and the last, when the software runs at the tick of a clock (Perros, 2003). In OBSim, the events that run on the simulator are messages. These may be sent by the edge nodes, or generated at a core node, as defined by the resource reservation protocols above presented.

Stochastic simulators, opposed to deterministic simulators, rely on random entities (usually random variables of numerical value) to simulate the randomness of real-life events. OBSim used the Java class Random, who generates pseudo-random values (of several types), using a congruential algorithm. Pseudo random variables must pass two tests that certify, first, the homogeneity of its distribution, and second, the independence of the generated values (Perros, 2003). Java class Random satisfies these conditions (Sun Microsystems Inc., 2003).

Symbolic simulators use some type of symbols to copy the behavior of real elements. In OBSim, these symbols are Java classes, which are instantiated as needed by the software, according to the input data provided initially by the user.

OBS Mesh Network Under Study

In this chapter, we use the example of an OBS core network (*Figure 1*), to support and present the discussed design and entities of OBSim simulator. Therefore, this sub-section shows the network topology and an example of its virtualization when an edge node sends a burst to another edge node. The network considered has six core nodes (the OXC and its corresponding signaling engine – or switch control unit –, numbered from 1 to 6) and nine links (data and signaling channels are shown separately).

Figure 2 plots a scheme and a unified modeling language (UML) class diagram that illustrates how a burst being sent from an edge node (edge node "a" connected to core node "3") to another edge node (edge node "b" connected to core node "6") deploys and uses a set of class instances in the simulator.

Burst Traffic Model

Concerning OBS networks, it is assumed that each OBS node requires (Teng & Rouskas, 2003, 2005): *i*) an amount of time, T_{OXC}, to configure the switch fabric of the OXC in order to set up a connection from an input port to an output port, and requires *ii*) an amount of time, $T_{Setup}(X)$ to process the setup message for the resource reservation protocol X, where X can be JIT, JumpStart, JIT$^+$, JET, and Horizon. It is also considered the offset value of

Figure 1. OBS network topology with 6 nodes and 9 links

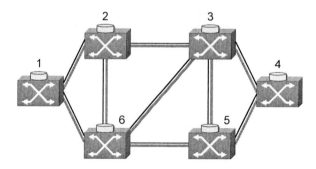

Figure 2. Classes instantiated when the edge node 3a sends a burst to the edge node 6b

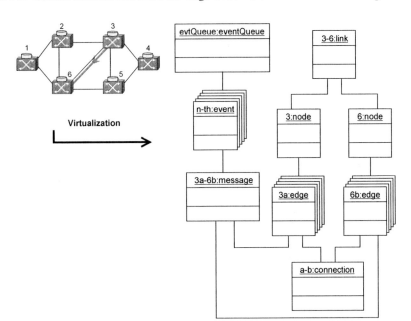

a burst under protocol X, $T_{offset}(X)$, which depends, among other factors, on the resource reservation protocol, the number of nodes the burst has already traversed, and if the offset value is used for service differentiation.

It is assumed that number of hops in the path of a burst for a given network is uniformly distributed from the ingress core node to the egress core node, and the burst offset time ($T_{offset}(X)$) is calculated by (1).

$$T_{offset}^{(min.)}(X) = k.T_{Setup}(X) + T_{OXC} \tag{1}$$

This study follows the simulation conditions presented in Teng & Rouskas (2003, 2005) in terms of traffic scenarios. In obtaining the simulation results, to estimate the burst loss probability, it was estimated 95% confidence intervals using the method of batch means (Perros, 2003; Schriber & Andrews, 1979). The number of batches is 30 (the minimum value to obtain the confidence interval) and each batch run lasting until at least 2,560,000 bursts are transmitted, assuming that each edge transmits, at least, 200 bursts and each batch contains 10 observations. The value of 2,560,000 bursts transmitted, at least, is obtained for a network with N=20 nodes, that is, 200 bursts transmitted x 64 edge nodes x 20 nodes x 10 observations = 2,560,000 bursts transmitted. It was found that the confidence intervals are very narrow. Therefore, in this chapter the confidence intervals are not shown in the figures with the objective to increase readability.

Concerning the number of data channels available per link, it is assumed that the number of channels connecting two nodes is $F+1$, where F is the number of data channels available in that link and the other channel represents the signaling channel. In this study F may have the value of 16, 32, 64 or 128 data channels per link ($F=2^n$, with $4 \leq n \leq 7$). Every channel is bi-directional.

In terms of setup message arrival process (and in consequence data bursts), it is assumed a Poisson point process with rate λ (where $1/\lambda$ is the mean duration of the burst inter-arrival time), such as in Chaskar, Verma, & Ravikanth (2000), Dolzer et al. (2001), Dolzer & Gauger (2001), Teng (2004), Teng & Rouskas (2003, 2005), Turner (1999), Verma, Chaskar, & Ravikanth (2000), Yoo, Qiao, & Dixit (2000). The Poisson arrival process may be justified considering that in current WDM switches, the variance and the mean of the switch processing and configuration times are of the same order of magnitude (Wei & McFarland, 2000). As in Battestilli & Perros (2004), Teng (2004), Teng & Rouskas (2005), Vokkarane & Jue (2003a, 2003b), and Xu, Perros, & Rouskas (2003a, 2003b), it is assumed that burst length, whether short or long, is limited (Qiao & Yoo, 2000) and follow an exponential distribution with an average burst length of $1/\mu$. Therefore, in OBSim, taking into account the average of burst length distribution ($1/\mu$) and the setup message arrival rate λ, the burst generation ratio is represented by λ/μ. It is also assumed that bursts are sent uniformly to every core node in the network, with the exception that a core node cannot send messages to itself and one core node may generate, at the most, one message by time-slot or time period. Edge nodes are responsible for the burst generation process, that is, neither the ingress core node nor any other core node processes the burst.

Another important issue with this study is related with the number of wavelength converters in each node. It is assumed that each OBS node supports full-optical wavelength conversion. In terms of number of edge nodes connected to each core node it is assumed that, for simulation effects, they are uniformly distributed and each core connects 64 edge nodes. Between core nodes it is assumed that the geographical size is large so the typical link delays are on the order of 10ms (Boncelet & Mills, 1992).

Other parameters included in this simulation tool are the propagation delay between edge and core nodes and between core nodes. The former refers to the propagation time delay when the message (and bursts) is in transit between the edge node and the ingress/egress core node. This parameter may be introduced in the input user interface. The latter is the propagation delay between core nodes. This parameter is defined in the text file with the definition of a network topology.

In the simulation, to select a free channel for an incoming burst (with equal probability), for JIT, JumpStart, and JIT$^+$, it is used the random wavelength assignment policy (Zhu, Rouskas, & Perros, 2000); whereas, for JET and Horizon, it is used the latest available unused channel (LAUC) algorithm (Xiong, Vandenhoute, & Cankaya, 2000).

Design of OBS Simulator

To study the problem and the characteristics of burst traffic in OBS networks, it is needed to evaluate the performance of different resource reservation protocols. This is achieved by studying its performance and behavior under different traffic conditions and network topologies. This tool simulates the behavior of a custom OBS network defined by the user. The simulator, called OBSim, allows to assess and compare the performance of one-way resource reservation protocols and load profiles to a given network topology.

Design Considerations

As it was needed a simulator independent of existing network data encapsulation protocols, OBSim was built from scratch. Java was the programming language chosen to build OB-Sim for several reasons, namely: *i*) the quality and ease of use Java available programming tools; *ii*) the robustness of Java in object and memory handling; and *iii*) the wide platform portability of the code.

Several assumptions were made while building OBSim. These assumptions occur in respect to the definition of the resource reservation protocols. Concerning network modeling, these are the following: *i*) all the nodes work in an independent and similar way; *ii*) all time scales are normalized in time-slots; *iii*) a path is used by a burst or by a setup message independently of the state of the network; *iv*) when a signal (burst or setup message) arrives to a node, it follows a predefined path calculated previously by Dijkstra's algorithm (Goldberg & Tarjan, 1996); and *v*) between two consecutive nodes, the wavelength (data channel) used is chosen by the algorithm defined by the user in the network topology definition file (*random* or *first-free*) (Li & Somani, 2000).

Figure 3. Nine of the 15 defined routes for topology presented in Figure 1

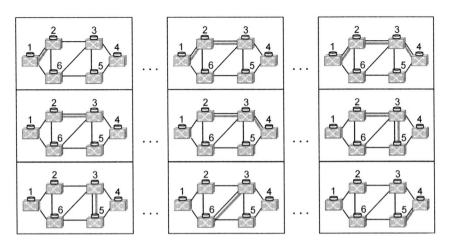

OBSim maintains an event queue that accepts and removes events, and forwards each event to its corresponding object (that with other objects compose the virtual network) so it can be processed.

Abstraction in OBSim is achieved by the behavior of the objects of the model. The OBS network wanted to simulate a set of defined real-world objects (eventually, real-world objects yet to be real-objects), each having its function and behavior, each interacting with the remain objects of the network according to a defined set of rules (e.g., the algorithms of the signaling engines). As an example, the network topology defined initially by the user is processed according to the Dijkstra's algorithm (Dijkstra, 1959), and for each pair of nodes, is defined a route. *Figure 3* shows the routes found by the algorithm.

Architecture of the OBSim Simulator

To simulate an OBS network several objects (Java classes) were defined, each having methods that may be activated by other objects. These mechanisms, common in object oriented programming (OOP) languages such as Java and C++, are, namely, inheritance, polymorphism, encapsulation, and also, dynamic instantiation and dynamic memory management. They allow one to create a working model that behaves like the real OBS network would. Time flow is simulated through a queue of events, ruled by a clock, and is introduced forward.

Figure 4 shows how a potentially existing OBS network is virtualized in Java classes. As shown, each link is composed of two nodes, and a node may belong to more than one link (e.g., node '3').

Figure 4. UML object diagram modeling an OBS network (presented in Figure 3)

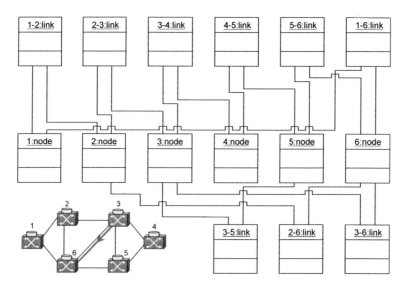

The main method is defined in *OBS class* and this calls several other objects, namely, the following:

- *NetworkBuilder*, that builds the network from the topology file; this process is the virtualization that builds the class Network;
- *Network*, composed of by Links and Nodes;
- *RouteBuilder*, that builds a route for any two nodes, and stores these routes in the *PathTable*; this class implements the Dijkstra's algorithm (Alavi et al., 1985; Goldberg & Tarjan, 1996);
- *PathTable* manages the paths defined by *RouteBuilder*;
- *Global* stores and manages the global constants and variables of the simulator;
- *Simulator*, which starts the burst request actions from each edge node at every core node;
- *EventQueue* manages all the *Events* according to the *Clock* class;
- And the classes *ErrorMsg* and *DebugMsg*, which manage the output of the debugging and error messages.

It may also be seen that the Node class is a generalization for classes *NodeJIT*, *NodeJET*, *NodeHorizon*, *NodeJumpStart*, and *NodeJITP*, which in turn, model the nodes of these resource reservation protocols. With this approach, adding a new protocol can be made by the definition of a new class and implementing it with its own set of specific algorithms.

NetworkObject is the generalization class for *Node*, *Edge* and *Event*. These are the main actors of an OBS network, and their behavior and interaction, as seen before, creates instances of classes like *Link*, *Connection*, *Event* and *Message*. In a non-parallel approach, classes *NetworkBuilder*, *Network*, *Simulator*, *EventQueue* (and *Clock*), *RouteBuilder* and *PathTable*, along with classes *ErrorMsg* and *DebugMsg* are instantiated only once. Classes *Node*, *Link* and *User* are instantiated as many times as defined in the network topology file. Other classes are instantiated as many times as needed, either by the stochastic workflow of the simulator or by running algorithms like Dijkstra.

Another important issue that may be described at this point of the explanation of OBSim simulator is the program execution. OBSim follows a linear functional execution scheme and performs five main tasks, which are the following: *i*) validate initial parameters defined by the user; *ii*) read network topology from network file; *iii*) build the network model for that network topology; *iv*) simulate traffic; and *v*) show simulation results.

Validation of initial parameters and the network topology allow the definition of the abstract model of a given topology to simulate. Each node of this network, according to their traffic load parameters, is instantiated to generate messages that are deposited in a queue. After reading network topology from the correspondent file, the abstract model of that network is created. To create this network model are instantiated the corresponding classes available on the simulator and they are initialized the correspondent data structures of each one. Following, `Routing` class executes Dijkstra algorithm for each node of the network. Every shortest path is saved in `Routing` class into a map. `Simulator` class initializes each core node and, for each one, starts their edge nodes (only for each edge that transmits bursts). To start each edge node, coded in `Edge` class, it is added a burst request in the event calendar (`EventCalendar` class). After every edge node that has permission to send bursts put their request of a burst, OBSim will run all events that are pending in the event calendar. This event running, that may be executed either in an instance of `Node` class or in an instance of `Edge` class, generates more events that will be added to the queue. Simulation finishes when there are no more events to perform in the queue. At this point, the method that calculates ratios in function of values stocked into several classes is called. Briefly, these values may be the number of bursts sent and lost in each core node and in each edge node, and the number of bursts sent and lost in each hop either per core node or per edge node (`Node` and `Edge` classes). Using these values it is calculated the burst loss probability per hop that is shown for the user.

Session Traffic and Scenario Generations

Traffic generation is an important issue in the model. As OBSim is an event driven simulator, initially there is the necessity to simulate the need to transmit bursts between nodes. As seen before, in a simulation, it is assumed that the bursts are sent evenly to every node in the network. Since every burst must be preceded by a setup message, and since edge nodes connected to core nodes send bursts at a random time, it is considered that time between these messages follow an exponential distribution, simple or with an offset (Paxson & Floyd, 1995; Schäefer, 2003). The traffic is then simulated when the OBSim starts to process the event queue, which was, at the start of the program, loaded with requests (messages) from the edge nodes. These requests, when processed, normally generate more messages that are added to the queue. Each time a message is added to the queue, the simulator timer generates a time interval according with the distribution defined by the user, and this time is added to the simulator clock, defining the time the event will be scheduled to happen.

The network scenario is read from a text file that defines the number of nodes (core nodes), the number of connections, the allowed attributes for each node, and the definition of the existing links in the network.

The text file that defines the network topology to be studied for the OBS network considered in this chapter (*Figure 1*) is illustrated in the *Figure 5*. Therefore, the next paragraphs present, in detail, the content of this file.

The first line of the text file defines the number of core nodes (6), the channel schedule algorithm used to assign the burst to a data channel on the outgoing link, and the number of links (9). For the channel schedule algorithm, one may use *0* or *1*, where *0* represents the use of random allocation channel algorithm and *1* represents the use of first free channel algorithm. Following, there are inserted data to define the parameters of each node:

- First two numbers represent the coordinates (*x* and *y*) of each core node in terms of two-dimensional localization;
- The third number represents that:

  0 – edge nodes of this core node cannot generate burts;

Figure 5. Text file with the definition of a network topology

```
609
01264
12264
22264
31264
20264
10264
1 2 10
1 6 10
2 3 10
2 6 10
3 4 10
3 5 10
3 6 10
4 5 10
5 6 10
```

1 – half of edge nodes of this core node can generate bursts;

2 – all edge nodes of this core node can generate bursts.

- The last number indicates the number of edge nodes per core node.

The following lines have three values each (*1 2 10*, for the example considered). The first two values identify the number of core nodes interconnected and the last number indicates the propagation delay of this link. This value may be defined in function of the link length, that is, in function of the real distance between the two nodes. In this study, the assumed geographical size is large so that the typical link delay is on the order of 10ms (Boncelet & Mills, 1992).

The creation of the network abstraction—the network model that supports the simulation—is accomplished by the classes defined in the program. These classes are responsible for the virtualization of the model.

Input User Interface of OBSim

The input user interface of the simulator allows the definition of several simulation attributes. The network is fully defined with these attributes and the attributes described for each node and each link in the above-mentioned text file.

The interface shown in *Figure 6* shows the parameters used to configure the model of the OBS network. These parameters are the following:

1.   **Resource reservation protocol:** This field registers the definition of the OBS resource reservation protocol to be used. The allowed values are JIT, JumpStart, JIT$^+$, JET, and Horizon.

2.   **Generation distribution function:** This field defines the statistic model to generate the time interval between events in the simulator. The allowed functions are the exponential and two state exponential.

3.   **Burst generation ratio:** This field admits a real value between 0 and 1, and it represents the burst generation ratio per node (λ/μ).

4.   **Available data channels per link:** This field allows numerical values between 1 and 2147483647 [Biggest allowed value for an int (integer) in Java] and represents the available number of data channels per link.

5.   **Setup message process time:** This field admits a decimal value greater than 0, and represents the amount of time that the OXC spends to process the setup message, defined as $T_{Setup}(X)$.

Figure 6. Input user interface of OBSim

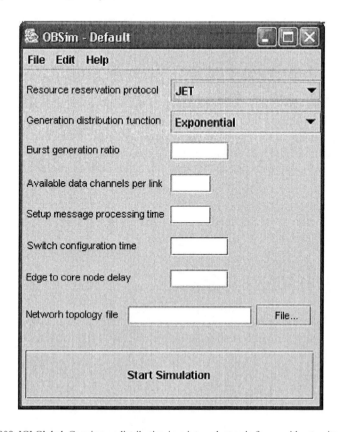

6. **Switch configuration time:** This field admits a decimal number greater than 0 and means the time the OXC takes to configure the optical switch matrix, after the setup message has been received and interpreted, defined as T_{OXC}.

7. **Edge to core node delay:** This is a numerical field greater than 0 and stores the time the message (and bursts) are in transit between the edge node and the ingress/egress core node.

8. **Network topology file:** This field stores the name of the file where the network is defined. It has an associated button (*file...*) that allows browsing the directory structure showing the available files.

Simulator Validation

Validation is a key issue to entrust the use of the results given by any simulator. Perros (2003) defines the validation of the model as the verification of the following five steps:

1. Check the pseudo-random numbers generator
2. Check the stochastic variable generator
3. Check the logic of the simulation program
4. Relationship validity
5. Output validity

In OBSim, the accuracy of the pseudo-random number generator is guaranteed by the Java language definition standards, and confirmed through the qui-squared test, and the independence test performed on the Java class random (Perros, 2003; Sun Microsystems Inc., 2003).

The stochastic variable generators have been separately validated by Ma and Ji (2001), Schäefer (2003), and Willinger and Sherman (1997).

The logic of the simulator and the validity of the relationships are inherent to the design of the resource reservation protocols, and to the Java programming environment, above mentioned.

The output validity has been achieved through comparison with the results of Teng and Rouskas (2005). For this purpose, a sample simulation was run considering a single OBS node, in isolation, for JIT, JET, and Horizon resource reservation protocols. It is assumed that (Teng & Rouskas, 2005): $T_{Setup}(\text{JIT})=12.5\mu s$, $T_{Setup}(\text{JET})=50\mu s$, $T_{Setup}(\text{Horizon})=25\mu s$, $T_{OXC}=10ms$, the average of burst length $1/\mu$ was set to 50ms (equal to $5T_{OXC}$), and burst arrival rate λ of *setup messages*, is such that $\lambda/\mu=32$, assuming 64 edge nodes per core node.

Figure 7 shows the burst loss probability as a function of the number of data channels per link for the OBS network presented above, given by OBSim and compared to the results presented in Teng and Rouskas (2005). As may be seen in this figure, the results obtained

Figure 7. Burst loss probability, as a function of number of data channels per link (F) (Rodrigues et al., 2004), in a single OBS node for JIT, JET and Horizon resource reservation protocols given by OBSim compared to results published in Teng and Rouskas (2005)

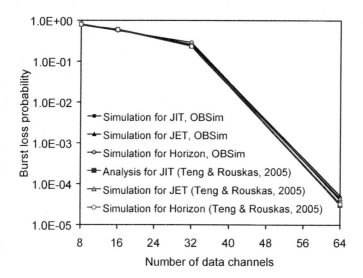

by OBSim are in a close range of those published by Teng and Rouskas (2005). The small variation perceived is expectable because of the stochastic nature of the events that are modeled.

Conclusion

In this chapter was presented the objectives, design, implementation and validation of a simulator for optical burst switched (OBS) networks, named OBSim. This simulator implements a model of OBS networks based on objects, which allows the estimation of the performance of a custom designed OBS network. The results of the simulator have been validated, and thus the simulator may be used as a tool to estimate the performance the OBS networks. Therefore, OBSim is used to as a simulation tool to support the performance studies of OBS networks.

References

Alavi, Y., Chartrand, G., Lesniak, L., Lick, D.R., & Wall, C.E. (1985). *Graph theory with applications to algorithms and computer science.* New York: John Wiley & Sons.

Baldine, I. et al. (2005). JumpStart deployments in ultra high performance optical networking testbeds. *IEEE Communications Magazine, 43*(11), S18-S25.

Baldine, I., Cassada, M., Bragg, A., Karmous-Edwards, G., & Stevenson, D. (2003, December 1-5). *Just-in-Time optical burst switching implementation in the ATDnet all-optical networking testbed.* Paper presented at the IEEE GLOBECOM 2003, San Francisco.

Baldine, I., Rouskas, G., Perros, H., & Stevenson, D. (2002). *JumpStart: A just-in-time signaling architecture for WDM burst-switched networks. IEEE Communications Magazine, 40*(2), 82-89.

Baldine, I., Rouskas, G.N., Perros, H.G., & Stevenson, D. (2003). Signaling support for multicast and QoS within the JumpStart WDM burst switching architecture. *Optical Networks Magazine, 4*(6), 68-80.

Bartford, P., & Landweber, L. (2003). *Bench-style network research in an Internet instance laboratory. ACM SIGCOMM Computer Communications Review, 33*(3), 21.

Battestilli, T., & Perros, H. (2004, October 25). *End-to-end burst loss probabilities in an OBS network with simultaneous link possession.* Paper presented at The Third International Workshop on Optical Burst Switching (WOBS 2004), San José, CA.

Bhide, N.M., & Sivalingam, K.M. (2000, January). *Design of OWns: Optical wavelength division multiplexing (WDM) network simulator.* Paper presented at the First SPIE Optical Networking Workshop, Dallas, TX.

Boncelet, C.G., & Mills, D.L. (1992, August 17-20). *A labeling algorithm for just-in-time scheduling in TDMA networks.* Paper presented at the Conference on Communications Architecture & Protocols - ACM SIGCOMM, Baltimore.

CACI Products Company. (2005). *About SIMSCRIPT II.5.* Retrieved July 26, 2005, from http://www.simprocess.com/products/simscript.cfm

Center for Hybrid and Embedded Software Systems (CHESS) University of California at Berkeley. (2005). *Ptolemy Project.* Retrieved July 28, 2005, from http://ptolemy.eecs.berkeley.edu/

Chaskar, H.M., Verma, S., & Ravikanth, R. (2000, January 5). *A framework to support IP over WDM using optical burst switching.* Paper presented at the IEEE/ACM/SPIE Optical Networks Workshop.

DAWN Networking Research Lab. (2004). *DAWN Research Lab.* Retrieved January 15, 2004, from http://dawn.cs.umbc.edu/

Dijkstra, E.W. (1959). A note on two problems in connexion with graphs. *Numerische Mathematik, 1,* 269-271.

Dolzer, K., Gauger, C., Späth, J., & Bodamer, S. (2001). Evaluation of reservation mechanisms for optical burst switching. *AEÜ International Journal of Electronics and Communications, 55*(1).

Dolzer, K., & Gauger, C.M. (2001, September 24-28). *On burst assembly in optical burst switching networks: A performance evaluation of just-enough-time.* Paper presented at the 17th International Teletraffic Congress, Salvador, Brazil.

Goldberg, A.V., & Tarjan, R.E. (1996). *Expected performance of Dijkstra's shortest path algorithm*. Technical Report TR-96-062. NEC Research.

Institute of Communication Networks and Computer Engineering. (2004). *IND Simulation Library*. Retrieved February 17, 2004, 2004, from http://www.ikr.uni-stuttgart.de/INDSimLib/.

Jeruchim, M.C., Balaban, P., & Shanmugan, K.S. (1992). *Simulation of communication systems*. New York: Plenum Press.

Jue, J.P., & Vokkarane, V.M. (2005). *Optical burst switched networks*. New York: Springer.

Lee, E.A. (2005). Absolutely positively on time: What would it take? *IEEE Computer, 38*(7), 85-87.

Li, L., & Somani, A.K. (2000). Dynamic wavelength routing techniques and their performance analyses. In S. Subramaniam (Ed.), *Optical WDM networks - Principles and practice*. Kluwer Academic Publishers.

Ma, S., & Ji, C. (2001). Modeling heterogeneous network traffic in wavelet domain. *IEEE/ACM Transactions on Networking, 9*(5), 634-649.

MathWorks. (1994-2005a). *MATLAB*. Retrieved July 23, 2005, from http://www.mathworks.com/products/matlab/

MathWorks. (1994-2005b). *Simulink*. Retrieved July 23, 2005, from http://www.mathworks.com/products/simulink/

McAdams, L., Richer, I., & Zabele, S. (1994). *TBONE: Testbed for all-optical networking*. Paper presented at the IEEE/LEOS Summer Topical Meetings.

Murthy, C.S.R., & Gurusamy, M. (2002). *WDM optical networks, concepts, design and algorithms*. NJ: Prentice Hall PTR.

The network simulator ns-2. (2002). Retrieved November 10, 2002, from http://www.isi.edu/nsnam/ns/

The network simulator ns-2: Documentation. (n.d.). Retrieved November 10, 2002, from http://www.isi.edu/nsnam/ns/ns-documentation.html

Nicol, D., Liu, J., Liljenstam, M., & Yan, G. (2003, December 7-10). *Simulation of large-scale networks using ssf*. Paper presented at the 2003 Winter Simulation Conference, New Orleans, LA, U.S.

OPNET. (n.d.). *OPNET Modeler*. Retrieved June 2004, from http://www.opnet.com/products/modeler/home.html

Optical Internet Research Center. (2005). *OIRC optical burst switching Simulator*. Retrieved July 25, 2005, from http://wine.icu.ac.kr/~obsns/

Paxson, V., & Floyd, S. (1995). Wide area traffic: The failure of Poisson modeling. *IEEE Transactions on Networking, 3*(3), 226-244.

Perros, H. (2003). *Computer simulation techniques: The definitive introduction!* Retrieved December 18, 2003, from http://www.csc.ncsu.edu/faculty/perros/

Qiao, C., & Yoo, M. (1999). Optical burst switching (OBS): A new paradigm for an optical

Internet. *Journal of High Speed Networks, 8*(1), 69-84.

Qiao, C., & Yoo, M. (2000). A taxonomy of switching techniques. In S. Subramaniam (Ed.), *Optical WDM Networks: Principles and practice* (pp. 103-125). Kluwer Academic Publishers.

REAL Network Simulator. (2002). Retrieved November 14, 2002, from http://www.cs.cornell.edu/home/skeshav/real/overview.html

Rodrigues, J.J.P.C., Garcia, N.M., Freire, M.M., & Lorenz, P. (2004, November 29 - December 3). *Object-oriented modeling and simulation of optical burst switching networks.* Paper presented at the IEEE Global Telecommunications Conference (GLOBECOM2004) - Conference Workshops, 10th IEEE Workshop on Computer-Aided Modeling, Analysis and Design of Communication Links and Networks (CAMAD2004), Dallas, TX, U.S.

Rodrigues, J.J.P.C., Freire, M.M., Monteiro, P.P., & Lorenz, P. (2005). Optical burst switching: A new switching paradigm for high-speed Internet. In M. Pagani (Ed.), *Encyclopedia of multimedia technology and networking* (Vol. II). Hershey, PA: Idea Group Reference.

Rosberg, Z., Vu, H.L., Zukerman, M., & White, J. (2003). Performance analyses of optical burst-switching networks. *IEEE Journal on Selected Areas in Communications, 21*(7), 1187-1197.

RSoft Design Group I. (n.d.). *LinkSim.* Retrieved May 6, 2003, from http://www.rsoftinc.com/linksim.htm

RSoft Design Group, I. (2002). *System simulation.* Retrieved May 8, 2003, from http://www.rsoftdesign.com/products/system_simulation/

Schäefer, A. (2003). *Self-similar network traffic.* Retrieved February 3, 2004, from http://goethe.ira.uka.de/~andreas/Research/Fractal_Traffic/Fractal_Traffic.html

Schriber, T.J., & Andrews, R.W. (1979). *Interactive analysis of simulation output by the method of batch means.* Paper Presented at the 11th Conference on Winter Simulation, San Diego, CA, U.S.

SimReal. (n.d.). *NCTUns Simulator.* Retrieved June 21, 2005, from http://nsl10.csie.nctu.edu.tw/

The SSFNet project. (n.d.). Retrieved July 2004, from http://www.ssfnet.org

Sun Microsystems Inc. (2003). *Class Math.* Retrieved February 23, 2004, from http://java.sun.com/j2se/1.4.2/docs/api/java/lang/Math.html/

Teng, J. (2004). *A study of optical burst switched networks with the Jumpstart just in time signaling protocol.* Raleigh: University of North Carolina State University.

Teng, J., & Rouskas, G.N. (2003, October 13-18). *A comparison of the JIT, JET, and Horizon wavelength reservation schemes on a single OBS node.* Paper presented at the First International Workshop on Optical Burst Switching (WOBS), Dallas, Texas, U.S.

Teng, J., & Rouskas, G.N. (2005). A detailed analysis and performance comparison of wavelength reservation schemes for optical burst switched networks. *Photonic Network Communications, 9*(3), 311-335.

Turner, J.S. (1999). Terabit burst switching. *Journal of High Speed Networks, 8*(1), 3-16.

Verma, S., Chaskar, H., & Ravikanth, R. (2000). Optical burst switching: A viable solution for terabit IP backbone. *IEEE Network, 14*(6), 48-53.

Vokkarane, V., & Jue, J. (2003a). Burst segmentation: An approach for reducing packet loss in optical burst-switched networks. *Optical Networks Magazine, 4*(6), 81-89.

Vokkarane, V., & Jue, J. (2003b). Prioritized burst segmentation and composite burst assembly techniques for QoS support in optical burst-switched networks. *IEEE Journal on Selected Areas in Communications, 21*(7), 1198-1209.

Wei, J.Y., & McFarland, R.I. (2000). Just-in-time signaling for WDM optical burst switching networks. *Journal of Lightwave Technology, 18*(12), 2019-2037.

Wen, B., Bhide, N.M., Shenai, R.K., & Sivalingam, K.M. (2001). Optical wavelength division multiplexing (WDM) network simulator (OWns): Architecture and performance studies. *SPIE Optical Networks Magazine, Special Issue on "Simulation, CAD, and Measurement of Optical Networks*, pp. 16-26.

Willinger, W.T., & Sherman, R.M.S. (1997). Self-similarity through high-variability: Statistical analysis of ethernet LAN traffic at the source level. *IEEE/ACM Transactions on Networking*, (5), 71-86.

Xiong, Y., Vandenhoute, M., & Cankaya, H.C. (2000). Control architecture in optical burst-switched WDM networks. *IEEE Journal on Selected Areas in Communications, 18*(10), 1838-1851.

Xu, L. (2002). *Performance analysis of optical burst switched networks*. Unpublished PhD Thesis, North Carolina State University, Raleigh.

Xu, L., Perros, H.G., & Rouskas, G.N. (2003a, August 31-September 5). *Performance analysis of an edge optical burst switching node with a large number of wavelengths*. Paper presented at the 18th International Teletraffic Congress (ITC-18), Berlin, Germany.

Xu, L., Perros, H.G., & Rouskas, G.N. (2003b, March 30 - April 3). *A queueing network model of an edge optical burst switching node*. Paper presented at the IEEE INFO-COM, San Francisco.

Yoo, M., Qiao, C., & Dixit, S. (2000). QoS performance of optical burst switching in IP-Over-WDM networks. *IEEE Journal on Selected Areas in Communications, 18*(10), 2062-2071.

Zaim, A.H., Baldine, I., Cassada, M., Rouskas, G.N., Perros, H.G., & Stevenson, D. (2003). The JumpStart just-in-time signaling protocol: A formal description using EFSM. *Optical Engineering, 42*(2), 568-585.

Zhu, Y., Rouskas, G.N., & Perros, H.G. (2000). A comparison of allocation policies in wavelength routing networks. *Photonic Networks Communication Journal, 2*(3), 265-293.

Chapter IV

Using Natural Language Modeling for Business Simulation

Peter Bollen, University of Maastricht, The Netherlands

Abstract

Natural language modeling (NLM) is a conceptual modeling language that can be used for requirements determination for business application systems. In this chapter we will show how the NLM methodology for requirements determination can be extended to serve as a blueprint for business (or management) simulation by providing an initial model for creating a business simulation. We will do this by defining the content of the communication documents for runtime management and we will subsequently show how this meta-UoD can be incorporated into an application UoD. This allows us to capitalize on conceptual models in a business that have been created for requirements determination by extending them with the conceptual model of runtime management. Subsequently, we will incorporate the simulation requirements into the latter UoD and we will give some guidelines on how the conceptual models for the "real life" runtime application can serve as a starting point for the conceptual sub-models for the simulation UoD.

Introduction

In the simulation literature, there's a growing number of articles that discuss the interdependence between a conceptual model (or operational model) of a subject area and a simulation model (de Swaan Arons & Boer, 2001; de Vreede, Verbraeck, & van Eijck, 2003; Floss, 1997; Gregoriades & Karakostas, 2004; Julka, Lendermann, Chong, & Liow, 2004). In parallel other authors indicate the interdependence between the simulation data sources and the operational business data sources that are contained in the corporate business systems (Perera & Liyanage, 2000; Robertson & Perera, 2002). These authors commonly identify the need for an integrated approach towards (the design of) business simulations and the (design of) operational information systems. Another stream of research is concerned with the definition of a methodology for business (process) simulation (Greasly, 2006; Hlupic & Robinson, 1998; Sol, 1982) that can be used for assessing potential "to-be" business process designs (Giagles, Paul, & Hlupic, 1999) and in which structural validation of a simulation with the actors in the system is advocated (Berends & Romme, 1999). Nidumolo, Menon, & Zeigler (1998) discuss the application of information systems architecture-based (ISA) approaches for object-oriented modeling and simulation (OOMS) in which they conclude that the strong roots of these approaches in static IS modeling, amongst other factors, limits them to be used for business process simulation (Nidumolu, Menon, & Zeigler, 1998). In this chapter we will challenge Nidumolu, Menon, and Zeigler's position and take the application of ISA approaches for business simulation one step further by providing an integrated framework for the three levels (information base, schema and meta-schema) and three perspectives (information, process and behavioural) for information systems (Nijssen, 1989). We will apply this framework on a subject area or Universe of Discourse (UoD), by defining the communication documents for runtime management and by incorporating the simulation requirements into the application requirements documents. A requirements determination approach in which a strong modeling methodology is contained that has been developed within the architecture of this framework is natural language modeling [NLM] (Bollen, 2002, 2004, 2005).

Chapter Structure

In this chapter we will first focus on the modeling constructs in NLM for the *information-oriented-, process-oriented perspective* and on the NLM modeling constructs and methodology in the *behaviour-oriented perspective* (Olle, Hagelstein, Macdonald, Rolland, Sol, van Asche, & Verrijn-Stuart, 1988). In the next part of this chapter we will show how the meta concept in the event perspective—*runtime management*—can be incorporated into the conceptual application models. In the next section we will specify how the derived documents for a "real" application subject area can be used as an initial model for the definition of a simulation model and how the elements of this initial model can be used to define a business simulation for the behaviour in an application area.

The NLM Constructs for Modeling the Information Perspective

The NLM approach is an evolution of the fact-oriented approach, which has evolved from the pioneering work of Abrial (1974) on the semantic binary relationship model, Falkenberg's object-role model (1976), the ENALIM methodology (Nijssen, 1977) and Control Data's (binary) NIAM (Verheijen & van Bekkum, 1982). In the late 1980s, binary NIAM evolved into N-ary fact-oriented information modeling (Nijssen & Halpin, 1989). The modeling constructs for the data perspective in NLM and its ancestor methodologies differ from (extended) entity-relationship ((E)E-R) approaches (Chen, 1976; Teory, Yang, & Fry, 1986) and the class diagrams in UML (Booch, Rumbaugh, & Jacobson, 2005; Rumbaugh, Jacobson, & Booch, 2004) in the way in which *facts* and *population constraint types* can be encoded. The most striking differences are in the absence of a modeling procedure in most (E)E-R approaches and the UML class diagrams and the presence of the attribute modeling construct in these approaches that can lead to modeling anomalies and modeling rework (Halpin, 2001, p.351). A survey of (extended) E-R literature shows that in Silva and Carlson (1995), the MOODD method is documented in which a restricted subset of English (RSL) can be used to express user requirements. However, the RSL specifies the possible outcomes of the requirements gathering process, but does not give explicit guidance for the analyst on how these outcomes are obtained in the requirements gathering process.

In this section we will define the set of modeling constructs of natural language modeling (NLM) for the information-oriented perspective.

The choice of names used in communication is constrained by the reference requirement for effective communication. For example, a chemical company will use a *customer code* for referring to an individual *customer*. The use of names from the name class *customer name* in the customer management registration subject area for referring to individual customers, however, will not lead to effective communication because in some cases *two or more* customers may be referenced by *one* name instance from this name class. This is one of the reasons why not all names can be used for referencing entities, things or concepts in a specific part of a real or abstract world. On the other hand it is evident that knowledge workers that are involved in activities in an application subject area have knowledge on the reference characteristic of the potential name classes for the different groups of "things" or concepts in their day-to-day business activities. This means that they should be able to tell an analyst whether a name from a specific name class can be used to identify a thing or concept among the union of things or concepts (in a specific part of a real or conceived world).

The main principle in NLM is the "natural language principle:" all appearances of verbalizable information (e.g., forms, note-books, Web pages) can be expressed as declarative natural language sentences. In every (business) organization examples of verbalizable information can be found. These examples can be materialized as a computer screen, a World Wide Web page, a computer report or even a formatted telephone conversation. Although the outward appearance of these examples might be of a different nature every time, their content can be expressed using natural language.

Figure 1. Customer Management Chemical International Corporation (CIC)

SAPclient 002					
Cust.No	Customer Name	Customer group	Group name	Int./ external	Sales Org.
abch567896	Jansen ag	INDU	Industry	internal	1001
567fhf67r7	Kaiser	RETA	Retail	external	1004

Example 1. We will give a common example (of verbalizable information) of a SAP R/3 business application (Curran & Ladd, 2000; Jacobs & Whybark, 2000) for customer management at a company called Chemical International Corporation (CIC)[a].

In *Figure 1* an example is given of a *customer management document.* In this example, the Chemical International Corporation (CIC) wants to record information about the (types of) customers, whether they are internal or external customers. Furthermore, CIC has defined customer groups for analytical purposes, for example, *private, retail, wholesale* and *industry.*

Algorithm 1. Verbalization

BEGIN VERBALIZATION (UoD$_i$, G) {UoD$_i$ is the universe of discourse that contains 1 or more 'real life'user examples. G is the group of users of the 'real life'examples in UoD$_i$}
WHILE still significant parts of user examples are not shaded
DO   let knowledgeable user (g□G) verbalize the next
      unshaded part from the significant[c] part of the
      UoD. The analyst will shade this part on the real-
      life example and he/she will add the verbalized
      sentences on the document verbalized sentences.
ENDWHILE
Replace dependent sentences by self-contained sentences.
{Reconstruction check}
Let the analyst recreate the original example documents by translating the verbalized sentences document onto the corresponding parts on the original document.
IF the recreated document is identical with the shaded part of the verbalized document
THEN {no information loss has occurred}
ELSE{information loss has occurred,
      VERBALIZATION(UoD$_i$,G)
{Have the user verbalize the example again, thereby using a different naming convention and/or verbalizations that refer to bigger parts on the example document}
ENDIF
END

It is assumed that a *customer number* can be used to identify a *specific* customer among the *union* of customers within the given *SAP client* that can be identified by a SAP *client code*. A customer furthermore can be assigned to a customer group. A *customer group* is identified among the union of customer groups by a *customer group code*. Furthermore, every *customer group* has a name and must have a *customer group status* that is either *internal* or *external*.

The application of the natural language axiom on the example of communication from *Figure 1* is embedded in step 1 of the NLM modeling procedure: verbalization[b]. We will give a detailed specification of this step in the NLM procedure here.

The *verbalization* transformation is essential for the implementation of the modeling foundation of *NLM* (the natural language axiom). The *verbalization* transformation is the *unification* transformation that uses *all* appearances of *declarative verbalizable* information as an input and will converge it into natural language sentences as an output.

The verbalization transformation will result in the following sentence instances (amongst others):

Figure 2. Example verbalizable communication documents

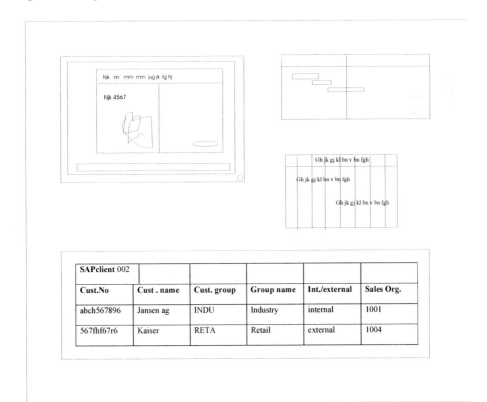

SAPclient 002					
Cust.No	**Cust . name**	**Cust. group**	**Group name**	**Int./external**	**Sales Org.**
abch567896	Jansen ag	INDU	Industry	internal	1001
567fhf67r6	Kaiser	RETA	Retail	external	1004

The customer abch567896 within client 002 has the name Jansen ag

The customer 567fhf67r7 within client 002 has the name Kaiser

The customer abch567896 within client 002 belongs to the customer group INDU

The customer 567fhf67r7 within client 002 belongs to the customer group RETA

We can now in principle verbalize all relevant examples of communication within this Universe of Discourse that contain the communication about the "real" world (see *Figure 2*).

If we analyze our four example sentences that have resulted (amongst others) from verbalizing the *customer management* example in *Figure 1*, we can divide them into two groups according to the type of verb that is common to the sentences in these sentence groups. This is the result of step 2 of the NLM modeling procedure: grouping [for a detailed specification of this modeling step see Bollen (2002, 2004)]. In Bollen (2002, 2004) a further specification of the remaining modeling step—classification and qualification—is given.

Sentence group 1:

The customer abch567896 within client 002 has the name Jansen ag

The customer 567fhf67r7 within client 002 has the name Kaiser

Sentence group 2:

The customer abch567896 within client 002 belongs to the customer group INDU

The customer 567fhf67r7 within client 002 belongs to the customer group RETA

If we focus on the first sentence group we can derive a sentence group template in which we have denoted the verb as text and the variable parts or *roles (or placeholders)* as text between brackets:

The customer <CUST> within client <CLIE> has the name <NAME>.

Figure 3 shows a graphical representation of all derived sentence groups from the CIC customer management example in *Figure 1*. Each role (or placeholder) is represented by a "box," for example, *CUST*. Each sentence group is represented by a combination of role boxes. Sentence group SG1 is represented by the combination of role boxes *CUST, CLIE* and *NAME*. For each sentence group one sentence group template is positioned underneath the combination of role boxes that belong to the sentence group. We will call the diagram in *Figure 3* an information structure diagram (ISD).

If we inspect *Figure 3* we see that a sentence group template must reveal additional information about the type of things that can be "inserted" into a *role* variable.[d] For example, the word "customer" specifies what type of thing (or concept) is allowed to play the role "CUST" in sentence group *Sg1* but also what type of thing (or concept) is allowed to play the role "CUST" in sentence groups *Sg2, Sg3, Sg4* and *Sg5*. We will call the "customer"

Figure 3. Communication and naming convention sentence group and sentence group template(s) plus population constraints for the customer management example

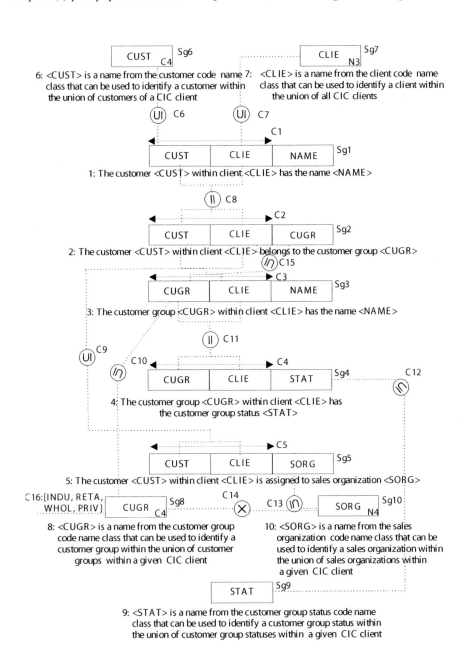

part in the verbs of sentence groups in *Figure 3* the *intention* of the role "CUST" in sentence groups *Sg1, Sg2* and *Sg5*.

Since we model the communication about some "real" or "constructed" world in NLM we need to incorporate a list of definitions of the *concepts* or *nouns* in the sentence group templates including the intentions. Such a list of concepts and their definitions must contain a definition for *each* intention in the UoD. The definition of an intention should specify how the knowledge forming the intention (definiendum) is to be constructed from the knowledge given in the definition itself and in the defining concepts (definiens). A defining concept (definien) should either be an intention or a different concept that must be previously defined in the list of concepts or it should be defined in a common business ontology or it must be a trivial generally known concept (for example, sun, moon). The list of definitions (in order of comprehension) for the CIC consumer management example is given in *Figure 4*.

Naming Convention Sentence Groups and Compound Referencing Schemes

In this section we will further formalize the outcome of the process of the selection of a name class for referring to things in a real or abstract world. The outcome of such a naming process will result in the utterance of naming convention sentences, for example the following sentences:

abch567896 is a name from the **customer code** *name class that can be used to identify a customer within the union of customers within a given CIC client.*

567fhf67r7 is a name from the **customer code** *name class that can be used to identify a customer within the union of customers within a given CIC client.*

Figure 4. List of definitions for CIC customer management example (in order of comprehension)

Concept	Definition
Customer	is a business partner against whom receivables are held for goods and services rendered.
Customer group	is a freely definable grouping of <customers> for pricing or statistical purposes.
Customer group status	is used to indicate whether or not the <customer group> consists of internal or external <customers>
Client	A self-contained unit in a R/3 system with separate master records and its own set of tables
Sales organization	is defined in order to structure the company with respect to its sales requirements. A sales organization is responsible for selling physical products and services

These sentences express that a certain name belongs to a certain name class and that instances of, for example, the name class customer code can be used to identify an instance of the intention customer within a given SAP client of the CIC corporation.

We note that the sentence group template for a naming convention sentence group (see sentence groups *Sg6, Sg7, Sg8, Sg9* and *Sg10* in *Figure 3*) contains the precise semantics of the naming convention. It is specified within what context the names of a name class can be used to identify instances of the intention for which the name class can serve as a naming convention. The relevance of this NLM modeling facility becomes clear when we consider an SAP application in which system codes are defined in many cases in the context of a pre-defined SAP client. In order to be able to make a distinction into "global" or "local" reference characteristics of a name from these name classes we must specify the context in which this name can be used to identify instances of an intention.

Population Constraints

After the sentence groups[e] and naming conventions have been derived for the application, we can define further population constraints for the sentences that might be contained in the application information base at any point in time. The types of pre-defined population constraints consist of (internal) *uniqueness constraints* (e.g., $c1$, $c2$ in *Figure 3*), *value instance constraints* (e.g., $c16$ in *Figure 3*) and *set-comparison constraints* (e.g., $c8$, $c13$, $c14$ in *Figure 3*). For an elaborate treatment of population constraints we refer to Bollen (2005). The accompanying steps from the NLM modeling procedure for deriving constraints of the types that are covered in this section are specified in detail in Bollen (2004).

We will call the combination of the ISD (sentence groups and population constraints) in *Figure 3* and the list of definitions from *Figure 4* the *NLM application information grammar* (AIG) for the consumer management business application area.

The NLM Constructs for Modeling the Process Perspective

In this section we will define the modeling constructs for the process-oriented perspective that can be applied within business subject areas. The process perspective in an enterprise subject area is concerned with "how" fact instances can be composed from other fact instances. In businesses we can consider facts as either an *outcome* of an enterprise activity or an *ingredient* for an enterprise activity. Enterprise activities are executed under the responsibility of a *user* from a *user group*. We will call a *user* that creates facts an *active user*. The border concept in the process perspective will show what *user groups* can be held responsible for the creation of fact instances in the UoD. We will call this border concept the *sphere of influence* (SoI) (Parker, 1982, p.116).

Definition of Conceptual Process Types

Consider two different examples of billing, for example, in a bistro and in a fast food restaurant. Although the *spheres of influence* are different in these examples, the description
of the informational activity that creates the *order total* on each order receipt in terms of
instances of fact types in the application data model is identical [i.e., it is "organization
independent," see Gorry & Scott-Morton (1971)]. This indicates the need for a theoretical
construct that abstracts from the concrete way in which a fact-creating activity is performed
(e.g., performed manually by a bistro waiter or automatically under the responsibility of a
fast food restaurant counter employee). This theoretical construct is the *conceptual process
instance*.

Definition 1. A *conceptual process instance* is the abstraction of an organizational activity
    that is responsible for the creation of (a) *fact instance(s)* by an *active user*.

Definition 2. A *conceptual process type* is the intention of a subset of the conceptual process
    instances that are responsible for the creation of fact instances of the same (set of) fact
    type(s) by active users in one or more user groups.

Example 2. This example UoD is an order- document in which the ordered quantities for
    the desired items are recorded, together with the item prices and the order total [see
    Figure 5(a)]. The NLM application grammar for example 2 is given in *Figure 5(b)*.

Figure 5. (a) UoD example 2 (b) NLM AIG for example 2

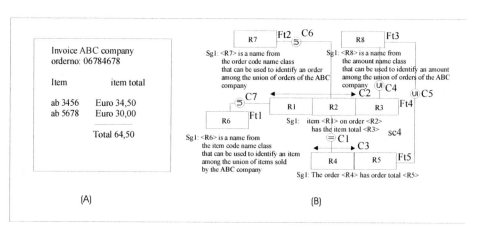

Derivation Process Types

The fact type(s) of the fact instances created in (an) instance(s) of a conceptual process type will be referred to as the *resulting fact type(s)* for the conceptual process type. The *ingredient fact type(s)* of a conceptual process type specif(y)ies what the fact type(s) is (are) for the fact instances that serve as an input for the creation of a fact in a process instance of a given conceptual process type. In case the "underlying mechanism" is a procedure or a derivation rule (see *Figure 6*) on a descriptive document that specifies for *all* instances of the conceptual process type how the resulting fact instance(s) (contained on a declarative document) can be derived from the ingredient fact instances we will call such a conceptual process type a *derivation process type*.

Definition 3. A *derivation process type* is a conceptual process type whose process instances create fact instances by applying the same derivation rule on instances of the same ingredient fact type(s).

Determination Process Types

Some facts will be created without a known (or existing) derivation rule: for example the creation of the first name of a newborn. However, in many cases the creation of such a fact is subject to constraints. In the example of the name assignment for a newborn, the following

Figure 6. Conceptual derivation process type

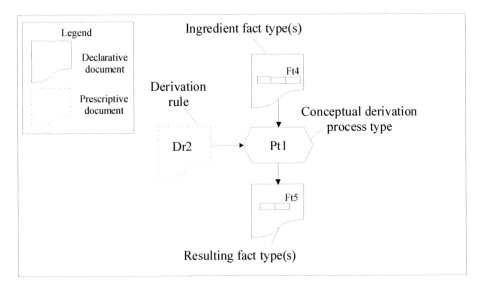

constraint exists: a baby of the female gender must be assigned a girl's name and a baby of the male gender must be assigned a boy's name (from a predefined list of names).

Definition 4. A *mixed determination process type* is a conceptual process type in which the active user uses instances of the same ingredient fact types (that are contained in the application's information grammar) in all process instances (see *Figure 7*).

The conceptual process that creates the names of a newborn baby: *We have decided to call you John. We have decided to call you Alice.* These examples do not involve any derivation rule or (formal) procedure, but it is assumed that ingredient fact instances exist, for example: *John is the name for a boy, Alice is the name for a girl. The child that should be named is a girl* must be known before a name can be created for a specific child. The way in which a name is assigned in a specific instance, however, can not be determined in advance. Some people might select the name of their own father or mother for their child. Others might choose the name of their favourite rock star. On a "process type" level, however, we can never know what selection criterion (or derivation rule), will be applied in a specific process instance. The same parent might use different criteria for every newborn.

In addition to *derivation* and *mixed-determination* process types we can distinguish conceptual process types which have no known and fixed set of ingredient fact type(s) and derivation rules: *strict-determination process types* (*Figure 8*). This type of process is used in managerial decision-making, for which, in some cases, *decision support systems* are employed: "The user may only need 40-100 data variables, but they must be the right ones; and what is right may change from day to day and week to week." (Sprague, 1980, p. 21).

Figure 7. Conceptual mixed-determination process type

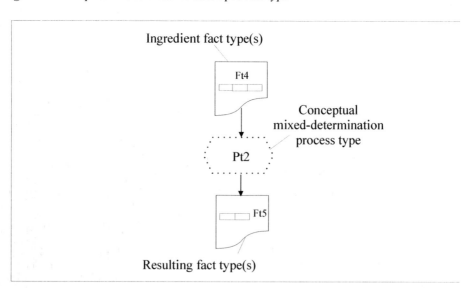

Figure 8. Conceptual strict-determination process type

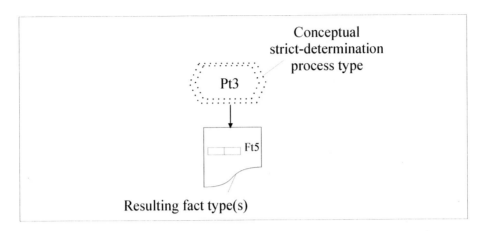

Definition 5. A *strict-determination process type* is a conceptual process type in which the active user does not use a known derivation rule all the time and the active user does not use instances of the same ingredient fact types (that are contained in the application's data model) in all process instances.

The Instantiation of Conceptual Process Types

We now take the *enterprise* or *application information base* (AIB) as a starting point and subsequently apply definitions 4 and 5, which tells us that every fact instance is created in a conceptual process instance. The collection of *conceptual process types* that are relevant for the enterprise subject area is called the *application process description* (APD).

For this process type instantiation we need parameters that tell us *what* fact instances will be the "tangible" end results of the execution of a conceptual process and what other values are needed for such a process execution. We will call such a set of parameters: the *conceptual process type argument* (see *Figure 9*).

Definition 6. A *conceptual process type argument* specifies the types of values that must be supplied for instantiating a conceptual process.

When we consider the derivation process type *create-order-total* in *Figure 9*, it will only create (a) fact instance(s) of fact type FT5 when at least one fact instance of fact type FT4 exists in the *AIB* in which the value for the role "order code" is equal to the value for the process argument "arg1". If we inspect the derivation rule for this conceptual process type

Figure 9. Instantiation of a conceptual process type via conceptual process type argument

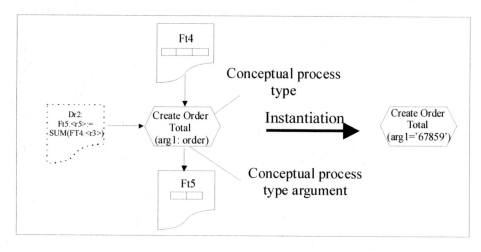

and the instantiation values for the *process type argument* it should be clear whether the execution of the process will lead to a result **before** the derivation rule is actually executed or fact instance(s) are determined by an active user.

The *pre-condition* for a conceptual process type serves as a checking mechanism for the instantiation of a process type (*Figure 10*). If the *pre-condition* is violated by the actual

Figure 10. Conceptual process execution: Pre-condition

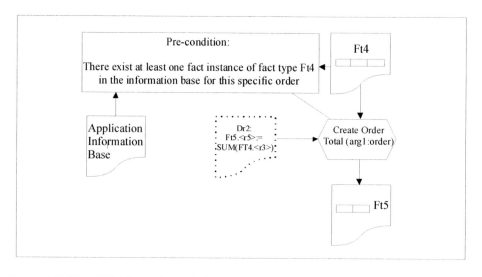

Figure 11. Conceptual process execution: Post-condition

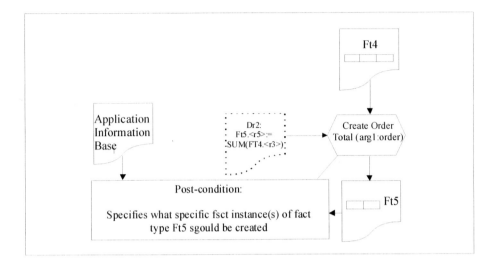

content of the *AIB*, the process will **not** be executed and (a) resulting fact instance(s) will **not** be created.

Definition 7. A *precondition* in a conceptual process type checks whether the required *input fact instances* for the *derivation process* or *the mixed determination process* exists in the *enterprise database.*

The post-condition specifies what the fact argument is for the facts that will be created in the conceptual process (see *Figure 11*). Furthermore, it is specified *how* the fact values that will be created in the conceptual process will be obtained. In case of a *derivation process* a reference is given to a derivation rule. In case of a *mixed-* or *strict-*determination process, it is stated that (a) fact(s) has (have) to be created (by an active user). This post-condition, furthermore, specifies how the resulting fact type(s) of the process type must be instantiated as a function of the instantiation values for the conceptual process type argument.

Definition 8. A *post-condition* of a conceptual process type specifies (parts of) the fact argument for the instances of the resulting fact type for the conceptual process. A *post-condition* in a conceptual process indicates that (a) fact value(s) ha(s)ve to be determined. A *post-condition* in a derivation process type, furthermore, specifies what derivation rule is used for the creation of the resulting fact instance(s).

Figure 12. (a) Textual process type representation and (b) Accompanying graphical representation

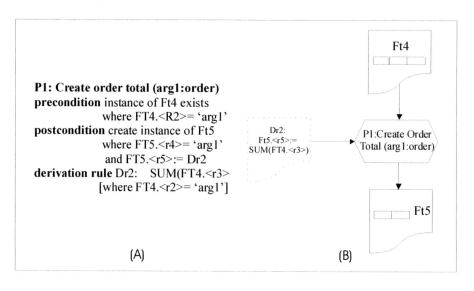

(A)   (B)

We will now simplify the specification of a conceptual process type by dividing such a specification in (at most) three parts. In the case of a *derivation process type* we will specify a *precondition*, a *post-condition* and a *derivation rule*. In case of a *mixed-determination process type* we will specify the *precondition* and *post-condition* and, finally, in case of a *strict-determination process type* we will only specify the *post-condition*.

For some fact types in an AIG, the conceptual processes that create fact instances are performed under the responsibility of users outside the application's sphere of influence. We will by default define such a fact creating process type as an *enter* process type. In *Figure 12* we have given the textual APD for example 2 together with a graphical representation. An elaborate description of the methodology for the process-oriented perspective can be found in Bollen (2006).

The NLM Constructs for Event Modeling

Although the execution of conceptual processes is constrained by the *pre-conditions, post-conditions* and *derivation rules* from the respective conceptual process types there still remain degrees of freedom with respect to *when* and in *what sequence* these conceptual processes can be executed. In some business situations the compliance to the *AIG* and the *APD* is sufficient for enforcing the business rules in the application area. In other application areas additional modeling constructs are needed that can specify **when** the instances of conceptual process types from the *APD* will be executed. For examples of workflow management see Reijers and van der Aalst (2005).

Example 3. Consider the following example of a workflow management application (see *Figure 13*).

Whenever an *insurance application* is created an instance of the process type *credibility checking* must be executed:

ON   insurance application is created

IF   credibility checking capacity is available

THEN   perform credibility checking

The description of the activity that is specified in the ON clause in the above example can be considered an event or occurrence of something that has happened in the application subject area.

Figure 13. AIG and APD for example 3

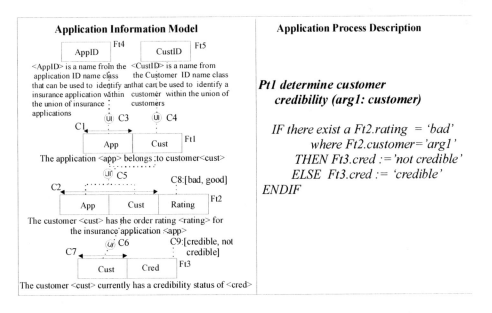

Event and Event Type

In the information systems literature numerous definitions of the *event* concept can be found: *"An event is an occurrence or happening of something in the environment under consideration"* (De & Sen, 1984, p. 182). *"An event is a noteworthy change of state; all the changes of state of objects are not events"* (Rolland, 1983, p. 34). *"When the environment does something to which the system must respond, an event is said to occur."* (Wieringa, 1996, p. 198). *"An event is a happening that changes the state of a model (or system)"* (Schriber & Brunner, 1997, p. 15). We will now give the following definition of *event*:

Definition 9. An event is an occurrence or happening in the application subject area that can lead to the execution of one or more conceptual processes within the application subject area's sphere of influence.

In the insurance application subject area we have to specify whose insurance application request is created. We therefore have to *qualify* the example event from the insurance application example—*Insurance application is created*—into the following event: *Insurance application with application number 45678 is created.* If we observe this application subject area over a certain period of time we can encounter also the following event instances *Insurance application with application number 45679 is created, Insurance application with application number 45680 is created.* We can conclude that the former verbalization of events can be further grouped and qualified into the *event type: Insurance application created (arg1: application number)*

Definition 10. An event type is a class of events that have the same intentions according to a user group *G* for an universe of discourse *UoD* and a sphere of influence *SoI*.

In order to make a distinction into fact types and event types we will model the *"role(s)"* in an event type as *event argument(s)*. The intention that plays such a role in an event type in principle is defined in the same way as the intention concept in the information perspective. A specific intention that is defined in an event type, however, does not necessarily have to be defined in the AIG of that subject area. We will now formalize the intentions(s) of the event type by structuring these(is) intention(s) into the *event type argument set*[g]. We will derive the set of arguments for the event type by *classifying* and *qualifying* a significant set of verbalizations of event instances, for example:

Insurance application 257892 is created.

Insurance application 28923 is created.

This will result in the following event type and its argument set:

Insurance application created (arg1: application)

Definition 11. An *event type argument set* of a given event type specifies all intentions, instances of which should be supplied at the occurence of an event instance of the event type.

Consider the following event type: ET1: *Customer at table wants to pay(arg1:customer)*

An *instance* of this *event* type is: *Customer at table wants to pay (arg1:'Piet Janssen').*

The conceptual process that should be instantiated as a result of this event is an instance of the following conceptual process type: *P1: determine order total (arg1:order).* The instances of the intentions in the argument set for the event type can be used for instantiating the process type (in an *impulse*). The modeling construct of *event* refers to an action that can occur, for example, *customer places an order*, or it can refer to a more "static" action, for example the start of a new day when *the clock strikes 12:00 P.M.* We can conclude that events can have different appearances and therefore we will use the *AIG* and the *APD* in combination as a starting point for "detecting" events that are relevant for the application subject area, which means that all events that do ***not*** potentially trigger a conceptual process will be left out.

Event Condition and Event Condition Type

An *event* can start the execution of a process (in some cases) under (a) condition(s) on the information base. In the population constraints from the AIG we have modeled the business rules that are always applicable (or invariant) in terms of the AIB: for example, the business rule that states that *each order has at least one orderline*. In the pre-conditions for the conceptual processes in the application process description, the business rules are modeled that specify what ingredient fact instances should be available in order to "compose" or "derive" the resulting fact instance(s) in the conceptual process. In the event perspective we will model the business rules that contain the knowledge under what condition (on the AIB) the occurence of an *event* of an *event type* will trigger a *process instance* of a specific *process type*.

An *event condition* is a *constraint* for the execution of a conceptual process that is "triggered" by a specific event. In addition to the conditions that are given in the *pre-condition* of the conceptual process or a condition that is enforced by the *population constraints* in the AIG, the condition that is specified in the event perspective is defined in terms of the *AIB* at that moment in relative time in which the *event condition* is checked.

Definition 12. An *event condition* is a proposition on the information base.

Example: $c_1: \exists f \in EXT(FT1)[f.<r2>= \text{'Piet'}]$

In most business application areas the occurence of an event instance at t_1 should lead to the execution of a conceptual process (see *Figure 14*). The occurence of the same event

Figure 14. Event, condition and process

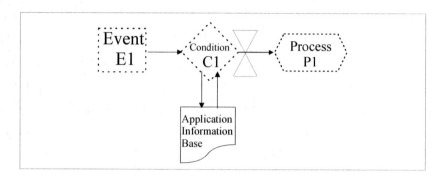

instance at another moment in relative time should not necessarily lead to the execution of a conceptual process because the *AIB* can be in a different state.

Example:

Event type *order request(arg1: customer, arg2:product)*

Event instance *order request (arg1:"Piet Janssen," arg2: "Bicycle")*

Under the condition that the customer has a satisfactory credit rating this event should lead to the instantiation of the following conceptual process type: *prepare order confirmation (arg3: customer, arg4: product)*. The instantiation will lead to the following process: *prepare order confirmation (arg3: "Piet Janssen," arg4: "Bicycle")*. A different event occurence of the same type is *order request (arg1: "Hans Koek," arg2: "Scooter")*. Given the fact that this customer in this case does not satisfy the credit rating, the event occurence should **not** lead to the instantiation of (a) conceptual process(es). In most business application areas it is possible to specify these conditions on a type level: *"If a customer wants to order a product he/she should have a sufficient credit rating."* A condition (instance) of the aforementioned condition type is: *"If Piet Janssen wants to order a product he/she should have a sufficient credit rating."* It can be seen that the instantiation values for the event type can be used in this case for the instantiation of a condition on the information base.

Definition 14. An *event condition type* is a set of propositions defined on the information base. An instance of a condition type: a condition can be derived whenever a(n) event instance is specified.

Example:

ET1: Customer orders an order (arg: customer)

CT: $\exists_x \in EXT(R_2) [x=ET1.arg]$

Impulse and Impulse Type

We will call the effect of an event occurence into the execution of one conceptual process (eventually under an event condition on the information base) an *impulse* (instance).

Definition 14. An *impulse instance* is the potential triggering of the execution of a *conceptual process instance(s)* from the *APD* dependent upon the condition in the *AIB*.

Example:(order 56 is delivered, (stocklevel <125), P_2)

Events that do not have the potential to "trigger" conceptual processes from the APD are not relevant for the description of the behavioural perspective in a given application subject area. We can now classify all impulses that have the same *event type*, the same *process type* and the same *event condition type* into a set of impulse instances that belong to the same *impulse type*.

Definition 15. The intention of an *impulse instance* is an *impulse type*.

We will now introduce a construct in the event perspective that enables us to derive the instantiation value for the process argument in an impulse whenever the values of the *event type argument set* are known. This is the construct of an *impulse mapper.* The impulse mapper is a mechanism that encodes the business rules in the subject area that specify *how* a conceptual process is instantiated when an event occurs and the condition on the information base (specified in the impulse) is satisfied.

An impulse mapper is a construct that transforms values of event type arguments and fact instances from the AIB into instantiation values for the *process type argument set(s)* for the process type(s) that will be potentially instantiated in the impulse.

Event type	*Et1: insurance application created (arg1: application).*
Process type	*Pt1: determine customer credibility (arg1:customer).*
Condition type	*Ct1: ET1.arg1 EXT(person3)*
Impulse type	*IT1:=<Et1, Ct1, Pt1>*
Impulse mapper	*Pt1.arg1:=Ft1.cust (where Ft1.app='Et1.arg1')*

We still need a modeling concept that allows us to express the "triggering" of conceptual processes whose types are **not** contained in any of the impulses of the *application event description* (AED) so far. This concept will be the *trigger-process* event type. An occurence of such an event type will immediately and unconditionally result in the instantiation of the conceptual process type that is specified. In the *trigger-process* event type the argument set of the event type is equal to the argument set of the process type. Including such a *non-*

Figure 15. The event modeling methodology for the behaviour-oriented perspective

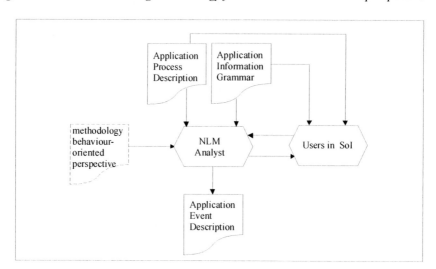

conditional and *unqualified* impulse in an event description implies that such a conceptual process can be instantiated at any time and it therefore implies that no constraints on the behavioural perspective exist for the instantiation of such a process type. In addition, we need two time variables that can encode the potential time delays between, on the one hand, the occurrence of an event and the check on the event-condition (in the impulse) and, on the other hand, the check on the condition and the execution (if any) of the conceptual process [see Bollen (2004, pp. 207-208)].

The NLM Event Modeling Methodology

In addition to the AIB and the AIG in the information perspective and the APD in the process perspective we will define the AED as the document that constrains the knowledge behaviour for an application subject area in the event or behavioural perspective. In Bollen (2004, pp. 162-163) we have specified the procedure that an analyst must apply in order to yield the AED. A more elaborate treatment of the modeling constructs and a description of the modeling methodology in the behaviour-oriented perspective or event perspective can be found in Bollen (2004, pp.161-165, pp. 199-209).

Extending the Subject Area with Runtime Management

The conceptual documents we have defined so far (see *Figure 16*) *intentionally* constrain the communication and "knowledge" behaviour within an organizational subject area. In this section we will introduce an additional *universe of discourse* and *sphere of influence* for the application areas in our architecture. This specific UoD and SoI will be needed for enforcing the *extensional* compliance to the *fact types* and the population *constraints* in the *AIG*, the *process types* in the *APD* and the *impulse types* in the *AED*. We will call this additional subject area *runtime management*. The sphere of influence of runtime management contains three user groups: the *event manager,* the *process manager* and the *information base manager*. We will define the UoD of runtime management as the union of "real life" examples that are used by these user groups.

The Event Manager User Group

We will now introduce a conceptual agent that will evaluate the occurence of events during runtime. This conceptual agent will detect the *occurence* of an event, determine its *event type*, evaluate the *event condition* in the *impulse* relative to the information base at the moment of evaluation and invoke the relevant *conceptual process types*. We will call this agent the *event manager*. The event manager uses the *AED*, the *APD* and the *AIB* as input

Figure 16. The four conceptual documents for an application subject area

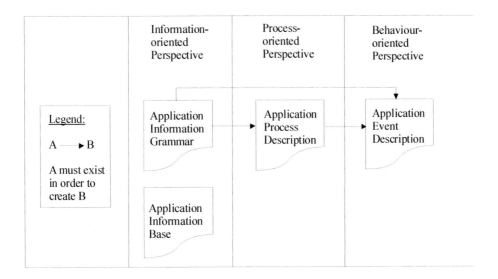

documents. The event manager can be considered an organizational function that "scans" the application subject area for event occurrences. Once the event occurrences are detected by the event manager it will determine to what event type the event instance belongs [*event recognizer*, see Wieringa (1996, p. 199)]. This is done by scanning the AED. Once the event type is determined, the impulse types that contain this event type will be selected. The event manager will now add the event occurence on the *event condition check list* (see *Figure 17*) together with the moment in relative time in which the condition according to the qualified impulse in the event description has to be checked.

The event condition check list will be evaluated by the *process trigger manager* (see *Figure 17*). The process trigger manager will check the conditions in the impulse at the moment in which the specified moment in time form the impulse will be equal to the current time. If the condition in the impulse is satisfied then the process that should be executed will be added to the *event condition check list* and in addition the trigger time for the conceptual process will be derived from *the condition-process trigger type* as specified in the impulse (type).

The *conceptual event processor* is a prescriptive document that specifies how to evaluate events and conditions.

It should be noted from *Figure 17* that the condition/process trigger time can be a function of the state of the information base at moment "now;" therefore, this expected process execution time should be recorded in this document[1]. The *event condition check list* serves also as an input document for checking the conditions if the relative time changes, for example, from t_1 to t_2. The processing and evaluating of the input document and checking the event occurence in the subject area will be performed by using the *conceptual event processor* as a prescriptive document. The results of the application of this procedure have to be recorded in order to instantiate the conceptual processes using the *impulse mapper* from the conceptual event description.

The event manager needs a document on which to record what process(es) need to be executed in future (relative) time: *The process execution list* (see *Figure 18*). This means that

Figure 17. Event condition check list[h]

EVENT CONDITION CHECK LIST at relative time: t0				
Event instance	**condition check on**	**condition type**	**process to be executed**	**at relative time**
(E1(arg1: Piet Jannsen),t-1)	now	CT1	p2	t1
(E1(arg1: Hans Loos),t-1)	now	CT2	p3	t2
(E2(arg2: July),t0)	t1	CT1	p3	t1
(E2(arg2:July),t0)	t1	CT2	p4	t2
(E3(arg4: 23567),t0)	t4	CT1	p2	t5

Figure 18. The process execution list

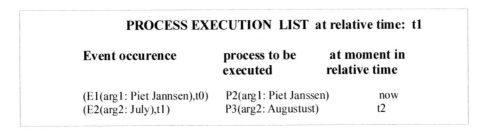

PROCESS EXECUTION LIST at relative time: t1		
Event occurence	**process to be executed**	**at moment in relative time**
(E1(arg1: Piet Jannsen),t0)	P2(arg1: Piet Janssen)	now
(E2(arg2: July),t1)	P3(arg2: Augustust)	t2

Figure 19. The event manager subject area

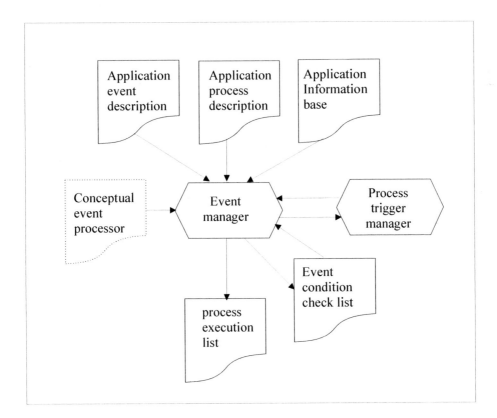

on the *process execution list* only those impulses will be recorded that satisfy the condition on the AIB.

In *Figure 19* the "real life" communication documents that are used and created by the event manager user group in the runtime management UoD are given.

The Process Manager User Group

The first task of the event manager is to evaluate all events and check whether the impulse-conditions are not violated at the moment in relative time of the event-condition trigger. Secondly the process types will be instantiated when their execution time is due. The second user group within runtime management we will call the *process manager*. This user group will be responsible for the process execution and it will use a *conceptual process processor* that will *instantiate, sequence* and *execute* triggered process types. We thereby use the document on which the future executions of conceptual process types are recorded by the event manager: the *process execution list* (see *Figure 18*) as an input. The *process execution list* will show the conceptual processes that are planned to trigger in the future. The *conceptual process processor* is the organizational function that takes care of the execution of multiple processes "at the same time" and makes it appear that a compound process is executed at a single moment in relative time. Furthermore, the process manager has to be able to sequence conceptual process executions within the same moment of time (e.g., an age can only be derived if the date of birth is known). We will call this function the *process inference manager.* This "conceptual" manager communicates the proper sequence of conceptual process instantiations for execution by inspecting their *pre- and post-conditions.* The successful execution of a conceptual process will lead to the generation of an instance of a post-condition for each fact that is created in a process *Create instance of fact type.* The primary task of the process manager is to generate either an INSERT(fact created) or UPDATE(fact created) request for each fact instance that is created in such a conceptual process instance[j]. The secondary task of the process manager is to inspect the *external update request list* and compare it with the *APD.* Only those fact instances can be contained on this external update request list that are created outside the sphere of influence (in an enter process type).

Definition 18. The *conceptual process processor* is a prescriptive document that specifies how to evaluate post- and pre-conditions of conceptual process types and how to sequence the execution of conceptual process at some relative time t.

Figure 20. The update request list

UPDATE REQUEST LIST at relative time: t2			
execution priority	update request	event that caused process	at time
1. INSERT(arg1: Piet Jannsen has a salary of $ 5000)		(E1(arg1: Piet Jannsen),t0)	now
2. INSERT(arg1: Jake Jones has tenure)		(E4(arg2: July),t1)	now
3. DELETE(arg1:Sylvia Fraser is enrolled in the marketing dpt.)			now

Figure 21. The process manager subject area

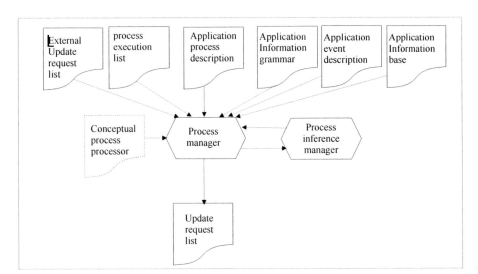

The process manager has an integrated document on which all IBMS events are recorded. We will call this list the *update request list* (see *Figure 20*).

The Information Base Manager User Group

The third user group within runtime management is the *information base manager* (see *Figure 23*). The information base manager takes the *update request list* as an input document. The information base manager will communicate further with another conceptual agent: *the information base update manager.* By checking the *information base before*, the *proposed information base after* and the *application information grammar* of the subject area, the information base manager will conclude in a dialogue with the *information base update manager* whether a proposed (compound) update request on the information base will be successful.

The *information base update manager* is a function that evaluates proposed information base updates at moment in relative time *t* and the information grammar and the (proposed) information base. The function of the *information base manager* is three-fold. Firstly, the *information base manager* checks whether adding, updating or removing fact instance(s) in (from) the information base is allowed according to the AIG. Secondly, in case this is allowed, (a) fact instance(s) will be added, updated or removed in (from) the information base. The *information base update report* (*Figure 22*) serves as the *proof of acceptance* of the facts that are negotiated or proposed in a conceptual process (or proposed outside the

Figure 22. Information base update report

INFORMATION BASE UPDATE REPORT for relative time: t1

Fact instance	Status
Jake was born on July 23, 1987	Added
The curent temperature is 23 degrees	Removed
Jim was born on July 29 1988	Rejected

Figure 23. The information base manager subject area

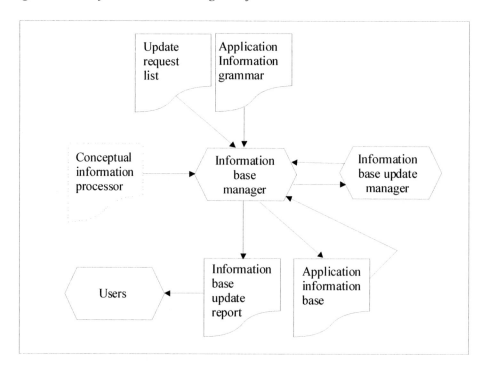

SoI) and in addition it informs the relevant environment of the application information system whether *update requests* have been *accepted* or *rejected*.

The *conceptual information processor* is a prescriptive document that specifies how to evaluate that the *AIB* will be in compliance with the *AIG* and how to evaluate the update request list and how to change the information base accordingly (Nijssen, 1981).

The messages that are sent whenever a state in the *information base* changes are, firstly, recorded in the *information base update report*, and, secondly, these messages are interpreted as potential events while creating the *AED*. An example of such an information base event is the following: *The fact instance "Jake was born on July 30, 1987" of fact type FT3 is added to the information base.*

Conclusions on Runtime Management

We can now extend the application universe of discourse with the following documents: *event-condition check list*, the *process execution list*, the *update request list* and the *information base update report*. Furthermore, we can integrate the *conceptual information processor* document, the *conceptual process processor* document and the *conceptual event processor* document into the *APD* and the *AED*.

We can conclude that the application independent *information meta grammar* is the meta concept in the *information-oriented* perspective, the application independent *information base management processes* is the meta concept in the *process-oriented* perspective and finally the application *independent runtime management* is the meta concept in the *behaviour-oriented* perspective.

Business Simulation and Regular
Application UoDs

We also need to create real life examples of the reports that we expect from a simulation. These reports will than subsequently be added to the application UoD, and the NLM modeling steps will be applied on this extended UoD. The most essential requirement for the conceptualization of a "simulation" UoD is that the *runtime management* must be part of the application UoD and in that case it can perform the role of "simulation executive" (Ball, 1996). The documents that contain the desired simulation output can be considered "real life" communication documents for a specific part of a business application. Adding these document to the "real life" application UoD, and applying the NLM methodology for conceptual modeling will result in an extended AIG, in which the simulation fact types are contained including the population constraints. With respect to the APD we can take the process description for the "real" UoD as a starting point. In order to derive a business simulation application UoD from this "real" application UoD, the determination process types have to be transformed into derivation process types. This will normally take place by introducing some form of "stochastic" process, in which a probability distribution must be selected first. Secondly, the parameters of such a distribution must be set or estimated [e.g., μ (distribution mean) and σ (standard deviation)], see for an example Sadoun (2005, p.659). Once this has been done, the resulting process can be considered to be a derivation processes in the context of the application business simulation UoD by incorporating a random generator. Finally, we need to add an environmental process that generates the "updates" from the (external) update request list.

Figure 24. Relevant model elements for each perspective in "real life" versus simulation

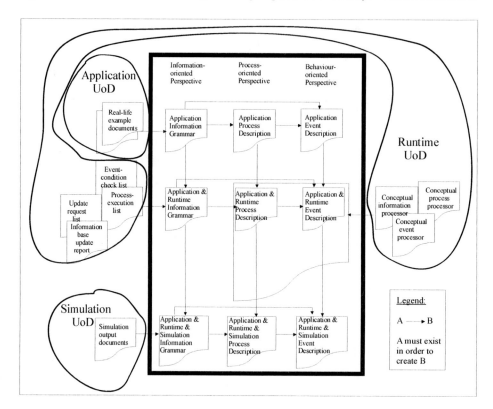

In *Figure 24* we have summarized how some of the elements in the conceptual schema of a general application UoD can be used as a starting point for defining the derivation processes and impulses for the "simulation" part of the application UoD. It follows, from this restriction to derivation process types and impulse mappers, that runtime management can act autonomously in the simulated application UoD.

The Citizen Service Case Study

In this section we will briefly discuss a simplified case study that illustrates the modeling concepts that we have introduced for simulations in this chapter. In a city hall of an average city, citizens are able to use the following services: register a newborn (NB), apply for a passport (PP), apply for citizenship (CS), apply for or renew a driver's license (DL) and ask for a birth-certificate (BC). The way that this service organization operates is as follows. Citizens enter the building and then have to apply for a counter-receipt. In this

Figure 25. Simulation output documents for city hall subject area

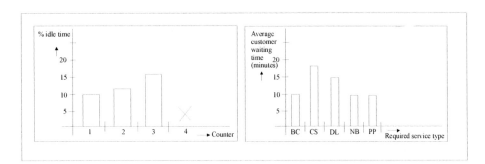

"transaction" the citizen's time of arrival is recorded. Furthermore, the requested service has to be specified. As a result the customer receives a ticket with the arrival ID (unique on any given day). At any time 0, 1, 2, 3 or 4 counters can be operational. In general, one or more services can be performed at any counter, depending upon the experiences of the civil servant. Every time a "customer" has been processed at a counter, the next in line will be announced on a central display. The central display lists the arrival ID plus the number of the counter where this client will be serviced (1, 2, 3 or 4). The AIG, APD and AED for this application subject area are straightforward. In this example we will focus on the simulation requirements. The management of the city hall is interested in the effects of opening/closing one (or more) counters on the (average) waiting times for the clients and the (average) idle time (for a given counter) for the civil servants. The simulation requirements, furthermore are based upon the assumption that customers for every service arrive at the city hall service counter according to a Poisson distribution and that the service times (for a specific service) follow a (negative) exponential distribution (Krajewski, Ritzman, & Malhotra, 2007, pp. 295-297). The example documents that are contained in this simulation requirements UoD are given in *Figure 25*.

Parts of the resulting AIG, APD and AED for the integration of the "operational" UoD and the "simulation" UoD are given in *Figures 26, 27* and *28*.

From the partial application and simulation process—and event—description in *Figures 27* and *28* we can clearly see how the application processes can be used to define the simulation processes.

Figure 26. Partial application and simulation information grammar

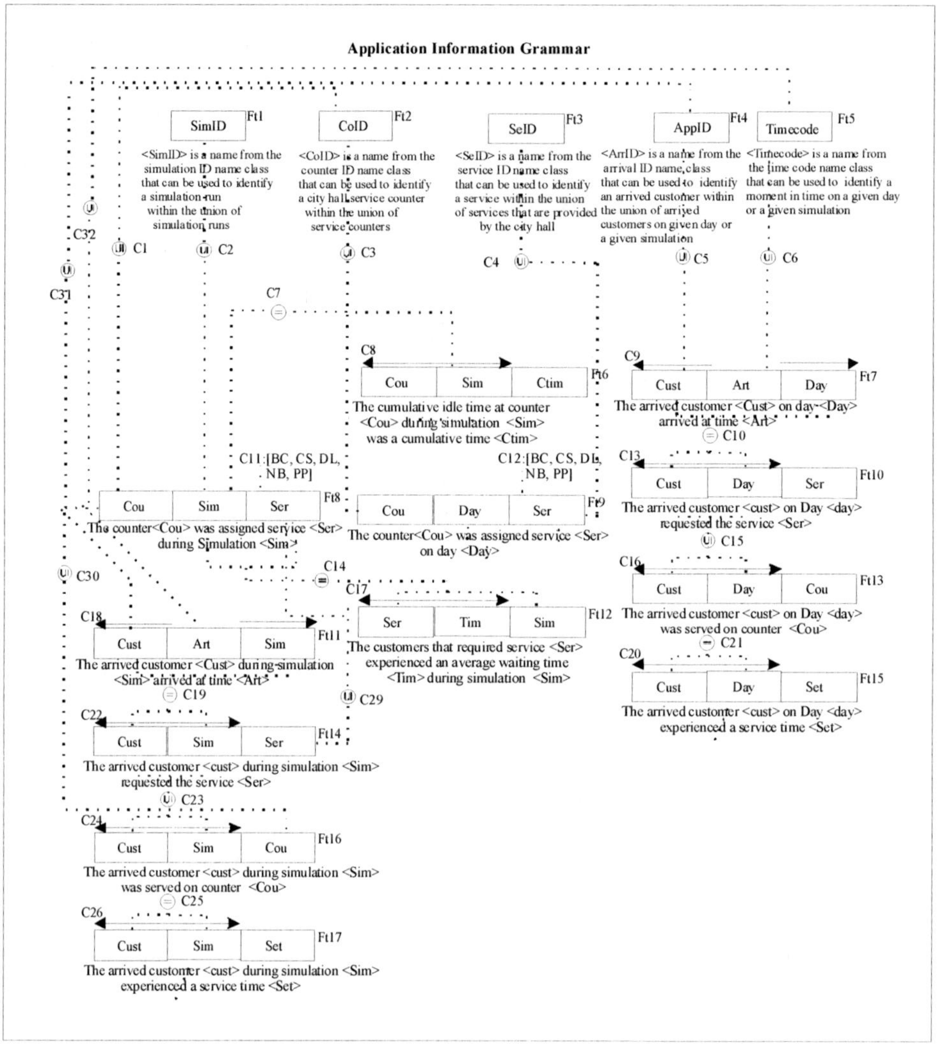

Conclusion

In this chapter we have introduced the NLM (NLM) methodology for conceptual modeling and we have shown how it can be used to model not only the "intentional" communication in an application subject area, but also how it can be used to define the prescriptive and declarative communication documents in the runtime management subject area. Finally, we have shown how the conceptual schema for a simulated application subject area can be incorporated into the "regular" conceptual documents for an application subject area by defining the "real life" communication documents for the simulation UoD and, by adding

Figure 27. Partial application and simulation process description

Pt1 determine service counter for arrived customer (arg1: customer, arg2: service)
IF there exists a FT10.cust where FT10.ser='arg2'
THEN SELECT from FT9.cou where FT9.ser='arg2'
          Such that EXPQUELENGTH('FT9.cou') is minimal
                    FT13.cou=MIN(FT9.cou) where FT13.cust='arg1'
Pt2 determine service counter for simulation customer (arg1: customer, arg2: service)
IF there exists a FT14.cust where FT14.ser='arg2'
THEN SELECT from FT8.cou where FT8.ser='arg2'
          Such that EXPQUELENGTH('FT8.cou') is minimal
                    FT16.cou=MIN(FT8.cou) where FT16.cust='arg1'

Figure 28. Partial application and simulation event description

ON Et1: customer arrives (arg1: customer, arg2: time, arg3:service)
DO Pt1:determine service counter for arrived customer (arg1: customer, arg2: service)
          Where pt1.arg1:=et1.arg1 and pt1.arg2:=et1.arg3
ON Et2: simulation customer arrives (arg1: customer, arg2: time, arg3:service)
DO Pt2:determine service counter for simulation customer (arg1: customer, arg2: service)
          Where pt2.arg1:=et2.arg1 and pt2.arg2:=et2.arg3

the simulation derivation process types and impulses leading to the integrated application information grammar, process description and event description of an application subject area including its simulation requirements. NLM turns out to be an approach that allows for a seamless integration between operational enterprise systems and data on one hand, and the simulation systems and data on the other hand, for the static models (information-oriented perspective) as well as for the behavioural models (process- and behaviour-oriented perspectives).

Future Trends

The NLM modeling constructs for the data, process and behavioural aspects in verbalizable application subject areas already contain the necessary modeling constructs for capturing domain and generic ontologies and the declarative modeling of knowledge and information processing. This trend does also apply to the subject area of simulation.

References

Abrial, J. (1974). Data semantics. In J. Klimbie & K. Koffeman (Eds.), *Data Base Management* (pp.1-59). Amsterdam: North Holland.

Ball, P. (1996). *Introduction to discrete event simulation.* Paper presented at the 2nd Dycomans Workshop on Management Control: Tools in Action, Algarve, Portugal.

Berends, P., & Romme, G. (1999). Simulation as a research tool in management studies. *European Management Journal, 17*(6), 576-583.

Bollen, P. (2002). *The NLM procedure. Lecture Notes in Computer Science 2382* (pp.123-146). Berlin-Heidelberg: Springer Verlag.

Bollen, P. (2004). *On the applicability of requirements determination methods.* PhD Thesis, University of Groningen. Retrieved January 21, 2005, from http://dissertations.ub.rug.nl/FILES/faculties/management/2004/p.w.l.bollen/

Bollen, P. (2005). NLM for business application semantics. *Journal of Information Science and Technology, 2*(3), 18-48.

Bollen, P. (2006). *Conceptual process configurations in enterprise knowledge management systems.* Paper presented at the 21st ACM symposium on applied computing (SAC 2006), Dijon, France.

Booch, G., Rumbaugh, J., & Jacobson, I. (2005). *The unified modeling language user guide* (2nd ed.). Reading MA: Addison-Wesley.

Chen, P. (1976). The entity-relationship model: Towards a unified view of data. *ACM Transactions on Database Systems, 1*(1), 9-36

Curran, T., & Ladd, A. (2000). *Sap R/3 business blueprint: understanding enterprise supply chain management* (2nd ed.). Upper Saddle River, NJ: Prentice Hall PTR

De, P., & Sen, A. (1984, September). A new methodology for database requirements analysis. *MIS Quarterly*, 179-193.

De Swaan Arons, H., & Boer, C.A. (2001). Storage and retrieval of discrete-event simulation models. *Simulation Practice and Theory, 8,* 555-576.

De Vreede, G-J., Verbraeck, A., & van Eijck, D. (2003). Integrating the conceptualization and simulation of business processes: A modeling method and arena template. *Simulation, 79*(1), 43-55.

Enter, N. (1999). *The semantics of the CIC SAP R/3 core.* Master thesis project, international business studies. Maastricht, The Netherlands: University of Maastricht, Faculty of Economics and Business Administration.

Falkenberg, E. (1976). Significations: The key to unify data base management. *Information Systems, 2*(1), 19-28.

Floss, P. (1997). *Requirements for transitioning business process simulation models to real-time operational systems.* Paper presented at the 1997 Winter Simulation Conference.

Giaglis, G., Paul, R., & Hlupic, V. (1999). Integrating simulation in organizational design studies. *International Journal of Information Management, 19,* 219-236.

Gorry,G., & Scott Morton, M. (1971). A framework for management information systems. *Sloan Management Review, 13*(1), 55-70.

Greasley, A. (2006). Using process mapping and business process simulation to support a process-based approach to change in public sector organization. *Technovation, 26,* 95-103.

Gregoriades, A., & Karakostas, B. (2004). Unifying business objects and system dynamics as a paradigm for developing decision support systems. *Decision Support Systems, 37*, 307-311.

Halpin, T. (2001). *Information modeling and relational databases*. San Francisco: Morgan Kaufmann Publishers.

Hlupic,V., & Robinson, S. (1998). *Business process modeling and analysis using discrete-event simulation*. Paper presented at the 1998 Winter Simulation Conference.

Jacobs, F., & Whybark, D. (2000). *Why ERP? A primer on SAP implementation*. Irwin McGraw-Hill.

Julka, N., Lendermann, P., Chong, C-S., & Liow, L-F. (2004). *Analysis and enhancement of planning and scheduling applications in a distributed simulation testbed.* Paper presented at the 2004 Winter Simulation Conference .

Krajewski, L., Ritzman, L., & Malhotra, M. (2007). *Operations management* (8th ed.). Upper Saddle River, NJ: Pearson Education.

Nidumolu, S., Menon, N., & Zeigler, B. (1998). Object-oriented business process modeling and simulation: A discrete event system specification framework. *Simulation Practice and Theory, 6*, 533-571.

Nijssen, G. (1977). *On the gross management for the next generation database management systems.* Paper presented at Information Processing 1977 IFIP. Amsterdam: North-Holland.

Nijssen, G. (1981). Databases of the future: Design and methodology, *Database Journal, 12*(1), 2-14.

Nijssen, G. (1989). *An axiom and architecture for information systems.* Paper presented at the IFIP Conference Information System Concepts: An In-depth Analysis. Amsterdam: North-Holland.

Nijssen, G.M, & Halpin, T.A. (1989). *Conceptual schema and relational database design: A fact oriented approach.* Upper Saddle River, NJ: Prentice-Hall.

Olle, T.W., Hagelstein, J., Macdonald, I.G., Rolland, C., Sol, H.G., van Asche, F.J.M., & Verrijn-Stuart, A.A. (1988). *Information systems methodologies: A framework for understanding.* Amsterdam: North-Holland.

Parker, M. (1982). Enterprise information-analysis. *IBM Systems Journal, 21*(1), 108-123.

Perera, T., & Lyianage, K. (2000). Methodology for rapid identification and collection of input data in the simulation of manufacturing systems. *Simulation Practice and Theory, 7*, 645-656.

Reijers, H., & Aalst van der, W. (2005). The effectiveness of workflow management systems: Predictions and lessons learned. *International Journal of Information Management, 25,* 458-472.

Robertson, N., & Perera, T. (2002). Automated data collection for simulation? *Simulation Practice and Theory, 9,* 349-364.

Rolland, C. (1983, Spring). Database dynamics. *Database,* 32-43.

Rumbaugh, J., Jacobson, J., & Booch, G. (2004). *The unified modeling language reference manual* (2nd ed.). Reading MA: Addison-Wesley.

Sadoun, B. (2005). Optimizing the operation of a toll plaza system using simulation: A methodology. *Simulation, 81*(9), 657-664

Schriber, T., & Brunner, T. (1997). *Inside discrete-event simulation software; How it works and why it matters.* Paper presented at the 1997 Winter Simulation Conference.

Silva, M., & Carlson, C. (1995). MOODD, A method for object-oriented database design. *Data & knowledge engineering, 17*(1995), 159-181

Sol, H. (1982). *Simulation in information systems development.* PhD Thesis, University of Groningen, The Netherlands

Sprague, Jr., R. (1980, December). A framework for the development of decision support systems. *MIS Quarterly,* 1-26.

Teory, T., Yang, D., & Fry, J. (1986). A logical design methodology for relational databases using the extended E-R model. *ACM Computing Surveys, 18*(2), 197-222.

Verheijen, G., & van Bekkum, J. (1982). *NIAM: An information analysis method.* Paper presented at the IFIP TC-8 Conference on Comparative Review of Information Systems Methodologies (CRIS-1). Amsterdam: North-Holland.

Wieringa, R.J. (1996*). Requirements engineering: Frameworks for understanding.* New York: Wiley.

Endnotes

[1]   Although the setting and documents in this example are fictive, they are based on a real life application of NLM in a large multi-national corporation that has implemented SAP (Enter, 1999).

[2]   The verbalization must lead to sentences that cannot be split up further without losing meaning (semantic irreducibility). A synonym for a set irreducible semantic equivalent sentences is a *fact.*

[3]   A significant part of a Universe of Discourse in this step of the information modeling methodology should be considered a part of that UoD that contains all possible variation in sentence types

[4]   If in the initial verbalization of a communication example the intention has been left out, it must be added in this step of the NLM modeling procedure.

[5]   Synonym for a set of semantically equivalent sentence groups is *fact type*.

[6]   We are only interested in conceptual process configurations in which an instance of a given fact type is created. Therefore, in our view only process configurations having a post-condition are relevant. Furthermore, we can only define a specific derivation rule when a post-condition exists.

[7]   For those intentions that have not been defined in the list of concepts that is contained in the AIG, a definition of such an intention will be added.

[8]   An event occurence is referenced by an event instance and the relative moment in time in which the event instance has occurred.

[9]   This means that the qualification of a condition-process trigger can contain a derivation rule that is defined on fact instances from the application information base.

[10]   To be able to decide between an INSERT or an UPDATE event for the creation of a fact instance, the process manager should at least have knowledge on the uniqueness constraints of the AIG, therefore the process manager needs access to the *AIG* (see *Figure 20*).

Chapter V

Relay Race Methodology (RRM):
An Enhanced Life Cycle for Simulation System Development

Evon M. O. Abu-Taieh,
The Arab Academy for Banking and Financial Sciences, Jordan

Asim Abdel Rahman El Sheikh,
The Arab Academy for Banking and Financial Sciences, Jordan

Jeihan Abu Tayeh, Ministry of Planning, Jordan

Abstract

This chapter introduces a suggested system development life cycle "relay race methodology" (RRM). The RRM is based on the philosophy of relay race, where each runner in the race must hand off the baton within a certain zone, usually marked by triangles on the track race. This chapter is comprised of nine sections. First, it shows the relationship between software and wicked problems. Then the chapter explains the motivation for analysis and design for simulation system projects. Furthermore, the chapter gives an overview of the reasons behind simulation project failures. Next, the chapter shows the sources for simulation inaccuracies and the project management issues in simulation. Subsequently, the chapter explains the proposed RRM life cycle. Consequently, the chapter presents the advantages and disadvantages of the proposed RRM life cycle, while relating the RRM to the risk factors. Finally, the chapter compares RRM to different life cycles.

Introduction

Software is defined by DeGrace and Stahl (1990) as a wicked problem; accordingly Rittell and Webber (1973) defined wicked problems as "problems that are fully understood only after they are solved the first time."

Moreover, according to Conklin (2003), the four defining characteristics of wicked problems are:

1. The problem is not understood until after formulation of a solution
2. Stakeholders have radically different worldviews and different frames for understanding the problem.
3. Constraints and resources to solve the problem change over time.
4. The problem is never solved.

Likewise, wicked problems, according to Horst and Webber, have ten characteristics:

1. There is no definitive formulation of a wicked problem.
2. Wicked problems have no stopping rule.
3. Solutions to wicked problems are not true-or-false but good-or-bad.
4. There is no immediate and no ultimate test of a solution to a wicked problem.
5. Every implemented solution to a wicked problem has consequences.
6. Wicked problems do not have a well-described set of potential solutions.
7. Every wicked problem is essentially unique.
8. Every wicked problem can be considered a symptom of another problem.
9. The causes of a wicked problem can be explained in numerous ways
10. The planner (designer) has no right to be wrong.

In the same token, Poppendieck (2002) proposed, in her article titled *Wicked problems* that "[t]here is nothing wrong with using adaptive processes to solve wicked software development problems. In fact, it is the problem cannot be tamed, this is the only 'good' choice" [Poppendieck, 2003]. Indeed, the author suggested *scrum* methodology in her paper. Moreover, Ambler suggested to "marry agility to the Unified Process (UP)" in his article *Unified and agile*.

Motivation for Analysis and Design in Simulation

Defining simulation in its broadest aspect as embodying a certain model to represent the behavior of a system, whether that may be an economic or an engineering one, with which conducting experiments is attainable. Such a technique enables the management, when studying models currently used, to respond appropriately and make fitting decisions that would further complement today's growth sustainability efforts, apart from cost decrease, as well as service delivery assurance. As such, the computer simulation technique contributed to cost decline; depicting the "cause & effect," pinpointing task-oriented needs or service delivery assurance, exploring possible alternatives, and identifying problems, as well as proposing streamlined measurable deliverable solutions, providing the platform for change strategy introduction, introducing potential prudent investment opportunities, and finally providing a safety net when conducting training courses. Yet, simulation development process is hindered due to many reasons. Like a rose, computer simulation technique does not exist without thorns; of which are the length, as well as the communication, during the development life cycle. Simulation reflects real life problems; hence, it addresses numerous scenarios with a handful of variables. Not only is it costly as well as liable for human judgment, but also the results are complicated and can be misinterpreted.

Within this context, there are four characteristics that distinguish simulation from any other software-intensive work. The following are the four characteristics, as discussed by Page and Nance (1997):

- *Time* as index variable: in simulation there is an indexing variable called *TIME*. In discrete event simulation (DES), this indexing variable "establishes an ordering of behavioral events" (Page & Nance, 1997, p. 91).

- *Correctness* of simulation software: in software engineering correctness is one of the objectives like reliability, maintainability, testability, and so forth. Yet the objective of correctness is very special to simulation software for a simple reason: how useful is a simulation program "if questions remain concerning its validity" (Page & Nance, 1997, p. 91).

- *Computational Intensiveness*: In their paper, Page and Nance (1997) clearly state that:

  The importance of execution efficiency persists as a third distinctive characteristic. While model development costs are considerable, as is the human effort throughout the period of operation and use, the necessity for repetitive sample generation for statistical analysis and the testing of numerous alternatives forces concerns for execution efficiency that are seen in few software-intensive projects (p. 91).

- *Uses of simulation*: the uses of simulation are not typical, in fact "[n]o typical use for simulation can be described" (Page & Nance, 1997, p. 91).

Why Do Simulation Projects Fail?

The arising issue of simulation projects falling short of being labeled as successful can be attributed to many reasons. An answer by Robinson (1999) had been put forth, as he listed three main reasons; the first being "poor salesmanship when introducing the idea to an organization" (p. 1702), which includes too much hope in too little time, while identifying the second reason as "lack of knowledge and skills particularly in statistics, experimental design, the system being modeled and the ability to think logically" (p. 1702), and pinpointing the third reason as "lack of time to perform a study properly" (p. 1702). Nevertheless, simulation inaccuracy has become a recurrent condition that instigated a thorough query in its sources.

Sources of Simulation Inaccuracy

There are three sources of inaccuracy in the simulation project that might be developed during the three major steps that are in the simulation life cycle, namely: modeling, data extraction, and experimentation (see *Figure 1*).

In this regard, the modeling process includes a subordinate set of steps, namely the modeler understanding of the problem, then developing a mental/conceptual model, and finally the coding. Noting that from these steps some problems might mitigate themselves, such as: (i) the model could misunderstand the problem, (ii) the mental/conceptual model could be erroneous, and (iii) the conversion from mental/conceptual model to coding could be off beam.

Furthermore, during the modeling process, the data collection/analysis is a key process element, particularly since the data collected is really the input of the simulation program, and if the data is collected inaccurately then the principle of *"garbage in garbage out"* is clearly implemented; likewise, the data analysis while using the wrong input model/distribution [see Leemis (2003)] is also a problem.

Last but not least, the third source of inaccuracies is experimentation, which is using the collected data used in the simulation system and comparing the end result to the real world, given that experimentation inaccuracies can result from ignoring the initial transient period, insufficient run-length or replications, insufficient searching of the solution space, or not testing the sensitivity of the results.

Within this context, ignoring the initial transient period, labeled as the first inaccuracy source during experimentation process, has been identified by Robinson (1999), when stating that: "Many simulation models pass through an initial transient period before reaching steady-state" (p. 1705). The modeler, suggests Robinson, can either take into account such a period or set the simulation system so that the system has no transient period. Moreover, insufficient run-length or replications, particularly as running the simulation system long enough is essential in order to reach the results that reflect real life; therefore, Robinson suggests two remedies; (i) run the simulation system long enough, or (ii) do may reruns (replications). In addition, insufficient searching of the solution space is the third and last source, which in turn would

Figure 1. The simulation modelling process (simple outline) (Robinson, 1999, p. 1702)

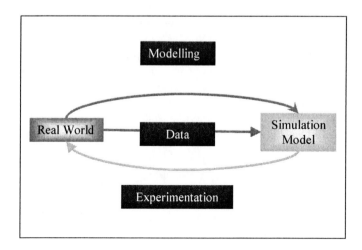

incite the modeler to "only gain a partial understanding of the model's behavior" (Robinson, 1999, p. 1702): such inaccuracy obviously leads to erroneous conclusion.

Indeed, errors in the simulation world do not originate only from one source, therefore validation, verification, & testing are not only considered a necessity, but also considered to be imperative and crucial. Therefore, Osman Balci, the well-known simulation scientist, declared 15 principles demonstrating the fundamentality and inevitability of conducting VV&T to the simulation world.

Project Management Issues in Simulation

In view of the aforementioned, a comparison is put forth between the simulation perspective and the software engineering perspective with respect to project management, project framework, and project life cycle.

When discussing project management issues in simulation, *Figure 2*, developed by Pidd, is taken into consideration. [Pidd, 1998] breaks any simulation project into two parts: simulation problem and simulation project. The simulation problem entails the structuring of the problem (specification) then modeling and finally the implementation. The second part, simulation project, reflects the management of the project. Both parts must work in parallel.

Moreover, because of the evolutionary nature of the simulation projects, the work of Sommerville on evolutionary development in software engineering is further considered. In *Figure 3* (Sommerville, 2001), evolutionary development, which is the second nature for any simulation project, is described. *Figure 3* (Sommerville, 2001) describes the process in two parts: activities and outcome of each activity, although the outcomes of each activity must inherently be supervised by project management.

In comparing both works, the resemblance of thought and work is apparent, as both break the problem into technical and management jobs, and both depict the jobs to be executed concurrently. Nevertheless, when taking a closer look at both figures, the only difference to

Figure 2. Solving problems and managing project (Pidd, 1998)

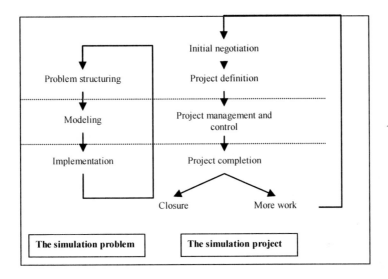

Figure 3. Evolutionary development (Sommerville, 2001)

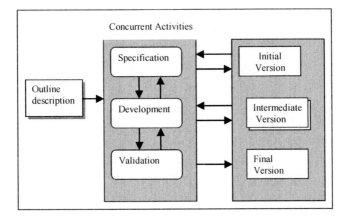

be noted would be the wording. As such, Pidd's problem structuring, as illustrated in *Figure 1*, is not different from Sommerville's specification in real life, and the same thing goes when Pidd tackles modeling and Sommerville tackles development. In fact, it is well established that this phase is, basically, designing the model and programming it.

Figure 4. Simulation framework (Paul & Balmer, 1993)

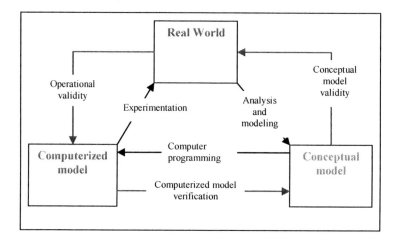

Figure 5. Software life cycle – waterfall model

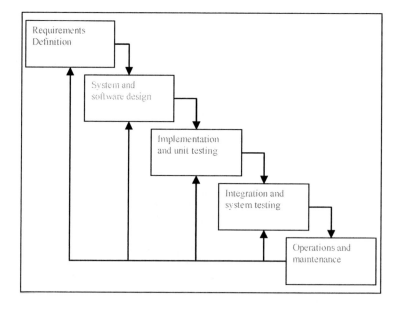

However, in view of the fact that the final stage represents basic implementation testing and ensuring that what was imagined would work as it was specified in the beginning (Banks, 2000O, and that it signifies implementation of the thought but before that the final product must be tested thoroughly and as an ongoing process, (Banks, 2000)"…whether the code reflects the description found in the requirement document" (McKay et al., 1986), as such, the final stage for Pidd is implementation and for Sommerville is validation.

In light of the aforementioned, the work of two famous simulation scientists, Paul and Balmer, are compared, in respect to the describing of simulation framework, to Sommerville's software life cycle. *Figure 4* [see Paul & Balmer (1993)] describes the framework of simulation. In *Figure 4*, one produces a conceptual model by analyzing and modeling the real world. The computerized model is produced by programming the conceptual model, then using experimentation to compare the computerized model with the real world.

Looking at the waterfall model (*Figure 5*), where the project is broken down to five phases and the product of each phase feeds as an input to the next, the five phases—requirement definition, system and software design, implementation and unit testing, integration and system testing, and operation and maintenance—are not different from the phases of *Figure 4*.

Analysis and modeling, in *Figure 3*, is not in nature different from requirement definition in *Figure 5*. The building of the conceptual model in *Figure 4* is not different from system and software design in *Figure 5*.

The two steps of computer programming and computerized model verification, in *Figure 3*, to build a computerized model are not different from Implementation and unit testing in *Figure 5*. Also, the two steps of experimentation and operational validity to compare computerized model to the real world are not different from system testing and operations and maintenance in *Figure 5*.

A third example of comparison is seen by looking at *Figure 6* and *Figure 7*. *Figure 6* has two figures. The first, *Figure 6.1* shows the basic process of computer aided simulation modeling (CASM) (Paul & Balmer, 1993). The second, *Figure 6.2*, shows a graphical simulation environment. *Figure 6* is system evolution (Sommerville, 2001).

Figure 6.1 shows the steps of the basic process in CASM. First the analyst produces a model logic, usually using activity cycle diagram notations. Then interactive simulation program generator (ISPG) produces a data file of the model logic. Another software subsystem will produce simulation model. *Figure 6.2* shows more detailed picture of *Figure 6.1*.

In system evolution, *Figure 7*, the four phases each produce an output that is used by the subsequent phase. The last phase feeds in to the first to show the evolution process.

Comparing the two figures, it is noted that building the model logic by the analyst (in *Figure 6.1*) is not different from "define system requirement" in *Figure 7*. "Assess existing system" is not different from "experimentation and operational validity" in *Figure 4*. Also, one must mention the cyclic nature in both *Figure 6.2* and *Figure 7*.

In conclusion, of all three comparisons, the main inference would be: simulation and software engineering both are using the same ideas but differ in the use of notations and vocabulary. Yet both can benefit from each other if both spoke the same language and agreed on the use of the same notations. For example, simulation is meticulous when discussing the testing phase. Simulation expresses the differences between validate and verify, which is understandable given the immense need for the testing phase. Whereas, simulation recapitulates the initial

Figure 6. CASM proposed life cycle

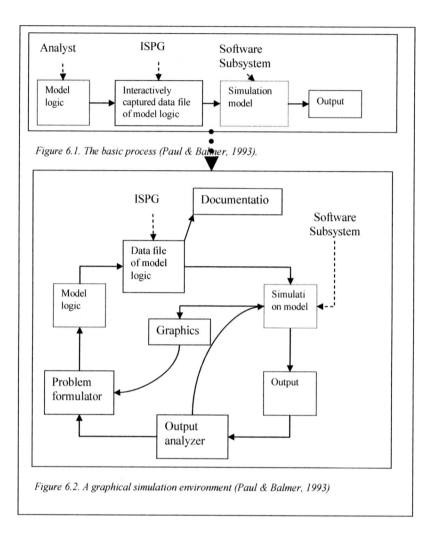

Figure 6.1. The basic process (Paul & Balmer, 1993).

Figure 6.2. A graphical simulation environment (Paul & Balmer, 1993)

phases: requirement definition through operations and maintenance into six process steps. Additionally, simulation and software engineering mix between product of a phase and the phase itself or the state of the project; for example, in *Figure 5* there are two jobs that must be done to reach from real world to conceptual model. In other words, what is considered a step of a process in simulation is a phase in software engineering *Figure 5*.

Furthermore, software engineering uses data flow diagrams (DFD), entity relations (ER), rational unified process (RUP) (Rational, 2004), and so forth in order to analyze the existing

Figure 7. System evolution from Sommerville (2001)

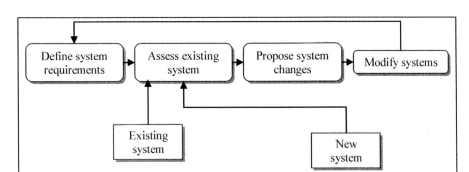

system, design and manage the software system, while simulation uses many tools among them activity cycle diagram (ACD) [see Paul (1993)].

"To be a good simulation modeler, you must merge good problem solving techniques with good software engineering practices" [see Shannon (1998)]. To bridge the gap between simulation and software engineering, the authors suggest a life cycle that captures the benefits of both worlds yet tries to use a language that is understandable to both.

RRM Life Cycle

As stated in the beginning of the chapter, RRM is based on the fundamentals of the relay race. In every phase a core team must attend: the core team is always composed of analyst, programmer, statistician, and user. In every phase, two of the quad-core team from previous phase must attend the subsequent phase. The idea of the relay race is apparent, still the reason is that some knowledge accumulated from the project in the team members can not be fully documented. Also the shared team members will be responsible for defending, explaining, or accepting changes suggested in the subsequent phases.

The use of case tools—planning toolkit, analysis toolkit, design toolkit, and coding toolkit—are essential in RRM. In addition, JAD sessions are highly encouraged, since part of their job is to clarify any ambiguity that may rise.

Next, the phases as seen in *Figure 8*, will be explained: start, requirements determination phase, and structuring system process requirements phase.

Figure 8. Proposed RRM

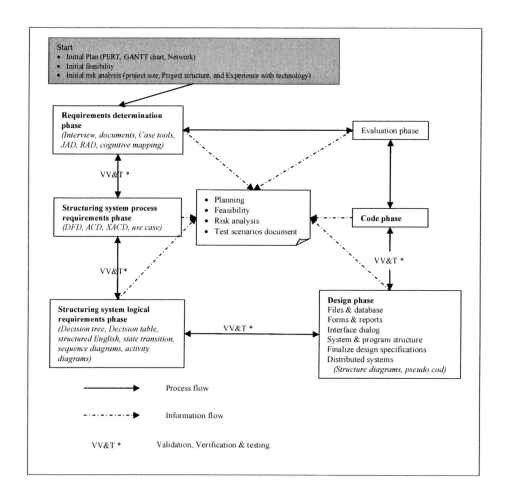

Start Phase

The start phase is the first phase in the life cycle. During start phase, three products are expected:

1. **Initial Plan:** There are many tools and techniques used to develop the plan document needed, that is, PERT, GANTT chart, network diagrams [best described in Hoffer (2005)]. The initial plan must be conducted from the start of the project and will be updated all during the project. Timesheets and milestone methods describe by Cadle

and Yeates (2004) are recommended for this document. Initial plan is a good tool to measure progress of the project.

2. **Initial feasibility study:** The team must conduct a feasibility study; Cadle & Yeats (2004) describe the contents of the initial feasibility study. Items like fixed cost, running cost, tangible and intangible cost and benefits. Many methods are used for estimation: analogy method, analysis method, programming method, direct estimation based on project breakdown, Delphi technique, constructive cost model (CoCoMo), and function point analysis. Such methods can be best understood by reading Cadle and Yeates (2004) and Hoffer (2005)

3. **Initial risk analysis:** There are many factors that effect the risk in the project including project size, project structure, and experience with technology [see Cadle & Yeates (2004) and Hoffer (2005)]. Initial risk analysis must be carried out for the simulation system in every step of the project.

All products are dependent on reading documents and past case studies. Start phase will produce the expected activity cycle diagram. Thus, the analyst will be prepared to interview the users in the subsequent phase.

Requirements Determination Phase

Requirements determination phase is the second phase. The phase shall entail individual interviews, which will produce more documents for the analyst to read. Towards the end of this phase, the analyst along with programmer will hold a JAD session. The JAD session will be held using RUP. Such an idea will help in eliminating contradictions.

The product of the phase is the requirements document.

Tools used in this phase include, but are not limited to: interview, documents, case tools, JAD, RAD, and cognitive mapping.

The requirements must be validated in a validation session or sessions. This phase is tightly linked to the subsequent phase. Part of the team from this phase must move to the next phase. In addition, moving forward in phase is a must; any alteration or amendments can be reflected in the appropriate documentation.

Structuring System Process Requirements Phase

Structuring system process requirements is the third phase in the RRM SDLC. The goal of this phase is to rid the system from any ambiguity and to build a model that reflects a real life. The core team will produce dataflow diagrams (DFD), activity cycle diagrams (ACD), or use case diagrams. The core team in a verification session will authenticate that the diagrams reflect the real life.

In case of any errors found in this phase that originate from the previous phase, a collective decision must be made to send the project to the previous stages. In such a case, the shared team member must either defend or explain the mishap.

Once this stage is completed, the project moves to the next stages, along with shared team members.

Structuring System Logical Requirements Phase

The fourth phase is structuring system logical requirements phase, which entails using the traditional tools and techniques like decision tree, decision table, structured English, state transition, sequence diagrams, activity diagrams.

The outcome of this phase will be the input of the design phase, but it must go through VV&T transition.

Design Phase

The design phase will produce a system/program design that can be converted to code. The goal of this phase is to build structure diagrams, with pseudo code that describe each unit. Also in this phase the interface dialog is designed, along with any files & database. Any design issues regarding the distribution of the system will be done in this phase.

Again, in case of any errors found that stem from previous stages a collective decision made to send the whole project to a previous stage. In such a case the shared team members of the two stages must explain, defend or accept such doing. In the case that the project is sent back to a previous stage the original team must be reinstated and must rectify the project. The project is not sent back to previous stages with out the proper reasons documented.

In case this stage is completed, the project moves to the next stage with, also, shared team members to carry the baton.

Code Phase

During the code phase all specification produced from design phase will converted to code. The use of code generation toolkit is essential in this phase given that the previous phases used the appropriate toolkits. Since RRM is iterative, and facing the fact of life, which is "nothing is engraved in stone," code generating tools are essential in this approach. Design is bound to change, therefore; the use of code generation tools like Designer 2000, Rational Rose will alleviate the frustration of the programmer and reduce errors that are sure to result.

The next stage in the code, which really runs in parallel with coding, is system integration. As the programmer is coding, the integration of the system will be taking place. Therefore, any errors will be discovered as soon as possible.

In case an error or mishap was found that stems from the previous stage, the project is sent to the previous stage. The shared team must explain, defend, or accept such a decision. The decision must be collective. In such a case the previous stage team must be reinstated and must rectify the mishap.

Evaluation Phase

In the evaluation phase, all the accumulated test cases & scenarios from previous phases will be tried out. In addition to matching the requirements, evaluation phase will include system test and installation. Due to the nature of simulation systems, testing the system does not mean the system is error free; therefore part of the system test is an initial installation, where the system will be tested using scenarios accumulated during the RRM life cycle.

Validation, Verification & Testing (VV&T)

No life cycle is complete without VV&T phase. Still, in this proposed life cycle VV&T transpires in every phase. Balci (1994), in his research, suggested a taxonomy of 45 VV&T techniques, moreover, in another paper, Balci (1997) defined 77 VV&T techniques and 38 VV&T techniques for object-oriented simulation. Abu-Taieh and El Sheikh (2006) accumulated most VV&T techniques and their uses in simulation.

The Team and Knowledge Management

People involved in the software development life cycle gain knowledge about the system at hand; the knowledge is stored in many ways yet the most of the knowledge stays in the head of the team member. In RRM, in order to avoid this leak of knowledge, the methodology insists on having part of the team that was involved in the previous stage to be involved in the next stage. As stated in the advice of Satyadas (2003) in the three key points to ensure the success of developing KM strategy: "A KM strategy should highlight and intertwine three areas: people, processes and technology" (Satyadas, 2003).

The team is made of four types of people: analyst /modeler, developer, customer, and statistician.

1. The analyst /modeler is the person that has the basic job of building the model to be developed later by the developer. Analyst/ modeler must very familiar with the system being built, whether by experience of by specialty, and must be familiar with analysis & design of information systems.

2. The user/customer is a team of people made with two tasks in mind: facilitate the job of the team and explain to the team what needs to be done. The user/customer team must have the knowledge and motivation to be involved in such a project. In order to achieve such goals there are two types of users needed: the high-end user like management, which is needed for commitment and to facilitate the work of the whole team, and the low-end user, which are the ones that know details of the system and usually have the knowledge.

3. The developer is the team of people that will convert the model to software program. Such team must be highly skilled in both ends of the project (simulation and animation).

4.   The statistician is the person who knows statistics related to simulation. S/he will be the expert that explains the statistics part of the simulation system to the rest of the team. Such person must at least have been involved with previous project.

The team works much like in relay race where each runner hands off the baton to the next runner. During the RRM life cycle phase, at least one of the team members will continue to the next phase. The baton in this case is the knowledge acquired during the previous phase

Each of the team members is required to spend some effort during all the phases, yet time spent will vary depending on the phase and the type of the team member. The analyst/modeler will have the lion's share in all the phase since s/ he has also the role of the project manager. The statistician will be involved in all the phases, but mainly in the design and code phases. In the previous phases the statistician will give as much as s/he takes. The team member of type developer will put in more effort, understandably, during the design and code phases; yet the developer must spend some time in the previous phases mainly to learn about the system to be developed. The user/customer must be available and participate in the all the phases, except for the coding phase. During the coding phase the user/customer will be less needed. especially if the user/customer spent enough time in the previous phases. *Figure 10* reflects the approximated time needed to be spent by the different team members.

Disadvantages of RRM

•   Relies heavily on highly skilled and experienced team members.

Figure 9. The team in proposed RRM

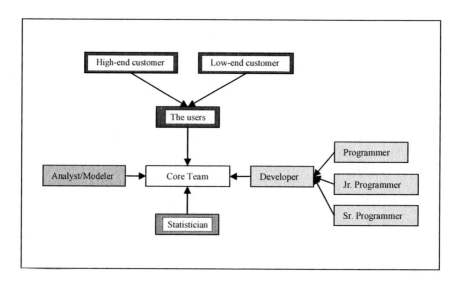

- User commitment is needed for final product; however simulation systems projects are long.

- Relies heavily on the use of case tools *Rational Rose and Designer 2000.* Such fact is not really a drawback. Case tools actually save much needed time and effort in simulation system project.

- RRM hugs the resources, which is translated to cost, yet one may defend such accusation by comparing the cost of finding the error before the end of the simulation system project or after.

Advantages of RRM

- RRM is flexible: from any phase one can go back to previous stage.

- RRM is adaptive.

- RRM has open channels of communication, which is essential to produce a correct simulation system.

- User involvement is a basic ingredient in RRM.

- Diagrammatic documentation: the use of graphic-based documentation.

Figure 10. The team involvement in the project phases

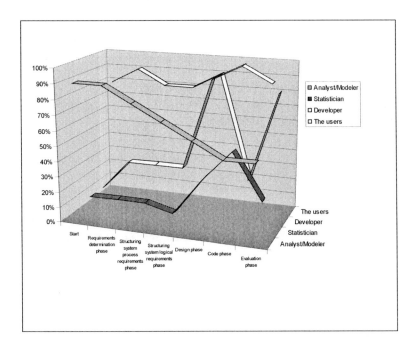

- There is initial business case or defined acceptance criteria, which facilitate the measurement of success/failure.

- RRM has discrete stages & defined structure, and addresses the quality management issues through validation and verification. Also covers the risk analysis issue.

Compare RRM to other SDLC

There are many traditional life cycle models: waterfall mode, "b" model, "v" model and incremental model. In addition, there are iterative or evolutionary life cycle models like Boehm's spiral model. RRM tried to take from both types the best traits.

RRM is like traditional life cycle models in the sequencing of the stages and addressing "the elements of quality management through verification and validation" (Cadle & Yeates, 2004). Yet RRM addresses risk factors as seen in the previous section, which are usual shortcomings in traditional life cycle models.

RRM is also like spiral model in some respects. Spiral model is a design for system where requirements are not well formed and /or understood by the user. Yet spiral model is evolutionary, in that it will increase the length of increases as the project evolves. Yet RRM insists on things that eluded spiral model: for example, RRM insists that to move from one stage to the next part of the team will move to the next stage also, since the accumulated knowledge is stored in the team member not on a storage device. In addition, RRM does not depend on evolutionary cycles; rather it depends on going back one-step forward/backward at the time. The team must move with the stage, for example, when going backward to a previous stage, the same team is reinstated and the reason for going back is resolved. The number of times of going backward to the same stage is limited, unlike spiral, which technically sweeps through the stages.

RRM and Risk Factors

Sumner summarized 17 risk factors in information systems projects, RRM tried to cover as many of those factors as possible. Each factor will be discussed next and how RRM counter-affected the risk factor:

The first factor, "Capabilities, failure of technology to meet specifications," was counter-affected by an iterative process of RRM. The second factor, "Lack of agreement on project goals," was countered by the open channel of communications. As for "Lack of technical expertise," RRM insists on the MIX of the types of people and also by insisting on the user involvement. The same goes for "Lack of application knowledge:" RRM insists on user involvement including management. The factors "Lack of user commitment, ineffective communications with users" and "Lack of senior management involvement" are also included in the RRM.

As for the "Application complexity (technical complexity)" factor, it is counter affected by use of code generation tools and insisting on the involvement of technical (programmers) from the start of the simulation project. The risk factor "Misunderstanding requirements, changes in requirements" was responded to in RRM by involving both the user and programmer from the start of the project. As for the risk factor "Organizational environment (resource insufficiency, extent of changes)," RRM insisted on the involvement of the management in all the phases of the project.

For the risk factor "Unrealistic schedules and budgets," RRM insisted on keeping the planning document, feasibility study, open for change in all the phases of the simulation system project. The risk factor "Lack of an effective methodology, poor estimation, failure to perform the activities needed" is dealt with in RRM by facing the fact of life that nothing is engraved in stone. The poor estimation is bound to become more accurate as the project progresses, yet an initial estimation is a must.

The risk factor listed by Sumner (2000), "Changing scope and objectives," is a nature of simulation system projects, yet since RRM is iterative and flexible. The involvement of the user offsets the risk factor "Conflicts between user departments;" also the user of JAD sessions. RRM insists on the type of the user and JAD sessions to avoid the risk factors: "Inappropriate staffing, personnel shortfalls" and "People and personality failures." RRM suggests many measurement techniques and systems to counteract the risk factor listed by Sumner (2000): "Lack of measurement system for controlling risk, inadequate project management and tracking."

References

Abu-Taieh, E., & El Sheikh, A. (2006). Discrete event simulation process validation verification and testing. In Dasso & Funes (Eds.), *Verification, Validation and Testing in Software Engineering*. Hershey, PA: Idea Group Inc.

Ambler, S.W. (2006, January). Unified and agile. *Software Development*. Retrieved January 2006, from www.SDmagazine.com

Balci, O. (1994). Validation, verification, and testing techniques throughout the life cycle of a simulation study. *Annals of Operations Research, 53*, 215-220.

Balci, O. (2003, October). Verification, validation, and certification of modeling and simulation applications. *ACM Transactions on Modeling and Computer Simulation, 11*(4), 352-377.

Balci, O. (1997, Dec 7-10). Verification, validation and accreditation of simulation models. In S. Andradóttir, K.J. Healy, D.H. Withers, & B.L. Nelson (Eds.), *Proceedings of the Winter Simulation Conference,* Atlanta, GA, U.S. (pp. 135-141).

Balci, O. (1995). Principles and techniques of simulation validation, verification, and testing. In C. Alexopoulos, K. Kang, W.R. Lilegdon, & D. Goldsman (Eds*.), Proceedings of the 1995 Winter Simulation Conference*. New York: ACM Press.

Cadle, J., & Yeates, D. (2004). *Project management for information systems*. Prentice Hall.

Conklin, J. (2003). *Dialog mapping: An approach for wicked problems*. CogNexus Institute.

DeGrace, P., & Stahl, L.H. (1990). *Wicked problems, righteous solutions*. Prentice Hall, Yourdon Press.

Hoffer, J. (2005). *Modern systems analysis & design* (4th ed.). Prentice-Hall.

Leemis, L. (2003, Dec 7-10). Input modeling. In S. Chick, P.J. Sánchez, D. Ferrin, & D.J. Morrice (Eds.), *Proceedings of the 2003 Winter Simulation Conference,* New Orleans, New Orleans, LA, U.S. (pp. 14-24). Winter Simulation Conference.

Page, H., & Nance, R. (1997, July). Parallel discrete event simulation: A modeling methodological perspective. *ACM Transactions on Modeling and Computer Simulation,* 7(3), 88-93

Rittel, H., & Webber, M. (1973). Dilemmas in a general theory of planning. In *Policy sciences* (Vol. 4) (pp. 155-169). Amsterdam: Elsevier Scientific Publishing Company.

Satyadas, A. (2003, January 13). *Growing with knowledge management*. Knowledge Discovery Business Leader. IBM Lotus Software. IBM/Lotus.

Sumner, M. (2000). Risk factors in enterprise wide information management systems projects. Evanston, IL: SIGCPR.

Wicked problems. (2002, May). Retrieved January 2006, from www.SDmagazine.com

Chapter VI

Information Feedback Approach for the Simulation of Service Quality in the Inter-Object Communications

R. Manjunath, UVCE, Bangalore, India

Abstract

Simulation of a system with limited data is challenging. It calls for a certain degree of intelligence built into the system. This chapter provides a new model-based simulation methodology that may be customized and used in the simulation of a wide variety of problems involving multiple source-destination flows with intermediate agents. It explains the model based on a new class of neural networks called differentially fed artificial neural networks and the system level performance of the same. Next, as an example, the impact of system level differential feedback on multiple flows and the application of the concept are presented, followed by the simulation results. The author hopes that a variety of real life problems that involve multiple flows may be mapped onto this simulation model and optimal performance may be obtained. The model serves as a reference design that may be fine-tuned based on the application.

Introduction

System simulation in the presence of limited data is challenging, especially when the real time outputs are to be generated. A system working with a limited data set is required to have built-in intelligence to take decisions and extrapolate the results. If there is a feedback from the output to input of the system, the characteristics of the system improve. When the feedback contains historical data or a set of previous outputs, it starts exhibiting interesting properties, such as self-similarity, abstraction, entropy maximization, and so forth, as explained in a future section.

In this chapter, a converse mapping is suggested to model the systems, that is, any system or a process where the indicated properties are observed, it may be simulated with a differential feedback model. It is meaningful since it is the differential feedback that imparts the indicated properties into the system. With this mapping, the other properties of the differentially fed system are explored. It provides better understanding of the system. For example, when the self-similarity of the nucleic acid sequences in a gene is modeled with a differentially fed system, it implies the maximization of entropy in the sequence. It finally leads to the fact that the nature maximizes the randomness of the genes for the better survivability.

In this chapter, the modeling and simulation example of a data network is discussed in depth. The network exhibits self-similarity. It is an ideal candidate to be modeled with a differentially fed system. Further analysis shows that the self-similarity originates as a result of usage of historical information in the network. One striking difference with this simulation model is that it can be made as a part of the real system generating the requisite signals to control the behavior of the system.

It has been shown that the entropy gets minimized with the differential feedback. The resource contention for communication is addressed and the solution is modeled.

The shifted feedback information may be used to achieve some additional quality of service deadlines, such as the absolute delay guarantee, fraction of the services lost, and so forth. The same would be agreed upon with the different communicating units well in advance.

Analysis shows that a system exhibits self-similarity to maximize the entropy (Manjunath, 2005). It is proved that, to maximize entropy, the system makes use of differential feedback of different degrees. They form various levels of abstraction and by and large carry redundant information (Manjunath, 2005). The self-similar property has been exploited here to maintain the quality of service constraints.

Because of abstraction and redundancy, even if a portion of the information is lost or if it is required to predict the future uncertainties with minimum available information, it can be repaired or re-synthesized using the available information. The self-similar property of the component induces interesting properties into the system. This property may be used as lead-lag components in controlling the information transfer over the network. Closed loop feedback is utilized to control the signals transferred over the network. Intermediate self-similar structures or switches may modulate feedback signals and control the system behavior.

The simulation techniques outlined here may be used for constraint bound communication between any pair of objects, be it the defense supply chain or the chemical reaction chamber.

Background

Model is the simplified representation of the system intended for the thorough understanding of the system. It is a symbolic representation of the working of the system. In this chapter, a DANN, to be introduced in the next section, would be the model for a complex system such as the Internet. A system may be expressible as a model in a variety of ways including mathematical, diagrammatic, and so forth. The mathematical models generally represent input-output relation in the form of equation. The equations could be simple linear equations or complex differential equations.

The technique of using the appropriate model to imitate the system and generating the desired input-output patterns refers to simulation. The power of simulation lies in the selection of the appropriate model. Hence a variety of models exist suitable for the different systems to be simulated.

The system and the model would produce exactly similar output for the similar inputs. Hence the models can safely drive the system to unexplored corners of the input. Simulation would find its real use in a random and evolving system. The model developed in one application area can be easily used in the other domain if the system behavior turns out to be similar. In this chapter, the similar behavior of a DANN and a data network are identified first. The other properties of the DANN are then extended to the data network. The simulation results prove the same.

The models are often inexpensive, reusable and safe to explore. For example, simulation model turns out to be safe and cheap to train a pilot before getting into the actual plane.

The analog systems are generally simulated with a differential equation. The equation relates the output and the input of the system. Further, Laplace transform is used to solve the equation and provide the output for the applied input. A discrete time system, on the other hand, used Z-transform in the place of Laplace transform. Depending upon the dependencies between the input and the output of the system, there are three categories of discrete time system models: autoregressive (AR), moving average (MA) and the auto regressive moving average (ARMA). In an auto regressive model, the present output happens to be a linear function of the applied input as well as the previous inputs. In a moving average system, it happens to be a function of previous outputs only. An ARMA is the combination of AR and MA models. It captures the correlations associated with the data and enhances the predictability. As a result, a smaller sample size is sufficient for simulation. One of the limitations of the model is that it is linear and unable to handle the non-linear relationship between the input and the output. The differentially fed neural networks introduced in the next section make use of ARMA as an integral component. The ARMA output would be passed through a non-linear function to cater for the non-linear relationship. The increased complexity of DANN relative to ARMA can be justified with the associated better error performance and the speed.

The simplest of all models that describe the relationship between a pair of variables is the linear or straight line model. The refined form of this linear model is the least square technique that finds the line minimizing the sum of distances between observed points and the fitted line. Other techniques of simulation include Monte Carlo simulation (Kreutzer, 1986), which combines the interaction among a large number of systems.

Finally, simulation may be applied on one system if it resembles another system. A good example is the pharmacological experiments on chimpanzees, as they share many characteristics of the biological system in common with the human beings. Similarly artificial neural networks model a general non-linear system. The differentially fed neural network can simulate all nonlinear systems especially where:

- There is a contention for resources
- There is self-similarity in the output
- There is long range dependency or past history in the output
- There is hierarchical representation of data with abstraction

This broad category makes them useful in the simulation of properties of genes (self-similarity) to packet transfer over computer networks (resource contention). The other areas include radio frequency channels (long range dependency), photosynthesis (self-similarity of plastid arrangement), human memory (hierarchical representation), and so forth. The usage of differential feedback model in these areas has thrown out interesting observations and results (Manjunath, 2005b). In this chapter the usage of differentially fed neural networks for resource allocation problem in a data network is discussed in detail. The techniques are applicable for similar situations including operating systems, supply-demand problems, and so forth.

Simulations can provide the system behavior by compressing the time and space. The universe model, for example, can provide the status of the universe at any point between fractions of a second to billions of years in no time. The other classification of model includes cycle-based and event-driven depending up on when the data gets updated.

Simulation Model

Model-based design approach is basically used to define design specifications, test systems and develop code for prototyping based on the reuse approach, that is, the entire project would be the amalgamation of a few reusable modules. Consistent usage of the same module tremendously helps in rapid prototyping. As seen, the DANN can be used to model almost anything. Thus a composite system may be modeled and built uniquely on one simulation platform.

The applications of model-based approaches for simulation has been given in the previous section. In this section an artificial neural network-based model is introduced. A variant of the model called differentially fed artificial neural network (DANN) will be explored for simulation application.

The model of a differentially fed artificial neural network is explained here. Artificial neural networks are use in different walks of life including finance, genetics, and so forth. The models based on neural networks learn the solution by experience. This unique feature helps in learning the trends of the system behaviour during simulation.

Figure 1. Artificial neural network

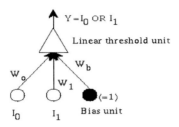

The problem of function approximation to map a set of data with another is challenging when they are non-linearly related. A variety of solutions based on neural networks are found in the literature. In this chapter, a new class of artificial neural networks with differential feedback is introduced. The different orders of differentials form a manifold of hyper planes (Manjunath, 2002).

Artificial neural network systems (ANN) are extremely helpful in modeling a system whose behaviour is unknown but for the enough non-linearly related input and output data. The structure of a simple artificial neural network with two inputs is shown in *Figure 1*.The network has two inputs and one output.

The output is given by 1 if $w_0 *I_0 + w_1 * I_1 + w_b > 0$ and 0 if $w_0 *I_0 + w_1 * I_1 + w_b <= 0$. The above network can learn Logical OR: output a 1 if either I_0 or I_1 is 1. The network learns or adapts by changing the weight by an amount proportional to the difference between the desired output and the actual output.

$$\Delta w = \xi.(error).i \tag{1}$$

Here i is the input driving the node, ξ is the learning rate. The network functions as follows: each neuron receives a signal from the neurons in the previous layer, and each of those signals is multiplied by a separate weight value. The weighted inputs are summed, and passed through a limiting function (non-linear in general) that scales the output to a fixed range of values.

A back propagation (BP) network learns by example. That is, a learning set has to be provided that consists of some input examples and the known-correct output for each case, like a look-up table. The network adapts for changes in the input. The BP learning process works in small iterative steps: first, one of the example cases is applied to the network, and the network produces some output based on the current weights (initially, the output will be random). This output is compared to the known-reference output, and a mean-squared error signal is calculated. The error value is then propagated backwards through the network, and small changes are made to the weights in each layer (proportional to the error). The whole process is repeated for each of the example cases, then back to the first case again, and so on. The cycle is repeated until the overall error value drops to a predefined value.

Differential Feedback Method

One of the major drawbacks of the conventional training methodology is that it is iterative in nature and takes a large number of cycles to converge to the pre-specified error limit. By intuition, if more information is made to be hidden with the data, it takes a less number of iterations to get stabilized to the pre-defined error limit. Auto regressive moving average (ARMA) model can be conveniently used in this direction. In a typical ARMA model, the output and input are related by:

$$y(n)=a_0*x(n)+a_1*x(n-1)+...+b_1*y(n-1)+... \tag{2}$$

The differentials of the output (& input) can be written in to similar linear combinations of present & previous outputs (inputs).

$$(dy/dt)=y(n+1)-y(n) \tag{3}$$

.

From the equations, it is clear that:

$$y(n)=f(x(n),x(n-1),..,dy/dt \,|t=n,d^2y/dt^2\,|t=n,..) \tag{4}$$

The ANNs are made to learn this function. Here the ARMA output is passed through a non-linearity for the same reason as in ANN that makes it possible to learn non-linear functions. The output y of a neural network but for the non-linearities can be written as:

$$y=\Sigma w_i x_i. \tag{5}$$

Where x_i are the inputs w_i, the corresponding weights. The space spanned by weight vector for different inputs is a hyper plane. The important factor is weight cannot span the entire input space, whatever may be the training mode. Again the linearity of the output may be viewed as a particular case of ARMA

$$y\,(n+1)=b_0 y(n)+b_1 y(n-1)+.....+a_0 x_n+... \tag{6}$$

Where b_0.. and a_0.. are constants. The auto regressive terms $b_0...b_n$ may be realized using an implied differential feedback (Manjunath, 2005).

Figure 2. Differentially fed artificial neural network

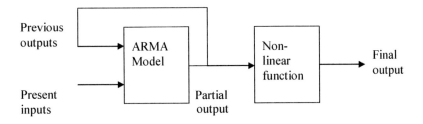

With differential feedback it has been found out (Manjunath, 2005) that the number of iterations required for training is reduced as shown in the Table I. XOR gate is considered for
simulation. Gaussian distributed random input with seed value 1000 is taken as the input.
The simulation is carried out in MTLAB version 6 release 11. The built-in XOR function
of MATLAB is used. A hard limiter is used as the non-linear function.

With I order different feedback, the output may be written as:

$$\Sigma w_i x_i + b_1 y_1 \qquad (7)$$

y_1 being the I order differential. This equation once again represents a plane parallel to $\Sigma w_i x_i$.
Thus the set of differentially fed ANNs form a manifold of parallel planes, with ∞ order
feedback being the plane with zero error. Also, simulation results of Table II show that two

Table 1. Performance with feedback

Order of differential	Square error	Iterations
No feedback	18	1156
I order	18	578
II order	18	289

Table 2. Performance with II order feedback

Order of differential	Square error	Iterations
II order Feedback	18	578
Equivalent Output	18	578

terms of II order differential feedback, that is, y_2-y_1 and y_1-y_0 can be replaced by a single equivalent plane represented by:

$$W_{eq}=(w_1* \text{ input } +w_2*\text{input1})/y_0 \qquad (8)$$

In II order differential feedback system, the two differential terms can be replaced by a single term. Extending this principle, the ∞ terms of ∞ order differential feedback can be replaced by a single term. This is termed as Eigen plane.

The differentially fed neural network exhibits other interesting properties (Manjunath, 2004). Specifically the following important properties will be used throughout the chapter.

1.   The different orders of differential feedback form a hierarchy of abstractions.

2.   The data corresponding to different hyperplanes, that is, the outputs for different orders of feedback are self-similar. That is, any of them can be obtained from the other one by convolving with a Gaussian pulse of appropriate scale factor. The scale factor decides the abstraction.

3.   Any of the hyperplanes can be expressed as the weighted sum of the other hyperplanes. If each of the hyperplanes represents the mathematical models, the weighted sum of the hyperplanes represents the ideal estimator. A set of ANN learning a set of rules may be replaced by a single DANN that can learn all the rules. This is possible by replacing piecewise learning by global learning.

The Concept of Differentially Fed Networks

When a differentially fed network is placed in a data network, the architecture consists of feed forward and feedback paths. The feed forward path consists of the actual information or data departing from the source. The feedback signal comprises of the position and status of the packet at the destination that has departed from the source. The differentially fed neural network sits as a controller, as a part of the loop comprising of the source, the forward path, the destination and the feedback path.

Since the differentially fed neural network is a part of the loop, its presence imparts interesting properties to the traffic in the loop. The DANNs make use of a large number of previous samples for decision-making. The output of such a system contributes to long-range dependency in the traffic. The abstract levels of hyper planes of DANN contribute to self-similarity of network traffic when observed over different time scales. The feedback information from the controller results in reduction/increase in the source operation rate, which in turn helps in proper scheduling. Based on the congestion status, different congestion control algorithms are used. Each one of them may be thought of as an estimator. A DANN works as an ideal estimator that happens to be the weighted average of these estimators.

In essence, insertion of DANN in the traffic loop makes the entire network behave as a differentially fed neural network, expressing all its properties. Hence DANNs play a role more than replacing the conventional ANNs in traffic shaping. The traffic shaping here refers to maintaining the schedules, reduction in the delays and reduction in stranded times or reschedules while maintaining the agreed service parameters. The notification signal or feedback signal is time shifted to get better performance. This algorithm is called random early prediction (REP) (Manjunath, 2004). In the subsequent sections, the effect of shift given to the feedback signal is analyzed.

The accurate traffic models are important in providing high quality of service. The network traffic does not obey Poisson assumptions used in queuing analysis. The complexity of traffic in the network is a natural consequence of integrating diverse ranges of members from different sources that significantly differ in their traffic patterns as well as their performance requirements over the same path. A long-range dependant or self-similar model fits with the observations

Impact of Shift Given to Feedback Signal on QOS Parameters

The presence of multiple data streams each with different priorities (such as real time video, non-real time file transfers) would have profound impact on the network resources. The streams are assigned quality of service QoS) parameters. The network is expected to maintain the absolute as well as the relative parameters with in a specified range. In a general scheduling algorithm, such as backlog-proportional rate (BPR) (Dovrolis, 2000) algorithm, the service rate allocations of classes dynamically get adjusted to meet proportional guarantees. The service rate allocation is based on the backlog of the classes at the scheduler.

The rates are proportional to the queue lengths of different members in the route. Such a scheme results in more absolute delays though the relative delay constraints are satisfied. This calls for a reduced queue length, that is, the absolute delay constraints are easily met in addition to the relative delay constraints if the queue lengths are reduced. The shifted or predicted feedback signal that is computed at the destination of the data actually reduces the queue size. The packet or cell drop ratio is used as the feedback signal. With the feedback signal in place, the data network exactly resembles a DANN. Hence the properties of DANN are expected to be reflected over the network.

The settling time of delay encountered in the network is reduced with feedback, that is, the time taken to stabilize the ratios of delays get reduced. This is because the queue flush time reduces as the queue size reduces with shift. The average delay also gets reduced.

The instantaneous and average delay metrics are suitable for a closed-loop control, characteristic of a reactive service rate allocation. Different from the above, the per-class delay metrics used in the case of a predictive rate allocation should attempt to measure the delay for the currently backlogged traffic.

With higher orders of feedbacks or larger shifts, the constraint is more easily satisfied as the instantaneous delay falls off with shifts. The same is true for the absolute loss rate. More rigorous proof is available in Manjunath (2004).

Knowing the network status well in advance with a differentially fed neural network allows taking corrective measure by incorporating appropriate shifts to the feedback signal. This ensures the absolute delay constraint (ADC) to be met in the future time.

The delay in the reception of the data is found to be having contributions from two components—one bursty or impulsive component and the other long-range dependent (LRD) component. The LRD component is auto regressive and the other part is impulsive. It remains independent of the shifts given to the feedback signal at any point of time. The LRD part is controllable with shifts or differential feedback. The analogy between the DANN model and the data network prompts to use the shifted feedback signal towards reducing the delay. The simulation results given in the next sections confirm the same. With shift given to the feedback signal, the queue length gets reduced while the input rate increases. The ADC is met easily with shifted feedback signal.

The proportional delay differentiation translates to relative delay constraint. To obtain a solution space, rather than a single solution for the service rate allocation, some slack in the expression of the relative delay constraints may be allowed. To satisfy proportional loss guarantees, traffic is dropped from classes according to a drop priority order.

For each class, the difference between the loss rate needed for perfect proportional differentiation and the observed loss rate defines an error. The larger the error of a class, the higher is its drop priority. For each class i, the controller stops dropping traffic when the loss guarantee L_i is reached or the path choking is resolved or a feasible rate allocation for absolute guarantees exists and there is no need for dropping the data or commodity anymore. The improved performance in handling multiple data streams is evident from the simulation results.

Applications

The properties of DANN may be used in a variety of applications ranging from molecular biology to data networks. The simulation model described above is generic and does not make any assumptions on the nature of the system.

As a stand-alone module, DANN may be directly used to extrapolate the data when the available data for simulation is insufficient.

DANN will be running as the background engine for the simulator. With this the following things are possible:

- Development of dynamic systems such as a data network
- Implementation of dynamic systems with the simulation models straightaway ported as working traffic shapers as in the network.
- Simulate linear, nonlinear, and discrete systems

In the products the DANN-based simulation module has to be integrated in to the actual system simulation environment. For example, it can sit as a proportional controller in a control system toolbox that is used to simulate control systems or as an optimizer in optimization toolbox or as a nonlinear ARMA in filters toolbox. The other applications include and are not limited to financial time series prediction, resource contention resolution, network traffic shaping, and so forth.

It can simulate linear, nonlinear, discrete and continuous systems. In heterogeneous simulation environment that integrates different simulators, a common DANN model may be used across.

Simulation

The DANN model basically supports any application listed in the above section. The system level simulation of a data network with regenerative feedback is discussed in this chapter. The resource status in a network may be simulated with the DANN included. The computer network shares many properties in common with a DANN. The autocorrelations of both the outputs would exhibit self-similarity property as a result of long-range dependency (Willinger, Paxson, Riedi, & Taqqu, 2001). In the case study of data networks, DANN is used as the appropriate model for the data network. The interest here is to study some important algorithms such as prediction-based scheduling of resources in the network. The properties of the model are mapped on to the network, that is, what is expected from the data network is also expected from the DANN model. For example, the output of a DANN consists of two parts -- a bursty part and a long-range dependant Gaussian part. It extends to the fact that the parameters in a network such as delay also consist of two components and the Gaussian part can be controlled through differential feedback. Simulation results prove the same. More details are provided in the subsequent sections.

The application of DANN component in network traffic shaping may be studied with this set up and the parameters of the DANN may be fine-tuned until the desired response is achieved. The simulations are basically used to analyze and tune quality of service metrics and allow exploring the different implementation options by conducting experiments. The simulation set up is as given below. Though the simulation is data network centric, the underlying methodology may be used for similar situations.

The superposition of the ON/OFF sources with heavy-tailed distribution results in self-similar aggregate traffic (Pruthi & Erramilli, 1995). Hence the same model is used here for data generation. *Figure 3* shows a source model to generate the self-similar aggregate traffic used in this section. To make a long duration of each period, the duration of each period is set to follow Pareto distribution with different values of α. The value of α has been taken as 1.5.

In this simulation study, we verify that the proposed active queue management scheme can keep an average queue length and the variation of queue length small when input traffic is generated by self-similar traffic sources. It is assumed that there is only one bottleneck switch in the network.

In the proposed scheme the predicted version of the probability of cell loss is given as feedback to the source. The DANN gets the training data from the background RED (random early detection) algorithm (Floyd & Jacobson, 1993). For some time, the DANN will be in learning phase. Then it predicts the data "k" steps in advance. This information is provided as the feedback to the source. The source then recomputes the transmission rate. It may be

Figure 3. Traffic model

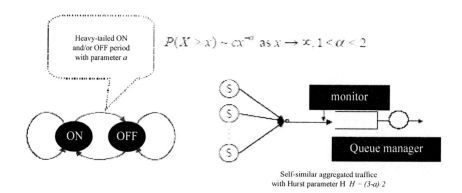

Self-similar aggregated traffice
with Hurst parameter H $H = (3-a)/2$

seen that the cell loss ratio has been reduced with feedback. In each case, 42 data points computed with RED are used for training. The input consists of 20 sources supporting the background traffic that exist over the entire simulation time. The maximum buffer size is 8000 and the cell size is 512.

The total cell loss ratio of an ordinary RED scheme is found to be 1.6. With a neural network prediction, it has been reduced to 0.05 and with the first order differentially fed neural network, it is reduced further to 0.04. A seven-step prediction has been used in the experiment.

It is desired to keep the variance of the buffer occupancy less. As explained in another section, the variance can be brought down with the increase of the differential feedback order. The use of differentially fed artificial neural network in Web traffic shaping has been explained at length in Manjunath (2004)

The prediction error for different orders of feedback is shown in *Table III*. As the order of differential feedback increases, the error reduces. Simulation time is set to 60 sec and 180

Table 3. Square error in training for different orders

No	Differential order	Square error
1	0	0.1943
2	1	0.1860
3	2	0.1291

samples are taken. Matlab version 6 and Simulink have been used to carryout the simulation. The proposed input rate prediction process captures the actual input traffic rate reasonably well.

General Characteristics

Figure 4 shows the instantaneous and mean queue length plot as a function of time. The buffer size of the gateway is 8000 packets and the target queue length is 3200 packets. 40 sources are considered in the experiment. As can be seen in the figure, in the steady state, the average queue size changes more slowly compared to the instantaneous queue size. This

Figure 4. Instantaneous and average queue

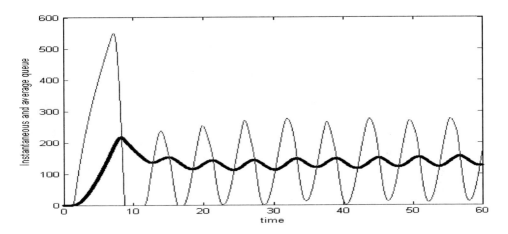

Table 4. Performance with RED and the proposed method for different sources

sl	Number of sources	Variance with RED	Variance with the proposed method	Maximum queue with RED	Maximum queue with proposed method
1	20	125.2279	111.5828	404	404
2	30	134.0159	126.0763	475	475
3	40	140.5793	129.6867	539	539
4	60	142.8687	111.8134	654	654
5	80	177.0254	126.0417	738	735
6	100	194.5093	138.2350	822	822
7	120	219.2376	151.1265	922	919

means that the proposed active queue management method is successful in controlling the average size at the router in response to a dynamically changing load and there is no global synchronization among the sources. The figure shows some large variation of mean queue length during the initialization phase. This is because it is assumed that the queue is initially empty, it takes some time for the proposed scheme to operate correctly. Because the proposed scheme randomly drops incoming packets according to the severity of the incipient congestion, there is no global synchronization. The large variation of the transient queue size is due to the bursty input traffic.

With the number of traffic sources that exist from the start to end of the simulation increased to 80, the total probability of cell loss or area under the error signal remains the same at 4.6. However, the variance of the queue length is considerably reduced with the proposed scheme. The results are summarized in the following tables. Here one-step prediction has been made use. The reduction in the variance is more pronounced with large traffic. The table for different prediction steps is shown below.

As seen in the tables, the variance will be reduced with increase in the prediction. This happens for some time. Again the variance shows upward trend. This is because the autocorrelation function of a long-range dependent series exhibits oscillations. As the number of sources increases, LRD is more pronounced with peaks of correlation separated far apart.

The above tables show the average queue length and its standard deviation for each method. RED detects incipient congestion by calculating the average queue size at each packet arrival. In other words, in RED, congestion detection and the packet drop are performed at a small time scale. On the contrary, in the case of the proposed scheme, the congestion detection and the packet drop is performed at a different time scale. When network congestion occurs, it takes some time for a traffic source to detect the congestion and reduce the sending rate to resolve the congestion situation. The shift, for which a further increase of shift shows a reverse trend in the variance, increases with increase in the load.

The proposed scheme can reduce the time for a traffic source to detect congestion, because the congestion detection is determined at a large time scale. In case that an incipient congestion is detected at a large time scale and a packet is lost within a sequence of packets, the

Table 5. Variance performance with RED and the proposed method for different sources

sl	Number of sources	K=1	K=3	K=4	K=6	K=8	K=10	K=14
1	20	111.5828	106.8508	106.5868	106.6906	NA	NA	NA
2	30	126.0763	120.8611	118.4315	119.2891	NA	NA	NA
3	40	129.6867	128.3137	128.5254	127.5494	126.1516	125.5521	130.2102
4	60	111.8134	110.6914	109.9463	108.7060	108.2325	108.2922	109.4781
5	80	126.0417	125.6543	125.5717	125.5579	125.7221	125.8909	NA
6	100	138.2350	137.1641	136.7088	135.3646	134.4704	134.2278	134.3958
7	120	151.1265	149.7342	149.0754	147.9885	146.8724	146.2454	145.8770

successfully delivered packets following the lost packets will cause the receiver to generate a duplicate acknowledgement (ACK). The reception of these duplicate ACKs is a signal of packet loss at the sender. So, the traffic sources can detect incipient congestion before it really occurs.

By using the proposed scheme the source can respond to the incipient congestion signal faster than in the case of using RED gateway. Therefore, both of the average queue length and the variation of the average queue length are smaller when the gateway uses the active queue management scheme proposed in this chapter rather than when the gateway uses RED. Because the proposed active queue management scheme can control the average queue size and the variation of the average queue size while accommodating transient bursty input traffic, the proposed scheme is well suited to provide high throughput, low average delay and low average delay-jitter in high speed networks.

The variance of the queue varies linearly with the shift. This confirms the existence of a constant term in the variance modulated by a variable term. The slope is independent of the load while the constant depends on it. Since the variance can be written as, $V = k_1 + k_2/shift$, k_1 and k_2 being constants, the variance varies linearly with the reciprocal of the shift. The constant variance term represents the bias. The second term happens to be the multi-resolution decomposition of the variance and is the sum of n terms, n being the order of the differential. By the weighted averaging of the hyper planes, the n terms in the second part can be replaced by a single term of highest degree or shift.

Congestion Control with Rate Feedback

The simulation is carried out in MATLAB version 6. The usage of the proposed model in the simulation of a simple switch is given here. A four-input four-output switch is considered for simulation. The traffic from these four inputs is destined randomly to any of the four outputs. Normal distribution with seed value 1000 has been used in the simulation. The addresses are uniformly distributed with seed value 1000. In the switch, the first input is given the highest priority while the fourth input the lowest priority. The cells from the first input need not be buffered. The cells from second input are buffered and the properties of this buffer are studied. The other cells are treated with low priority and dropped during the contention.

The growth of buffer size is a reflection of the efficiency algorithm followed in routing the cells. When differential feedback concept is applied, the maximum buffer occupancy gets reduced. Here the permission to pass or stop a cell at any instant is decided by an ANN with inputs being the previous permission vectors. Hence it results in a crisscross ANN with the output of one serving as an input for the other. The architecture is as shown in *Figure 5*.

Thus permission vector is a function of (previous) permission vector and differentially fed ANN can conveniently implement it. With this, the switch performance is found to be better as evident from the table. Table 6 compares the performance of the system with both these methods.

An ANN is used to compute the permission to transmit cells from a certain input, based on the previous permission vector. This makes the output dependent on the past history. In case 1, the decision is made at two levels. The first ANN, based on instantaneous maximum

Figure 5. Criss-cross architecture

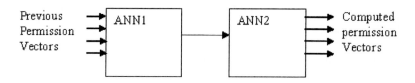

Table 6. Buffer utilization

Method	Total cycles to flush all the cells	Maximum buffer utilization
ANN without differential feedback method	953	599
ANN with differential feedback method	952	587

buffer utilization selects one of the rules. The ANNs are used to learn the rules, which are as follows:

Rule1:

Permission=1 if there is no blocking head. For input1 there is no blocking head. For input2 cells of input1 are the blocking heads and so on.

Rule 2:

Same as the rule 1 with input 2 given the highest priority.

Here, ANNs are trained separately according to the rule and then used. In an alternate architecture, a single differentially fed ANN is used to learn both the rules. It is clear from the simulation results that the differentially fed ANN can efficiently merge the two ANNs reducing the hardware as well as improving the switching efficiency in routing the cells. It may be observed that the differentially fed ANN based estimator outperforms the multiple scheduling schemes. Here the ANN is used to compute the permission to transmit cells from a certain input based on the previous permission vector, making the output dependent on the past history. With this, the switch performance is found to be better in terms of maximum buffer size and the total time to flush the cells.

The permission to pass or stop a cell (present permission vector) at any instant is decided by an ANN with the inputs being the previous permission vectors. This way the current output is made as a function of the previous outputs and a sort of long range dependency is given. Thus permission vector is a function of the permission vector, forming a loop, and differentially fed ANN can conveniently implement it. With this, the switch performance is found to be better as evident from the table. *Table 6* compares the performance of the system with both these methods.

At the system level, a neural network basically sits as a QoS monitor adjusting the rate of the source transmissions by controlling the flow of the cells that convey the feedback information in the feedback path. The QoS is guaranteed for the fixed network resources with the control of rate of transmission from the source. A set of ANNs, each learning a different scheduling algorithm, may be used. Any of them may be triggered based on the cost function, once again decided by another ANN. Each scheduling algorithm is an estimator. Finally, a single differentially fed ANN, as in the above experiment, may be used since all Bayesian estimators can be replaced by an ideal estimator.

Here, differential feedback has been tried at the system level over the feedback RM cells. Simulation results show that the resources consumed with such a feedback are arbitrarily small attributing smoothening of the traffic with differential feedback.

Performance with GREEN

GREEN is another standard scheduling algorithm (Feng, Kapadia, Thulasidasan, 2002) like RED. The proposed algorithm of random early prediction (REP) works well with Green. In the simulation, carried out with simulink model as shown in *Figure 7*, 40 greedy background sources are considered. The simulation lasts for 40 ms. The average loss probability and variance for different shifts is shown in the following tables.

After a shift of 8 the status reverses indicating gain can be achieved only up to a certain shifts because of the long-range dependency (LRD) or periodic nature. Also, for small shift,

Table 7. Cell loss for GREEN

SI	Shift	Average Probability of Loss (GREEN)	Average Probability of Loss (proposed method)
1	1	7.4%	6
2	2	7.4%	5
3	4	7.4%	0.725
4	6	7.4%	0.55%
5	8	7.4%	0.55%
6	10	7.4%	0.875

Table 8. Queue Variance for GREEN

SI	Shift	Variance of Q (GREEN)	Variance of Q (proposed method)
1	1	136.8258	133.3290
2	2	136.8258	133.9068
3	4	136.8258	135.5478
4	6	136.8258	136.1364
5	8	136.8258	136.3639
6	10	136.8258	135.7300

variance is reduced with not much gain in cell loss. It is reversed for large shifts. As the LRD increases with increase in the source, the gain in the variance from prediction will be more, but last for small shifts.

The instantaneous value falls as the sources go off. When more sources become on, with shift, it builds slowly and thus reduces the rapid fluctuations of the queue size. When the queue changes rapidly, queuing models do not work well. In the above experiment, the number of background sources is 20. Another 20 on-off sources start at t=0 to 70 other 20 from t=120 onwards.

The same set up may be used to demonstrate the usage of the model for relative quality of service (QoS) constraints. The simulation has been carried out in Simulink of MATLAB version 6 release 11. The network operations are modeled as on-off traffic sources. Two traffic classes or varieties are assumed. The length of each experiment is 70 ms of simulated time, starting with an empty system. In all experiments, the incoming traffic is composed of

Figure 6. Queue behavior versus time in Green

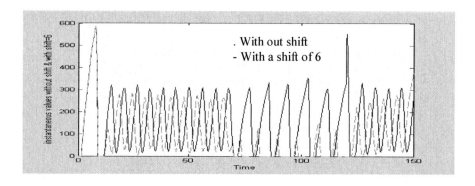

Figure 7. Simulink model

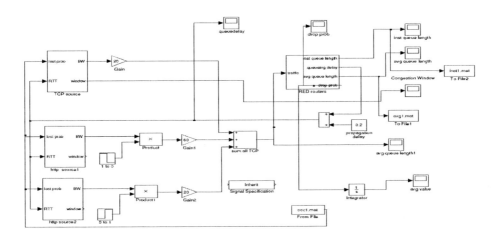

a superposition of Pareto sources with $\alpha= 1.2$ and average interarrival time of 300 μs. The shape parameter α of the Pareto distribution essentially characterizes the burstiness of the traffic arrivals. Smaller the α, burstier will be the traffic. The number of sources active at a given time oscillates between 200 and 550, following a sinusoidal pattern. Here relative loss rate constraint (RLC) is given as 4. The loss rate or the feedback signals are generated in the same ratio. The relative delay gets stabilized faster if more advanced signal is given as the feedback. The results are shown in *Figures 8* and *9*.

When the optimization problem does not yield a solution, meaning that it is impossible to satisfy all service guarantees simultaneously, some of the QoS guarantees are selectively ignored, based on a precedence order specified a priori. Due to the form of the constraints, the optimization problem is a non-linear optimization, which can only be solved numerically.

The computational cost of solving a non-linear optimization upon each arrival to the link under consideration may be prohibitive to consider an implementation of an optimization-based algorithm at high speeds. It has been proved in a set of simulation experiments that the REP algorithm is effective at providing proportional and absolute per-class QoS guarantees for delay and cancellation/reschedule. The plots show that the closed-loop algorithm reacts immediately when the routes are going from under load to overload and reacts swiftly when the routes go from overload to underload.

This indicates that the delay feedback loops used in the closed-loop algorithm are stable. Proportional delay differentiation does not match the target proportional factors $ki = 4$ when the route is underloaded, due to the fact that the algorithms are work-conserving, and therefore cannot artificially generate delays when the load is small.

Results for the ratios of delays indicate that proportional loss differentiation, that is, schedule cancellation is achieved when the outbound route is overloaded and the traffic is dropped. However, it is not met in any of the algorithms when the queue falls to 0. This implies that the algorithms basically manipulate the queue of the flow members and scheduling of the

Figure 8. Relative delay versus time plot for a shift of 0 & 10

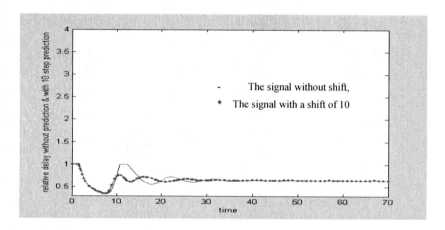

Figure 9. Instantaneous queue versus time plot for a shift of 0 & 10

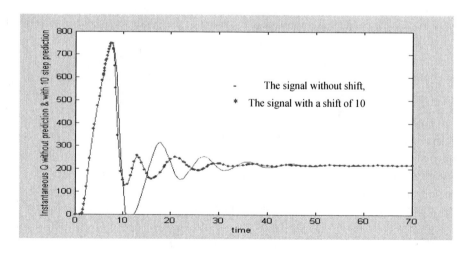

members to meet the relative delay and loss guarantees. From this simulation experiment, one can conclude that:

- The REP feedback loops used in the closed-loop algorithm appear to be robust to variations in the offered load.
- The results of the REP closed-loop algorithm are found to be better than the one without any shift.

The delays and losses experienced by classes are monitored, which allows the algorithm to infer a deviation compared to the expected service differentiation. The algorithm then adjusts service rate allocation and the drop rates to attenuate the difference between the service experienced and the service guarantees.

A non-linear control technique is used in the design of the algorithm. In particular, a prediction-based feedback control is proposed to achieve proportional delay differentiation. Absolute differentiation is expressed in terms of saturation constraints that limit the range of the controller. The control loop around an operating point is made stable through differential feedback and a stability condition is derived on the linearised control loop. The stability condition gives useful guidelines for selecting the configuration parameter of the controller.

Simulation results indicate that the proposed closed-loop algorithm is an effective approximation of the optimization-based algorithm, and that the feedback control is stable. Here two different classes of the network members or flows are considered. It may be generalized for multiple classes. The relative service parameters come into picture when the different classes of the flow contend for the common resources that tend to get choked and required to maintain a fixed ratio of the flows.

Issues and Solutions

The major limitation of the simulation models such as game theory is the availability of the data set. They do not provide an accurate model of the system when the sample size is small. The problem is handled by adding a bit of intelligence to the model. The ARMA model can work with reduced data size. They however fail to capture the non-linearities of the system. A neural network with a single hidden layer can capture all forms of non-linearities. They however require more data and time for training. A DANN can combine the features of both of these models.

One of the issues with model-based simulation is the scalability. The DANN architecture is scalable and extensible. The additional inputs may be separately passed through an ARMA and then added to the main branch before passing through the non-linear function.

Choice of simulation parameters largely influences the results. The uniformity and reproducibility of the results may be achieved with the help of bench marking results. Choice of the simulation model would affect the performance of the model. For example, where there is no feedback from the output to the input, selection of AR model would turn out to be wrong. In such cases, if the ARMA model is used, the AR parameters are set to zero.

The time taken for simulation is required to be arbitrarily small. The massive parallel structure of artificial neural network and DANN ensure that the computations and hence the simulation can be carried out fast.

The presence of multiple modeling techniques is a major drawback for the philosophy of modeling. Each model would throw out its own output. However, a DANN can merge these outputs in to a meaningful value (Manjunath & Gurumurthy, 2005a).

Future Trends

The properties, in a stand alone DANN, or in an embedded part of a large system may be used in various composite system simulations. Only a few properties of the DANN are explored in this chapter. Behaviour of the system in the presence of noise and capability to handle noisy data in the simulation has to be explored. The model tuning and the parameters are highly specific to the problem in hand. A network system problem is concerned about service quality whereas a flow management problem simulation is interested in the length of the stranded queue. More and more applications may be tried in the future exploiting all the properties of DANN. By fine-tuning this model various applications may be simulated in future. All these properties may be considered together and a universal system level simulator platform may be developed on a large scale.

The following variants of the simulation can make use of the DANN backbone.

Agent-Based Simulation

Agent is basically an independent entity that is programmed as a set of behavioral rules. A DANN agent allows these rules to be fine-tuned adaptively during the course of learning. An agent-based simulation model is different from the conventional one in the sense that the agent is decentralized. The agents are well-modeled with a feedforwared neural network that is again realizable with DANN. The interaction among the agents may be optimized by modeling them with DANN.

Web-Based Simulation

Web-based simulation is one of the areas where the simulation and the Web technologies meet. It includes (but is not limited to) the distributed simulation or simulation models distributed across the Web. It calls for reduced delay during simulation time to make it effective. The differential feedback provided in the Web or data network can reduce the transmission as explained in the previous sections and provides a better simulation environment on top of the best.

Collaborative Simulation

It calls for the integration of various distinct simulation tools that are time-shared. The duration of allocation for each application may be varied as per the priority calling for a quality-based service. This service may be made more realistic with DANN. The process of maintaining relative QoS parameters constant with the constraint of minimum burden to the resources is explained in Manjunath (2004) in the context of data networks.

Object-Oriented Simulation

The hierarchical representation and abstraction of the objects map onto the abstractions of the data. The REP algorithm described in the previous sections call for temporal shifts to the feedback signal. This temporal shift gets converted to spatial abstractions of the transferred data (Zeigler, 1995).

Summary

Availability of the right data for simulation is often a problem during the simulation of a system. The problem gets compounded when the data is available from multiple sources or when the sample size is inadequate. In this chapter, an artificial neural network-based solution is sought for the problem. The model developed is generic and may be put into a diverse class of applications including black hole collision, genome structure and mutation, database organization, and so forth. In this chapter, the example of Internet traffic control is considered with this simulation model.

References

Dovrolis, C. (2000). *Proportional differentiated services for the Internet*. PhD thesis, University of Wisconsin-Madison.

Feng, W-C., Kapadia, A., & Thulasidasan, S. (2002). *GREEN: Proactive queue management over a best-effort network*. Globecomo2.

Floyd, S., & Jacobson, V. (1993). Random early detection gateways for congestion avoidance. *IEEE/ACM Transactions on Networking, 1*(4), 397-413

Kreutzer, W. (1986). *System simulation: Programming styles and languages*. Reading, MA: Addison Wesley Publishers.

Manjunath, R., & Gurumurthy, K.S. (2002). Information geometry of differentially fed artificial neural networks. In *Proceedings of the 2002 IEEE Region 10 Conference on Computers, Communications, Control and Power Engineering (TENCON'02)*.

Manjunath, R. (2004). *Compact architecture for the analysis and processing of subnet signals using differentiators as building blocks.* Unpublished doctoral dissertation, University of Bangalore, India.

Manjunath, R., & Gurumurthy, K.S. (2005). *Hyperplanes Generation THROUGH Convolution With Gaussian Kernels,* IICAI.

Manjunath, R., & Gurumurthy, K.S. (2005a). *Bayesian estimator with differential feedback,* MWSCAS.

Manjunath, R. (2005b). *Continuous creation of matter across the black holes,* The Third 21COE Symposium: Astrophysics as Interdisciplinary Science Waseda University, Japan.

Pruthi, P., & Erramilli, A. (1995). Heavy-tailed on/off source behavior and self-similar traffic. In *Proceedings of the ICC 95* (pp. 445-450).

Willinger, W., Paxson, V., Riedi, R.H., & Taqqu, M.S. (2001). Long-range dependence and data network traffic. In Doukhan, Oppenheim, & Taqqu (Eds.), *Long Range Dependence: Theory and Applications.*

Zeigler, D.P. (1995). Object Oriented Simulation with Hierarchical, Modular Models: Selected Chapters Updated for DEVS-C++, Originally published by Academic Press, 1990.

Chapter VII

Model-Based Simulation to Examine Command and Control Issues with Remotely Operated Vehicles

Sasanka Prabhala, Intel Corporation, USA

Subhashini Ganapathy, Intel Corporation, USA

S. Narayanan, Wright State University, Ohio, USA

Jennie J. Gallimore, Wright State University, Ohio, USA

Raymond R. Hill, Wright State University, Ohio, USA

Abstract

With increased interest in the overall employment of pilotless vehicles functioning in the ground, air, and marine domains for both defense and commercial applications, the need for high-fidelity simulation models for testing and validating the operational concepts associated with these systems is very high. This chapter presents a model-based approach that we adopted for investigating the critical issues in the command and control of remotely operated vehicles (ROVs) through an interactive model-based architecture. The domain of ROVs is highly dynamic and complex in nature. Hence, a proper understanding of the simulation tools, underlying system algorithms, and user needs is critical to realize advanced simulation

system concepts. Our resulting simulation architecture integrates proven design concepts
such as the model-view-controller paradigm, distributed computing, Web-based simulations,
cognitive model-based high-fidelity interfaces and object-based modeling methods.

Introduction

Advances in technology, software algorithms, and operations research methods provide new opportunities for effectively building interactive models to support the study and analysis of human-computer issues in the operation and control of remotely operated vehicles (ROVs). The domain of ROVs is highly dynamic and complex in nature. Hence, a proper understanding of the simulation tools, underlying system algorithms, and user needs is critical to realize advanced simulation system concepts. With the increased interest in the overall use of pilotless vehicles in the ground, air, and marine domains for both defense and commercial applications, the need for the development of high-fidelity simulation models for testing and validating the operational concepts of these systems is very high.

ROVs are mobile systems controlled by human operators from a remote location. The utility and subsequent exploitation of ROVs have developed and diversified over the years. ROVs have been and continue to be used in a variety of ways, such as: decoys against enemy action or as friendly targets, scouts, reconnaissance and surveillance platforms, highly maneuverable bombs, and as transports (Christner, 1991). Incorporating these ROVs, such as unmanned aerial vehicles (UAVs), unmanned combat aerial vehicles (UCAVs), space maneuverable vehicles (SMVs), and unmanned emergency vehicles (UEVs), into military missions as required have been successful, although each present a real challenge for command and control station designers.

ROV operators have a wide range of responsibilities associated with multiple ROV coordination, handling multiple targets within an area and/or multiple target areas, detecting targets using a variety of sensor information, identifying targets sufficient for possible attach, planning routes given some set of targets (or locations), destroying targets in the case of armed ROVs, and the timely return of the ROVs to base. Operator responsibilities include supervisory control during normal operations, making minor flight or path adjustments when necessary, and overriding automated systems when abnormal situations occur, such as remotely versus automated landing of a UAV. Successful completion of any mission depends on an operator's ability to perform the control task(s) as well as maintain awareness of the task(s) performed autonomously by the ROV.

The dynamic operating conditions, the complexity of the fielded systems and the overwhelming amount of data processed by the human ROV operators are making automation a critical part of the planning, decision-making, and execution process. Since automated systems are capable of performing fast, accurate calculations, they can help improve performance in high-risk, time-sensitive and/or accuracy-critical situations. Although automation often improves overall performance in dynamic and complex systems by reducing workload and human errors, the advent of automation brings new problems associated with human automation interaction (Endsley, 1996; Wiener, 1995; Parasuraman, Molloy, & Singh, 1993). Bainbridge (1983), in describing the ironies of automation, points out that automation, designed to help reduce operator workload, sometimes increases workload. Automation also fails in many ways. First, it can fail to produce an outcome or a signal required for subsequent actions. Second, it may have sub-optimal accuracy causing wasted resource usage and potentially causing mission degradation. Finally, automation may work very accurately overall but not always at the right time. Each of these automation failures can be difficult for the operator to monitor and recognize. Wiener and Curry (1980) first described new errors and problems associated with increased automation and the guidelines for handling those problems. Current research investigating automation problems span a wide array of topics such as vigilance decrements (Ruff et al., 2004; Skitka, Mosier, & Burdick, 2000), out-of-the-loop performance problems (Kaber & Endsley, 2004; Endsley & Garland, 2000; Barnes & Matz, 1998), trust biases (Vries, Midden, & Bouwhuis, 2003; Mosier & Skitka, 1996), complacency (Prinzel, 2002; Moiser, Skitka, & Korte, 1994), loss of mode awareness (Mumaw, Sarter, & Wickens, 2001; Sarter & Woods, 1995), skill degradation (Mooij & Corker, 2002; Hopkin, 1995), and attention biases (Mosier et al., 2001; Mosier & Skitka, 1996).

As illustrated in *Figure 1*, many significant factors have been found to affect the interaction among humans, automation, and system performance. Many of the major factors that affect the use of automation and system performance are described in the next section.

Factors Affecting Automation

Rarely are systems classified as having purely zero automation or purely total automation. Various studies have attempted to classify the amount of automation into various levels based on functional allocation (Endsley, 1987; Sheridan, 1980), dynamic workload assessment

Figure 1. Major factors that affect automation use and system performance

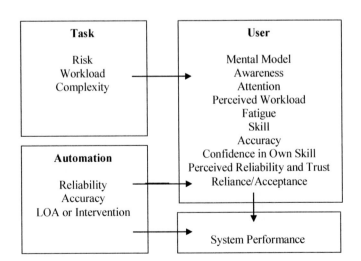

(Wickens, Lee, Liu, & Becker, 2004), and psychophysiological measures (Kramer, Trejo, & Humphrey 1996). The levels of automation may range from automation that simply suggests options for the user, to systems in which the user has the ability to override an automation decision or action, to the extreme wherein the system only provides the user the option to terminate the task or shut down the entire system. The problem with a system providing too little room for operator intervention is that this approach leads to operator complacency and out-of-the-loop performance problems. The problem with a system requiring too much operator intervention is that this approach leads to fatigue and workload problems.

Focusing on salient or readily available information (cues) may lead to loss of situational awareness (SA). The resulting penalty is that the actions taken, which are based on a faulty assessment, can lead to operator errors (Mosier & Skitka, 1996). Three major factors of automation that directly impact SA are: (1) changes in vigilance and complacency associated with monitoring, (2) assumption of a passive role instead of an active role in controlling the system, and (3) changes in the quality or form of feedback provided to the human operator. Each of these factors can contribute to the out-of-the-loop performance problem. In addition, automated systems, by nature of their complexity, also challenge the higher levels of SA (comprehension and prediction) during ongoing system operations (Endsley & Kaber, 1999).

Both Endsley (1996) and Sarter and Woods (1994) demonstrated that with highly automated and/or complex systems a loss of operator awareness of the system status leads to overall system and operator performance degradations. Endsley and Kiris (1995), in a decision-making task study, found that immediately following an automation failure, low operator SA corresponded with out-of-the-loop performance decrements in decision time. They also argue that this problem can be directly linked to a lower level of SA that exists when people function as monitors of automated systems (as opposed to active controllers of the system). This out-of-the-loop performance problem is worse when operators are neither updated nor alerted regarding the state of the automation, and changes to those states. Like out-of-the-loop performance problems, mode errors occur because the operator is not aware of what mode that the automation is in (Sarter & Woods, 1994) within the system.

The human operator and the automated portion of the human-computer system must be capable of communicating information, strategies, and commands to each other. Many of the awareness problems created by automation are due to the user not having current knowledge on what the automation has done, their not knowing what the automation is currently doing, and/or the operator not anticipating future automation actions or events that might arise due to the automation. The user must maintain constant diligence to understand the system state and anticipate events that will affect the system state (Sarter & Woods, 1994). Given proper feedback, much of the information arising from this work could be inherent in the display/interface, and thus require minimal cognitive processing. Providing proper feedback will increase operator awareness and reduce their workload. Determining feedback requirements for high workload automated systems is the challenge of automation research today.

Workload has been a key reason for introducing automation to systems. In an air traffic control experiment, Hilburn (1996) demonstrated that as traffic density increased so did the controllers' workload. Hilburn also showed that the use of automation worked as a workload reduction aid. System designers are not the only ones that chose to employ automation techniques to reduce workload. In a second experiment Hilburn showed that when users were given the ability to turn automation on and off, the use of automation was triggered by workload (traffic density). Sarter, Woods, and Billings (1999) noted that workload was a major factor in a pilot's choice of automation use inside the cockpit. Indeed human operators often cite excessive workload as a factor in their choice of automation in an attempt to maintain or improve performance.

An interesting irony is that automation does not seem to reduce operator perceived workload. Metzger and Parasuraman (2001) found that though automation improved performance, the perceived workload did not seem to be reduced. Endsley and Kiris (1995) and Billings (1991) found that subjective workload was no different for users from manual operation for automation monitoring, at any level of automation. These researchers argued that some of the operator workload might have shifted to the monitoring and planning portion of the task. Cummings and Mitchell (2005) stated that the automation designed to reduce workload sometimes increases it. Wiener and Nagel (1988) surveyed commercial pilots and found that one third disagreed with the statement, "automation reduces workload," and only one third agreed with the statement.

When the user must perform manual tasks while monitoring automation the potential for error increases significantly. Endsley and Kaber (1999) and Parasuraman, Molloy, and Singh (1993) found that users, while monitoring automation, had more difficulty detecting automation failures when they were responsible for other manual task(s). In such situations, Moiser and Skitka (1996) argue that operators over-rely on automation to perform the automated task(s), perhaps to compensate for the high workload involved with the manual operation of the manual task(s). Researchers even suggest that perhaps operators assume that automated aids may have better skills when under time pressure.

Given the issues associated with the interactive modeling of human-centered automation, such as operator trust in automation, the passive role of the human operator, out-of-the-loop performance, and situational awareness of human operators, there remains a pressing need for studies into and theories regarding the coordination between human operators and automated controllers in the remotely operated vehicle domain. Most of the current U.S. Air Force acquisition programs' efforts in unmanned vehicles are either too "vehicle centered" or rely heavily on available commercial, off the shelf technology to reduce costs and time. Particularly deficient are systematic approaches to allocating functions between humans and automation, studies of coordination between remote operators and intelligent controller nodes in the unmanned vehicles under realistic scenarios, and systematic studies of human/system interface development in remote command and control. We present a model-based approach that was adopted, for investigating the critical issues in the command and control of ROVs, through interactive model-based simulation architecture.

Interactive simulations support human interaction during the execution of the simulation model (McGregor & Randhawa, 1994). The primary purpose of an interactive simulation architecture is to study a human controller's behavior with the system and as such is designed to support active human interaction such that the operator interface displays the current system state and also facilitates the change of the simulated system state during execution. In interactive simulations, the simulation clock must be either scaled or must function in real time. The human operator must also be able to alter the simulation experimental parameters (e.g., simulation clock speed) or system dynamics (e.g., adding a constrained resource at run time). For example, in the context of the modeled UAV/UCAV domain of interest in our work, human operators interacting with the simulation during a mission must be able to modify UAV flight waypoints that define the UAV route or provide the simulated remote vehicle commands that initiate the actions to destroy a target such as when examining a UCAV mission.

An interactive simulation architecture was developed to facilitate concurrent, multi-user communication with ROV models. The architecture is called UMAST (uninhabited aerial vehicles modeling and analysis simulator testbed) and contains Java classes to model UAVs, the terrain, targets, and Web-based user interfaces (Ruff, Narayanan, & Draper, 2002; Narayanan et al., 1997). The architecture incorporates concepts from object-based modeling, Web-based simulations, and the model-view-controller paradigm, and can be used to rapidly instantiate event-driven interactive simulations in the UAV/UCAV domains. The major features of the UMAST architecture include support for interactive modeling, concurrent multi-user connectivity, reconfigurability of user interfaces, and modularity and reusability of software abstractions in the infrastructure. The requirements of the modeling architecture along with the design and implementation of UMAST are described in the following section.

UMAST Architecture

One of the primary purposes of the UMAST architecture is to study a human controller's behavior within the UAV/UCAV command and control domain. Thus, the architecture must support active human interaction where the visual interfaces display the state of the underlying simulation of the UAV/UCAV system and also facilitate the user changing the simulated system state during execution. Examples of interactive simulations discussed in the literature include Hurrion's (1999) work in job shop scheduling, Bell and O'Keefe's (1987) work in rail locomotive systems, Tai and Boucher's (2002) work in flexible manufacturing systems, and Narayanan et al.'s (1997) work in airbase logistics.

O'Keefe (1987) outlines the different views of interactive simulations in the analysis of complex dynamic systems: statistical analysis, decision support, and simulator. Under the statistical analysis view, there is little user interaction with the simulation model during program execution; simulation system interfaces are for post-simulation animation or statistical simulation output performance analysis. The more common applications of interactive simulations are for the last two views: decision support and as simulators. In the decision support view, the user can perform what-if analysis of scenarios immediately through interaction with the simulation model, with the interaction user-initiated or model-driven. Under the simulator view, interactive simulators can provide a high-fidelity synthetic environment for training human operators in a system. These interactive simulations are run in either a scaled time view (usually reduced time) or in a real-time view.

The UAV/UCAV domain features a high degree of complexity in entity interaction, a dynamically changing environment, and uncertainty in the outcomes of the scenario. It is typical in practice to simultaneously have multiple human operators controlling several UAVs. Concurrently, other human operators may be controlling the targets that may either be stationary or mobile providing a two-sided, dynamic gaming environment. The architecture must be designed such that multiple human operators can selectively control selected entities in the domain. *Figure 2* illustrates the overall framework for the UCAV domain with two UCAV controllers and a target controller connecting to and communicating through an executing system simulation model. The connection can be made through a Web browser or through a client computer. The interfaces presented to the controllers display appropriate system status and can accept relevant user commands to update the executing system's model. The behavior of system entities can be updated endogenously through transactions with other system elements or exogenously through dynamic data updates from real world sensors connected to the UMAST architecture.

The software abstractions employed to facilitate user interaction with the simulation architecture must be separate from the software abstractions used to represent the simulated world and its associated entities. The application of the model-view-controller (MVC) paradigm, where the representation of the application domain (i.e., model), the graphical interface specification (i.e., view), and the user interaction specification (i.e., controller) are partitioned provides a suitable framework to provide the requisite user interface. When the user interface is implemented as a separate process from the simulation model process, multiple distinct views of the underlying simulation can be attached to the same underlying simulation model. In the UCAVs domain, UCAV controllers and target controllers have different interface displays and different information displayed in the interface, each updated by the state changes from the same simulation model. For example, the UCAV controller would be able to view the location of the different UCAVs, but will see a target only when that target is within the sensing range of an air vehicle. Similarly, the target controller will have visibility into the target assets within the simulation scenario, but will view the UCAV state only when that UCAV is within the range of the target sensing capability. Interactive simulations that are built using the MVC framework allow interfaces tailored to the types of users and tasks for which the users need to interact with the system model.

Figure 2. A computational architecture for concurrent multi-user interaction in the UCAV domain (Adapted from Narayanan et al., 1999)

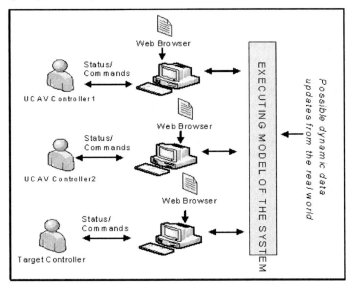

The development of the UMAST architecture is driven by the desire to simplify the process of creating models and interactive simulations for conducting studies pertinent to command and control issues within the UAV domain. The architecture is developed with modular and reusable software abstractions that can be assembled rapidly for implementing specific simulations of command and control scenarios and for prototyping user interfaces with which controllers connect to the implemented interactive simulations of the target study scenario. The application of object-oriented programming with a detailed domain analysis can facilitate the development of modular and reusable software for modeling and simulations (Narayanan et al., 1998). Through object-oriented programming, software abstractions in simulations can be developed that have a direct correspondence with real world objects, thus facilitating model development and validation (Bodner & McGinnis, 2002). Also, object-oriented programming with principled design methods can facilitate the development of a hierarchy of classes to permit modeling at different levels of abstraction in a domain. In the UAV domain, the software abstractions must be developed such that a specific UAV system can be sub-classed from a generic UAV class and instantiated through data input files that provide the specifics of the UAV system of interest.

The requirements and goals outlined above influence the design and implementation of the UMAST architecture. The goal of modularity and reuse of software abstractions is a natural fit to the use of object-based programming. Since the fundamental role of the UMAST architecture is to study the effects of complex decisions, whether those decisions are made by a human controller or some automated control process, partitioning the decisions, the data used in the decision-making process, and the results of the decision implementation on the physical system modeled in the interactive simulation in a structured manner facilitates analysis of the simulation model output. This approach of decomposing physical-control-information (PCI) entities has been found effective in other studies involving object-based simulations in manufacturing and airbase logistics simulation modeling domains (Narayanan et al., 1998; Mize et al., 1992). The PCI decomposition approach was used extensively in analyzing the UAV domain and developing the UMAST architecture to accommodate the requirements of the domain.

To meet the goal to develop interactive simulations with high-fidelity user interfaces, Java was chosen as the programming language for the development of UMAST. Java has become the object-oriented programming language of choice for Web-based systems development. Java supports object-oriented programming concepts, such as data encapsulation, inheritance, and polymorphism, and contains several packages for general data structures, file input/output, graphical user interfaces, and for creating programs on the Internet through applets (Niemeyer & Peck, 1996). Java is also multithreaded and suitable for distributed computing. One of the packages in Java is java.net, which provides Internet communication services through a native Java implementation of *Berkeley sockets* and a set of classes to download and manipulate objects from locations on the Web (Orfali & Harkey, 1997). Thus, Java has built-in support for developing simulations, user interfaces, and inter-process communication, all necessary in the development of tailored interactive simulations.

To incorporate capabilities of multi-user concurrent interaction and separate high-fidelity interfaces from common underlying simulations, mechanisms to facilitate inter-process communication and capability to support virtual reality modeling language are important. The requirement to support multi-user interactions through Internet browsers also influences mechanisms for inter-process implementations due to security considerations. In the development of UMAST, we developed our own tailored version of the inter-process communication module built on Java sockets that provides connectivity to a simulation server with interfaces implemented as applets on the Web or standalone Java applications.

The classes and the hierarchy of the software abstractions in UMAST are categorized into simulation/communication classes, domain-related classes, and the interface-related classes. An overview of the classes is provided in the following subsections.

Simulation/Communication Classes

The underlying inter-process communication between the simulation, the user interfaces and the knowledge structures needed to capture the state of the simulated system are facilitated through a package containing the class SimComm developed in UMAST. *Figure 3* provides a conceptual illustration of the role of SimComm. A SimComm object handles all network connections during the execution of the simulation model. A SimComm object must be instantiated on the same machine as the main simulation controller and must function as a server. In *Figure 3*, a simulator process is instantiated on a server computer. Three client computers containing interfaces to two UAV controllers and a target controller are connected to the server computer. On every other computer involved in the application, typically these additional computers are running a tailored interface to UMAST simulations, a SimComm object is initiated as a client. The server object has the ability to store the identity of the clients using the name of the client computer. The client process can connect to a server, register, and disconnect itself at any time while the server process is executing on a computer. The lower level communication between SimComm objects is enabled using

Figure 3. Inter-process communication through SimComm in UMAST (Adapted from Narayanan et al., 1999)

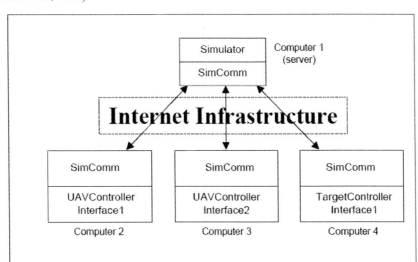

buffered reader/writer stream sockets, which are currently more efficient than other Java socket variants (Narayanan et al., 1998). The communication between both the simulation and interface process, however, is seamless to the simulation modeler. The interfaces are easily implemented through a Web browser or as standalone applications.

Domain-Related Classes

The domain-related classes in the UMAST architecture are categorized into physical, control, and information objects. The abstract physical clustering encapsulates the physical manifestations of the simulation entities, or the methods that indicate actions exhibited by the object. The control clustering encapsulates decision-making logic that results in actions on the physical object. The information clustering contains the data and parameters that are available to a particular object at a given instance within the simulation. Therefore, a complex object such as a UAV can be viewed as an aggregation of a set of physical, control, and information objects.

Figure 4 shows a partial hierarchy of the domain-related classes implemented in UMAST. As is typical in object-oriented programming, classes at the lower level are specializations of the upper level parent classes. The UAVPhysical and TargetPhysical classes share the characteristics of the parent Vehicle class, but have their own unique manifestations. In representing a specific UAV object such as a DarkStar high-altitude UAV, a class would be assembled with a DarkStarPhysical, DarkStarInfo, UCAVController object to represent the PCI aspects along with support information objects, such as WayPointInfo and so forth, used by the UCAV's physical and control objects. During instantiation of the specific Dark-StarUAV, data associated with the waypoints, physical, and information characteristics are read from data files. The control or decision-making behavior of the UCAV is embedded in the methods of the UAVController class. This control or decision-making behavior code is written by the analyst or by the architecture developer based on guidelines from the analyst. Alternate decision-making strategies are readily tested by altering the decision-making methods implemented in the UCAVController class. Similarly, different information and data input are easily evaluated using specific data file input.

UMAST contains Java classes to model UAVs, the terrain associated with the scenario examined, and the targets defined for the scenario. Targets can be stationary or mobile and can be armed; for instance a mobile target may contain surface-to-air missiles (SAM). Regardless of how many locations are visited during the UAV mission, each UAV has a route it must follow during the mission. Waypoints specify a set of locations along the route traversed by the UAV. These waypoints are defined in terms of the x, y, and z coordinates within the simulation scenario. The terrain is the geographical area traversed by the UCAV or UAV in which the targets are typically present. The performance measures for UCAV controller may include targets located, targets destroyed, time taken to complete mission, fuel consumed, UCAVs lost, or missiles expended. For the target controller, the performance measures may include UCAVs destroyed, missiles designated or fired, or resources used.

The UCAV controller and the TargetController are either human users of UMAST or computer agents within UMAST that monitor the situation evolving within the interactive simulation and then interactively provide control commands to the simulation. Such control commands may include alternate waypoints within the UAV route, identify targets pres-

Figure 4. A partial class hierarchy of domain-related objects in UMAST (Adapted from Narayanan et al., 1999)

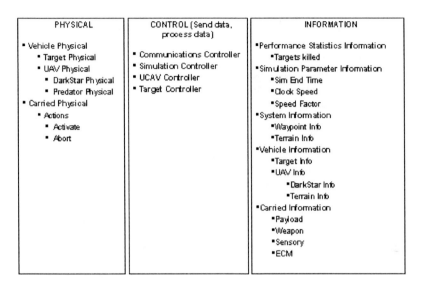

ent in the UAV field of view, or shoot targets within the UAV range. Thus, user interfaces need to display the state of system from the perspective of either an UCAVController or a TargetController and provide interface objects that can be used to accept input commands from the human controller.

Interface Classes

The abstractions in the interface infrastructure of UMAST have the capability to interconnect with the domain-related classes as well as human operators through the communication infrastructure provided by the SimComm class package. The fundamental building blocks in the interface include applets, drop-down menus, push buttons, check boxes, labels, panels, canvas, and scrolling lists, many of which are subclasses of the elements of the abstract windowing toolkit that is part of the Java programming language. Capabilities to communicate commands accepted from the human operators by the simulation interface to the underlying simulation model, as well as to parse messages from the simulation process into meaningful representations on the displays are also implemented in the UMAST interface infrastructure.

Applications

UAVs are employed in a plethora of domains, ranging from benign roles such as target drones to extremely active roles such as armed search and destroy platforms. The techno-logical promise of UAVs to serve across the full range of military missions, in areas such as communications relay and suppression of enemy air defenses (SEAD) mission, are quickly being realized. The objective of the SEAD mission is to neutralize, destroy, or temporarily degrade surface-based defenses by destructive and/or disruptive means (Flack et al., 1998). UCAVs are now quite likely to serve in such SEAD missions. UAVs and UCAVs have several advantages over manned systems including increased maneuverability, reduced development and operating costs, reduced radar signatures, longer endurance missions, and less risk to the aircrews operating the vehicle. Even though the human may be removed from the direct control of the aircraft, the human is typically involved in the process as a supervisor controller. The human supervisor charged with operating the UAV must be pre-sented with information necessary for tactical and strategic decision-making and provide the operational system the input necessary to augment any autonomous control of the vehicle. More specific instances describing the application of UMAST architecture to prototypical military missions are discussed below.

SEAD Mission

The specific scenarios modeled using UMAST can be described in terms of the UCAV capabilities and behavior, target capabilities and behavior, terrain characteristics, human operator interactions, and interface specifications. Assume there are four stealth-equipped UCAVs, each with four high-speed anti-radiation Missiles (HARM). These UCAVs have a maximum endurance of 8 hours. Each UCAV can fly at a maximum altitude of 25,000 feet with a maximum speed of 450 knots and a 14 cruise speed of 350 knots. The UCAVs have electro-optical (EO), infrared (IR), and synthetic aperture radar (SAR) with ground moving target indicator (GMTI) sensory systems with global positioning system (GPS) and inertial navigation system (INS) for navigation. The guidance and control systems for the UCAVS are preprogrammed, but can be overridden by a human operator from their remote location. Each of the UCAVs follow a pre-determined search path defined by multiple waypoints. In this scenario, a waypoint is the set of coordinates along with an action to be taken by the UCAV upon reaching that waypoint. While traveling the route, UCAVs attempt to locate targeting radars. Once a targeting radar is detected, the UCAVs split paths and attempt to precisely locate the enemy site use multiple lines of position (LOPs). Once the target is located, the UCAVs fire weapons at the target while evading any SAMs launched by the enemy. The UCAVs communicate with each other to share sensory data to each other and to collaborate on the task at hand.

To develop interfaces that are meaningful to the UCAV controllers, a detailed task analysis and developed human operator model called operational functional model (OFM) was performed. The OFM is defined in terms of a network of finite state automata (Mitchell, 1996). OFMs have been applied to describe and prescribe operator activities in supervisory control including monitoring, planning, and troubleshooting (Cohen et al., 1992). The OFM is a network in which nodes represent operator activities. Activities are structured hierarchically, representing operator goals or functions at the highest level and individual actions at the lowest level. Actions can be physical or cognitive. An OFM model represents how an operator might decompose a complex system into simpler parts and coordinate control actions and system configurations so that acceptable overall system performance is achieved. OFMs represent ideas of knowledge representation, information flow, and decision-making in complex systems.

The structure of the model is a heterarchical/hierarchical arrangement that structurally accounts for the coordination of operator functions and the focus of operator attention. Hierarchically in the network, the relationships among information needs, operator actions, tasks, sub-functions, and functions are represented. The top-level operator functions define the heterarchic level of the network, which has the operator goals and primary functions. Control actions represented in the OFM may take one of two forms. The first form is that an operator action can be manual, as in pushing a button to fire a missile. The second form of control action representation is the operator's cognitive actions, as in assessing the system state based on feedback presented on the interface. Cognitive actions require information gathering, integration or processing, and decision-making. Cognitive actions do not necessarily require any operator physical movement and do not usually alter the system state or

Figure 5. Partial operator functional model for UCAV controller in the SEAD mission scenario (Adapted from Narayanan et al., 1999)

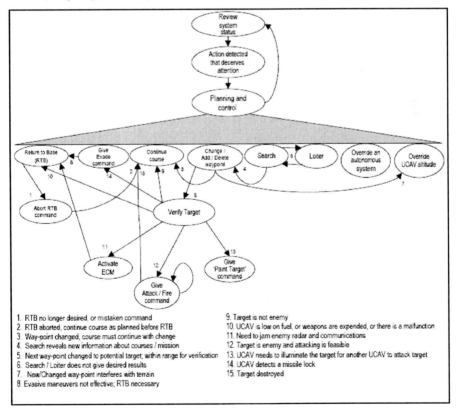

1. RTB no longer desired, or mistaken command
2. RTB aborted, continue course as planned before RTB
3. Way-point changed, course must continue with change
4. Search reveals new information about courses / mission
5. Next way-point changed to potential target; within range for verification
6. Search / Loiter does not give desired results
7. New/Changed way-point interferes with terrain
8. Evasive maneuvers not effective; RTB necessary
9. Target is not enemy
10. UCAV is low on fuel, or weapons are expended, or there is a malfunction
11. Need to jam enemy radar and communications
12. Target is enemy and attacking is feasible
13. UCAV needs to illuminate the target for another UCAV to attack target
14. UCAV detects a missile lock
15. Target destroyed

configuration. Application of the OFM for purpose of UCAV simulation interface design follows the operator actions. Nodes in the model denote operator functions, while the arcs linking these nodes represent data or contexts that drive the course of control actions. The arcs can also be described as system-triggering events that initiate or terminate operator activities. *Figure 5* presents a partial OFM for a UCAV controller within a SEAD mission scenario. Numbers on the arcs correspond to the numbered descriptions at the bottom of the graphic.

The results of this OFM study reinforce a human-in-the-loop approach to complex system design. The resulting interactive simulation architecture facilitated the backbone for the empirical analysis of the performance of the system based on different levels of automation. Results indicate that an automation level incorporating an intermediate level of automation had better performance than the more autonomous and less autonomous levels of automation.

Search and Rescue Mission

The next domain discussed is a simulated human survivor search and rescue mission performed by multiple robots supervised by a single human operator. The domain description is based on an international aerial robotics competition held in summer 2000. The overall goal of the robot-based search system is to locate and rescue human survivors in an urban catastrophe scenario and to determine current environmental conditions. The simulated robots are independent, intelligent, and cooperative software agents that can sense, plan and act/react semi-autonomously. The simulated robotic system is capable of deriving different problem solving strategies and then implementing them depending on the level of automation available. A human operator supervises the activities of the robots and is responsible for assessing, analyzing and determining the appropriate response of the system to operate the system at an optimal level, given the resources and constraints.

This simulation component of the system consists of simulated robots that navigate through an urban catastrophe scenario. A snapshot of the human operator console is displayed in *Figure 6*. The upper-left area of the display represents the map of the search area. The robot's physical activities are also displayed in this area. The search and rescue efforts in this particular domain are restricted to roadways. An aerial robot (i.e., a UAV) and several ground-based robots search the area; the UAV travels faster than the ground robots and thus initially surveys much of area before the ground robots have fully covered a portion of the area. As survivors are located they are categorized based on their perceived mobility and their proximity to hazardous conditions. The ground robots are then assigned to rescue them based on their analysis of time, distance and tasking information. The distance to each survivor from the present locations of the robots is calculated using an "A*" search algorithm. The ground robots are then assigned to rescue the survivors, using an assignment process that utilizes "Flood's Algorithm" to determine an optimal assignment (Kolman, 1980).

The simulation starts using predetermined search routes for the ground robots as well as a predetermined search pattern for the UAV. The routes employed by the ground robots change when the robots are tasked with rescuing survivors or returning to their default search routes once a rescue tasking is complete. The simulation terrain also presents obstacles in the form of low-level and high-level obstacles causing the robots to react to the situation appropriately.

Robot reactions include retracing its route on the road, re-planning its route to the survivors or slowing down through the obstacle area in order to circumvent it. The lower left corner contains select simulation system status information. The sensory information is presented in the lower middle part of the figure. The lower right portion of the display will contain information regarding the robot-to-survivor assignment process, along with selection buttons. The upper right area is designated for two different presentation concepts (means-end hierarchies and finite state diagrams), described below.

The simulation has two types of simulated robots: GBots, whose traversal routes are restricted to roadways only, and UAVs, which can hover over any area. The function of these UAVs is to detect and locate survivors, whereas the GBots can also physically move survivors to the centralized recovery center.

The simulation also allows incorporating different techniques to visualize automation behavior. Current implementations include a goal–tree and finite state diagram. The visualization techniques help the operator determine and analyze automation responses to different situations. For example, a derived response to an obstacle in the path may be to re-route via another path. A goal-tree structure is a structure that exhibits means-end hierarchy and addresses the critical question of why-what-how a goal can be achieved. Specifically, a goal-tree consists of top-down connected action bubbles where the preceding bubble represents a higher-level goal for which the current goal needs to be achieved, as shown in the right side of the screen depicted in *Figure 6.*

Figure 6. Snapshot of operator interface to search-and-rescue simulation scenario

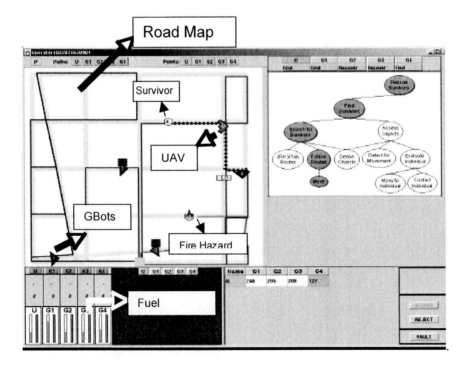

This architecture led to the development of a knowledge-based system that mimics supervisory control tasks (Dave et al., 2004). The simulation architecture was developed to facilitate human interaction with the system at different levels of abstraction. The human operator is capable of supervising the system ranging from total automation to situations where the operator forms an integral part of the decision-making process. This spectrum of choices allows testing to determine the adequate tasking level for humans as compared to automation. The simulation also allows incorporating different techniques to visualize automation behavior.

Summary

UMAST was described in the context of a prototypical military mission involving lethal SEAD mission scenarios. Future studies could include systematic evaluation of the extensibility of the UMAST architecture for modeling a class of similar problems and human-centered automation aspects of controlling uninhabited aerial vehicles. The UMAST architecture integrates concepts such as model-view-controller, distributed computing, Web-based simulations, cognitive model-based high-fidelity interfaces, and object-based modeling methods. The architecture supports the modeling requirements for analyzing uninhabited aerial vehicles including interactivity, multi-user connectivity, reconfigurable user interfaces, and modularity of software abstractions in the simulation infrastructure. While technological advances continue to be made in the computing and communications field, critical issues on the role of the human within these complex systems and the effective ways of supporting the human operator in these applications continue to be of paramount importance. This chapter provides potential answers to the critical issues associated with ROVs through interactive model-based simulation architecture.

References

Bainbridge, L. (1983). Ironies of automation. *Automatica, 19*(6), 775-779.

Barnes, M.J., & Matz, M.F. (1998). Crew simulation for unmanned aerial vehicle (UAV) applications: Sustained effects, shift factors, interface issues, and crew size. In *Proceedings of the Human Factors and Ergonomics Society 42nd Annual Meeting* (pp. 143-147).

Bell, P.C., & O'Keefe, R.M. (1987). Visual interactive simulation: History, recent developments, and major issues. *Simulation, 49*(3), 109-116.

Billings, C.E. (1991). *Human-centered aircraft automation: A concept and guidelines* (NASA Technical Memorandum 103885). Moffet Field, CA: NASA.

Bodner, D.A., & McGinnis, L.F. (2002). A structured approach to simulation modeling of manufacturing systems. In *Proceedings of the Institute of Industrial Engineers Annual Conference.*

Christner, J.H. (1991). Pioneer unmanned air vehicle accomplishments during operation desert storm. *Proceedings of the International Society for Optical Engineering (SPIE)*, *1538* (pp. 201-207).

Cohen, S.M., Mitchell, C.M., & Govindaraj, T. (1992). Analysis and aiding the human operator in electronics assembly. In M. Helander & M. Nagamachi (Eds.), *Design for manufacturing* (pp. 361-376). London: Taylor & Francis.

Cummings, M.L., & Mitchell, P.M. (2005). Managing multiple UAVs through a timeline display. In *Proceedings of the American Institute of Aeronautics and Astronautics Information Technology* (pp. 1-13).

Dave, R., Ganapathy, S., Fendley, M., & Narayanan, S. (2004). A knowledge-based system to model human supervisory control in dynamic planning. *International Journal of Uncertainty, Fuzziness, and Knowledge-Based Systems*, *1*(2), 1-14.

Endsley, M.R. (1987). The application of human factors to the development of expert systems for advanced cockpits. In *Proceedings of the Human Factors Society 31ˢᵗ Annual Meeting* (pp. 1387-1392).

Endsley, M.R., & Kiris, E.O. (1995). The out-of-the-loop performance and level of control in automation. *Human Factors*, *37*(2), 381-394.

Endsley, M.R. (1996). Automation and situation awareness. In R. Parasuraman & M. Mouloua (Eds.), *Automation and human performance: Theory and application* (pp. 163-181). Mahwah, NJ: Lawrence Erlbaum Associates.

Endsley, M.R., & Kaber, D.B. (1999). Level of automation effects on performance, situation awareness and workload in a dynamic control task. *Ergonomics, 42*(3), 462-492.

Endsley, M.R., & Garland, D.J. (2000). Pilot situational awareness in general aviation. *Proceedings of the Human Factors and Ergonomics Society 44ᵗʰ Annual Meeting.*

Flack, J., Eggleston, R., Kuperman, G., & Dominguez, M. C (1998). *SEAD and the UCAV: A preliminary cognitive systems analysis.* United States Air Force Research Laboratory (contract #: F41624-94-D-6000).

Hilburn, B. (1996). Dynamic decision aiding: The impact of adaptive automation on mental workload. *Engineering Psychology and Cognitive Ergonomics*, *1*, 193-200.

Hopkin, V.D. (1995). *Human factors in air-traffic control.* London: Taylor & Francis.

Hurrion, R. (1999). Discovering system insights using visual interactive meta-simulation. In *Proceedings of the 32ⁿᵈ Hawaii International Conference on System Sciences* (Vol. 6). IEEE Computer Society.

Kaber, D.B., & Endsley, M.R. (2004). Out-of-the-loop performance problems and the use of intermediate levels of automation for improved control system functioning and safety. *Process Safety Progress*, *16*(3), 126-131.

Kolman, B. (1980). *Elementary linear programming with applications.* New York: Academic Press.

Kramer, A.F., Trejo, L.J., & Humphrey, D.G. (1996). Psychophysiological measures of workload: Potential applications to adaptively automated systems. In R. Parasuraman & M. Mouloua (Eds.), *Automation and human performance: Theory and applications* (pp. 137-162). Mahwah, NJ: Lawrence Erlbaum Associates.

McGregor, D.K., & Randhawa, S.U. (1994). ENTS: An interactive object-oriented system for discrete simulation modeling. *Journal of Object-Oriented Programming, 5*(8), 21-29.

Metzger, U., & Parasuraman, R. (2001). The role of the air traffic controller in future air traffic Management: An empirical study of active control versus passive monitoring. *Human Factors, 43*(4), 519-528.

Mitchell, C.M. (1996). GT-MSOCC: Operator models, model-based displays, and intelligent aiding. In W. B. Rouse (Ed.), *Human-Technology in Complex Systems* (pp. 67-172). Greenwich, CT: JAI Press.

Mize, J.H., Bhuskute, H.C., Pratt, D.B., & Kamath, M. (1992). Modeling of integrated manufacturing systems using an object-oriented approach. *IIE Transactions, 24*(3), 14-26.

Mooij, M., & Corker, K. (2002). Supervisory control paradigm: Limitations in applicability to advanced air traffic management systems. In *Proceedings of IEEE Digital Avionics System Conference*, IC3.1- IC3.8.

Mosier, K., Skitka, L.J., & Korte, K.J. (1994). Cognitive and social psychological issues in flight crew/automation interaction. In M. Mouloua & R. Parasuraman (Eds.), *Human Performance in Automated Systems: Current Research and Trends* (pp. 191-197). Hillsdale, NJ: Lawrence Erlbaum Associates.

Mosier, K.L., & Skitka, L.J. (1996). Human decision makers and automated decision aids: Made for each other? In M. Mouloua & R. Parasuraman (Eds.), *Human performance in automated systems: Current research and trends* (pp. 201-220). Hillsdale, NJ: Lawrence Erlbaum Associates.

Mosier, K.L., Skitka, L.J., Dunbar, M., & McDonnell, L. (2001). Aircrews and automation bias: The advantages of teamwork. *International Journal of Aviation Psychology, 11*(1), 1-14

Mumaw, R.J., Sarter, N.B., & Wickens, C.D. (2001). *Analysis of pilot's monitoring and performance on an automated flight deck*. Paper presented at the 11th International Symposium on Aviation Psychology. Columbus, OH.

Narayanan, S., Schneider, N.L., Patel, C., Reddy, N., Carrico, T.M., & DiPasquale, J. (1997). An object-based architecture for developing interactive simulations using Java. *Simulation, 69*(3), 153-171.

Narayanan, S., Bodner, D.A., Sreekanth, U., Govindaraj, T., McGinnis, L.F., & Mitchell, C.M. (1998). Research in object-oriented manufacturing simulations: An assessment of the state of the art. *IIE Transactions, 30*(9), 795-810.

Narayanan, S., Malu, P., Krishna, A.P.B., DiPasquale, J., & Carrico, T.M. (1998). A Web-based interactive simulation architecture for airbase logistics systems analysis. *International Journal of Industrial Engineering, 5*(4), 324-335.

Narayanan, S., Narasimha, R. E., Geist, J., Kumar, P. K., Ruff, H. A., Draper, M., & Haas, M. W. (1999). UMAST: A Web-based architecture for modeling future uninhabited aerial vehicles. *Simulation, 7*(73), 29-39.

Niemeyer, P., & Peck, J. (1996). *Exploring Java*. O'Reilly & Associates.

O'Keefe, R.M. (1987). What is visual interactive simulation? (and is there a methodology for doing it right?). In *Proceedings of the Winter Simulation Conference* (pp. 461-464).

Orfali, R., & Harkey, D. (1997). *Client/server programming with Java and CORBA*. New York: John Wiley & Sons.

Parasuraman, R., Molloy, R., & Singh, I. (1993). Performance consequences of automa- tion- induced "complacency." *The International Journal of Aviation Psychology*, *3*(1), 1-23.

Prinzel, L.J. (2002). *The relationship of self-efficacy and complacency in pilot-automation interaction*. Report No: NASA/TM-2002-211925.

Ruff, H.A., Narayanan, S., & Draper, M.H. (2002). Human interaction with levels of automa- tion and decision-aid fidelity in the supervisory control of multiple simulated unmanned air vehicles. *Presence: Teleoperators and Virtual Environments, 11*(4), 335-351.

Ruff, H.A., Calhoun, G.L., Draper, M.H., Fontejon, J.V., & Guilfoos, B.J. (2004). Explor- ing automation issues in supervisory control of multiple UAVs. In *Proceedings of the Human Performance, Situation Awareness, and Automation Technology Conference* (pp. 218-222).

Sarter, N.B., & Woods, D.D. (1994). Pilot interaction with cockpit automation II: An experi- mental study of pilots' mental model and awareness of the flight management system (FMS). *International Journal of Aviation Psychology, 4*(1), 1-28.

Sarter, N.B., & Woods, D.D. (1995). "How in the world did we get into that mode?" Mode error and awareness in supervisory control. *Human Factor, 37*(1), 5-19.

Sarter, N.B., Woods, D.D., & Billings, C.E. (1997). Automation surprises. In G. Salvendy (Ed.), *Handbook of human factors and ergonomics* (pp. 991-925*)*. New York: Wiley.

Skitka, L.J., Mosier, K., & Burdick, M.D. (2000). Accountability and automation bias. *International Journal of Human-Computer Studies, 52*(4), 701-717.

Sheridan, T.B. (1980). Computer control and human alienation. *Technology Review, 10*, 60-73.

Tai, T., & Boucher, T.O. (2002). An architecture for scheduling and control in flexible manufacturing systems using distributed objects. *IEEE Transactions on Robotics and Automation, 28*(4), 452-462.

Vries, P., Midden, C., & Bouwhuis, D. (2003). The effects of errors on system trust, self- confidence, and the allocation of control in route planning. *International Journal of Human-Computer Studies, 58*(6), 719-735.

Wickens, C.D., Lee, J.D., Liu, Y., & Becker, S.E. (2004). *An introduction to human factors engineering*. New York: Longman.

Wiener, E.L., & Curry, R.E. (1980). Flight-deck automation: Promises and problems. *Er- gonomics, 23*(10), 995-1011.

Wiener, E., & Nagel, D.C. (1988). Cockpit automation. In E. Wiener & D.C. Nagel (Eds.), *Human factors in aviation* (pp. 433-461). San Diego, CA: Academic Press.

Wiener, E.L. (1995). Debate on automation and safety. In *Proceedings of the Human Factors and Ergonomics Society 39ᵗʰ Annual Meeting.*

Chapter VIII

Simulation Modeling as a Decision-Making Aid in Economic Evaluation for Randomized Clinical Trials

Tillal Eldabi, Brunel University, UK

Robert D. Macredie, Brunel University, UK

Ray J. Paul, Brunel University, UK

Abstract

This chapter reports on the use of simulation in supporting decision-making about what data to collect in a randomized clinical trial (RCT). We show how simulation also allows the identification of critical variables in the RCT by measuring their effects on the simulation model's "behavior." Healthcare systems pose many of the challenges, including difficulty in understanding the system being studied, uncertainty over which data to collect, and problems of communication between problem owners. In this chapter we show how simulation also allows the identification of critical variables in the RCT by measuring their effects on the simulation model's "behavior." The experience of developing the simulation model leads us to suggest simple but extremely valuable lessons. The first relates to the inclusion of stakeholders in the modeling process and the accessibility of the resulting models. The ownership and confidence felt by stakeholders in our case is, we feel, extremely important and may provide an example to others developing models.

Introduction: Challenges in Modeling Healthcare Decision Making

In recent years the healthcare sector in the UK and around the world has witnessed a wave of organizational reforms and changes, which mirror trends in the business sector. As a result of such changes there has been an increase in the level of constraints against health services provision. This has meant that health managers and professionals are increasingly faced with difficult decisions (Delesie, 1998), trying to balance the competing pressures of meeting ever-increasing needs and demands of the population whilst keeping the cost of treatment under strict control (Lagergren, 1998). In this environment, professionals are constantly faced with situations where they have to make immediate decisions, which, in many cases, can affect the quality of life of patients.

To support decision-making, we often seek to develop models of the decision space, which will allow us to gain insight into the likely outcomes of different decisions that we may take. There are several reasons, however, that make this modeling difficult in the health-care domain. Most contemporary healthcare systems are very complicated in terms of their structure, operation, and the diversity of people involved. Complexity often stems from the problems in healthcare lacking well-defined corners (Delesie 1998), with many interdependent entities competing for limited resources. This complexity can make it very difficult to predict the behavior of a healthcare system, since prediction is based on the application of collected historical data. Where a healthcare system is complex, it can be both difficult to determine the factors on which data collection should focus and challenging to define effective and appropriate data collection mechanisms. As a result, modelers tend to resort to assumptions in order to define the basic features of the problem, particularly when using mathematical models to represent the system. Another problem with healthcare systems is that they are multifaceted with different interdependent components. For example, in making decisions about patient treatment, the clinician has to look at the patient's profile alongside the cost of different treatments and the availability of resources to support such treatments. The involvement of different professionals in the healthcare system—such as clinicians and health managers—also makes modeling difficult (Delesie, 1998). These different stakehold-ers can have different views about the problem, expect different outcomes, and may make different decisions based on any model developed. This usually leaves the modeler with a major problem in integrating (and sometimes resolving) the different stakeholder perspec-tives involved in the problem.

The potential importance of modeling in supporting decision-making and the complexity of the situations that arise in healthcare suggest that studies of modeling approaches are worthwhile. The main purpose of modeling is to present an abstract picture of the real system and to examine the system's responses to different levels of inputs (Pidd, 1996). Many types of modeling are already used in healthcare problems. Lagergren (1998), for example, reports on a range of applications of modeling: in epidemiology for predicting future incidence, prevalence and mortality for broad sets of chronic diseases or for different specific diseases; for the evaluation of intervention strategies or disease control programs; in healthcare systems design, where the main concern is designing healthcare systems and estimating future resource needs as a training instrument for health managers; in healthcare systems operation, where the main objective of modeling is to improve performance by offering techniques for analyzing how existing resources could be used more efficiently; and in medical decision-making, where models are developed as support for analysis and decision-making in medical practice.

Given our emphasis on complexity and unpredictability, it is important to identify and study a case where these factors are prominent. A suitable area is randomized clinical trials (RCTs), which usually involve patients with a range of clinical needs, complex treatment pathways and states, and different medical professionals. Data collection also tends to be constrained by restrictions on time and cost, with an RCT based on random sampling of observations from different cohorts of patients. Determining the variables on which data is to be collected is extremely difficult, since it is likely that only a limited amount is known about the healthcare system that is to be studied in the trial. This will, however, be guided by the broad purpose of the RCT, such as the focus on cost-effectiveness of different treatments.

As well as defining a suitable application area, our interest lies in exploring the usefulness of a specific modeling approach: discrete event simulation. This chapter examines the suitability of the modeling approach in identifying the critical variables ahead of conducting the RCT, providing an insight into the role of simulation in countering the complexity and unpredictability issues and supporting medical decision-making. The chapter also shows how simulation can help the stakeholders—health economists in this particular example, since we are interested in this RCT from cost-effectiveness point of view—understand the healthcare system by allowing them increased access to the model. This includes allowing users to flexibly change the model's structure and study the implications of such changes.

The next section discusses our main field of study in healthcare—randomized clinical trials (RCTs)—and the use of modeling techniques alongside them. This will highlight limitations of these modeling techniques with respect to healthcare decision-making. We will then go on to look at the use of discrete event simulation in healthcare decision-making and the ways in which it might overcome such limitations. To illustrate our position, we will then present an RCT case study, starting with the users' requirements and moving into the development and appraisal of a simulation model to support decision-making. We conclude by considering the lessons of transferable value learned from this study and the limitations that should be considered.

Randomized Clinical Trials (RCT) and Modeling

RCTs represent one of the most complex classes of systems in healthcare. RCTs are used to examine new treatments based on randomization of treatment pathways or patterns of care for a sample of patients. The evaluation of treatments is usually based on the quantity and quality of life for each cohort in terms of cure, side effects, and relapses. Target data is collected throughout the course of trial for subsequent analysis to evaluate the efficacy of introducing a new treatment from clinical and economic perspectives.

A number of modeling techniques have been used for designing and analyzing RCTs, most commonly Markov modeling (Hillner et al., 1993; Sonnenberg & Beck, 1993). However, Markov modeling has characteristic restrictions: model cycles are limited to fixed time intervals, which means that patients can only change their health state at the end of each period in time; transition probabilities are time independent and are not influenced by previous health states experienced by the patient; Markov models only represent aggregate levels of the system without providing a view about individual cases or rare events; and Markov models require specialism, which means that they are usually inaccessible to healthcare professionals.

To support decision-making in RCTs more effectively, an alternative modeling approach that does not display these characteristics would be helpful. Such an approach should: offer the flexibility of dealing with systems at individual as well as aggregate levels; allow patients to experience events at any point of time after the previous event; support the recording and retention of the patient's history throughout the course of the trial so that it can be used to influence the pathways through the model; and allow the recording of other information about a patient, such as cost and quality of life effects associated with the events undergone. In summary, a modeling technique is required to follow patients individually as well as in aggregate through the model and decide their next states based on their history and current state.

Simulation Modeling in RCTs

We see discrete event simulation as a candidate for use in modeling RCTs. Simulation as a modeling technique has been widely used as an aid for decision-making in many health areas (Pidd, 1996; Klein et al., 1993). It has been used in diverse healthcare application areas, from predicting the increase or decrease of certain illnesses then making decisions about their corresponding treatments (Davies & Flowers, 1995; Davies & Roderick, 1998), through understanding the appointment systems in outpatients clinics in order to make decisions about scheduling strategies (Paul, 1995), to supporting strategic planning given the limited resources available (Pitt, 1997). A growing use of simulation in healthcare is in supporting economic evaluation (Halpern et al., 1994)—a key issue in RCTs. Where simulation *is* employed, a common objective is to use the models to provide some answers about the problem in hand. This reflects a tendency to see simulation as a way of deriving future outcomes and determining the effect of different model configurations on system behaviors, thus subsequently enabling decision-makers to appraise the implications of their decisions.

However, we see simulation as an approach that is not just for deriving answers. We are also looking at simulation modeling as a process to support problem understanding. To this end, we can identify two central capabilities of simulation modeling. Firstly, simulation provides a systematic debating vehicle between the different stakeholders who will contribute to decision-making in the healthcare system. Secondly, simulation offers the flexibility to accommodate as many changes as possible in the system model, either in aggregate or individually, to enhance the understanding of the system.

In this chapter we aim to demonstrate the suitability of simulation modeling as a decision-making aid for designing RCTs by identifying the key variables. The use of modeling in economic evaluation alongside clinical trials is not new (Schulman et al., 1991; Buxton et al., 1997), but its use to identify key variables for further data collection has thus far been under-recognized. The main purpose here is to understand the structure of the trial and identify the main variables for data collection. To achieve this we make use of the two central capabilities of simulation modeling identified above: the model may be built in a way that makes it easier to understand the system thus enhancing the level of debate between the different stakeholders; and the model is designed to provide flexibility for the users, allowing them to vary the simulation configurations and input factors in order to identify the main variables in the RCT.

The RCT Case Study

The particular RCT we modeled concerns the economic analysis of adjuvant breast cancer (ABC) treatment. The problem owners (or users of the model) are health economists who are interested in the economic assessment of the treatments, and clinicians who are expert in the way that the treatments are followed and the different possible side effects and relapses that may occur. From the perspective of our modeling, the health economists were our primary stakeholders, since they were our "customer" (as this was an advocacy modeling exercise). The trial structure may be thought of as representative of a typical healthcare system in terms of complexity and the large number of variables that are included. The remainder of this section provides a detailed description of the ABC RCT as background to the later discussion of the development of the simulation model in the fifth section. This background includes relevant clinical details of the RCT and of the basics of economic evaluation for the RCT, and provides justification for the use of a modeling technique to support decision-making for the RCT's design.

The ABC trial is a national collaborative RCT undertaken in the UK. The main objective of adjuvant therapy for early breast cancer is to prolong survival while maintaining a high quality of life. The principal aim of this trial is to assess the value of combining alternative forms of adjuvant therapy for early breast cancer. More formally, the trial aims to assess the value of adding cytotoxic chemotherapy and/or ovarian suppression to prolonged adjuvant Tamoxifen in order to treat pre/perimenopausal women with early breast cancer, and cytotoxic chemotherapy to prolonged adjuvant Tamoxifen in order to treat postmenopausal women with early breast cancer. The possible treatment alternatives are listed in *Table 1*.

At the beginning of the trial path, patients are routed based on their menopausal state. Patients who are pre/perimenopausal may be treated with Tamoxifen, Tamoxifen with chemotherapy, Tamoxifen with ovarian suppression, or Tamoxifen with both ovarian suppression and chemotherapy. Patients who are postmenopausal may only be treated with Tamoxifen or Tamoxifen with chemotherapy. After the menopausal classification, clinicians specify treatment paths that are randomly selected. All random alternatives are based on a simple rule:

Table 1. The randomization options for the ABC RCT

	Treatment plan	Randomization
Pre/perimenopausal women	Tamoxifen+OS	CT
	Tamoxifen+CT	OS
	Tamoxifen	CT and OS
Postmenopausal women	Tamoxifen	CT

Key: OS=ovarian suppression; CT=chemotherapy

for each level or treatment a patient is randomized between having a treatment (adjuvant) or not having that particular treatment (control). Then patients are sent to the next level. A patient who is prescribed Tamoxifen and ovarian suppression may be randomized for receiving chemotherapy or not. A patient who is prescribed Tamoxifen and chemotherapy may be randomized for receiving ovarian suppression or not. A patient who is prescribed Tamoxifen may be randomized for receiving chemotherapy and/or randomized to have ovarian suppression. Generally, each individual is allocated to the randomization option most suitable for her. Based on a patient's condition, clinicians may decide to allocate, or not allocate, a patient to specific treatment in any way and then randomize for the rest of treatment. It must be noted that all pairs of alternatives have equal probabilities (i.e., probabilities of 50-50 for each arm in a pair).

Table 2. Comparison for the clinical endpoints

Pre/perimenopausal women		
1	Randomized to +OS, regardless of CT	Randomized to -OS, regardless of CT
2	Randomized to +CT, regardless of OS	Randomized to -CT, regardless of OS
3	Tamoxifen+CT(given) +OS(randomized)	Tamoxifen+CT(given) -OS(randomized)
4	Tamoxifen –CT(not given) +OS(randomized)	Tamoxifen-CT(not given) -OS(randomized)
5	Tamoxifen+OS(given) +CT(randomized)	Tamoxifen+OS (given) -CT(randomized)
6	Tamoxifen-OS(not given) +CT(randomized)	Tamoxifen-OS (not given) -CT(randomized)
Postmenopausal women		
7	Tamoxifen+CT(randomized)	Tamoxifen-CT(randomized)

Key: OS=ovarian suppression; CT=chemotherapy

The ABC RCT aims to include four thousand pre/perimenopausal and two thousand post-menopausal women. The clinical endpoints of the trial are overall and relapse-free survival for five years. Health economists have the task of evaluating the economic implications of the ABC RCT to determine the cost effectiveness of the various adjuvant treatment combinations by comparing the additional resource use with the survival gains and quality of life effects.

Comparisons for pre/perimenopausal women are based on those randomized to ovarian suppression with those randomized not to have ovarian suppression, regardless of whether they have had chemotherapy or not. Also, an aim is to compare all those who are randomized to chemotherapy with those who are randomized not to receive chemotherapy, regardless of whether they have had ovarian suppression or not. For postmenopausal women, the comparison will be between those randomized to receive chemotherapy and those randomized not to receive chemotherapy. Comparison of the clinical endpoints is presented in *Table 2*.

Economic Evaluation Alongside RCT

Assessment of cost-effectiveness of the different treatments in the RCT is undertaken by an economic evaluation technique called cost-effectiveness analysis (CEA). This technique is actually concerned with the systematic comparison of alternative treatment options in terms of their cost and effectiveness (Drummond et al., 1997). CEA is the most commonly used form of economic evaluation and it is particularly used as an aid for decision-making. For detailed accounts of the role and recommendations for conducting CEA in healthcare the reader is referred to Russell et al. (1996) and Weinstein et al. (1996).

The results of the economic evaluation for the ABC trial are to be expressed in terms of the difference in costs, the difference in life years and difference in quality adjusted life years (QALYs) for the comparison groups. A QALY, as defined by health economists involved in the trial, is a duration of one year where the patient is in perfect health. Where appropriate the cost per additional life year gained and additional cost per QALY gained are to be estimated.

The Role of Modeling in the ABC RCT

Given the complex nature of the ABC RCT and the fact that it is time limited, those responsible for the trial felt that data collection for the economic evaluation should be kept to a minimum. It was suggested that data collection should be focused on the key variables or important factors that have direct effects on the objectives of the study. A pragmatic approach to data collection, without employing some form of modeling to identify key variables, was thought to be in danger of not identifying relatively infrequent but large resource-consuming activities. Hence the first phase of the evaluation was to conduct a modeling exercise using existing data and "expert opinion" prior to any primary data collection, in order to determine the key parameters influencing the cost-effectiveness of adjuvant therapies for early breast cancer. Further data collection could then be concentrated on those key parameters, thus ensuring that data collection during the trial was minimized whilst retaining the stakeholders' confidence in the data being collected.

The Modeling Process

Our basic objective was to produce a simulation model that enabled the health economists to understand the ABC RCT and establish the major variables that might affect the behavior of the cost-effectiveness of the adjuvant therapy. In this case modeling aims to provide facilities for flexible input alteration, which enables stakeholders to "play" with the model in order to establish the pattern of the different outputs. Another use of the model is as a medium of communication between health economists and clinicians. This meant that the model had to be built in a way that made it easily understandable by both parties.

As we noted earlier, the complicated nature of the trial with so many randomization options, treatment options, side effects, and relapses made discrete event simulation an appropriate modeling technique (Paul & Balmer, 1993; Davies & Davies, 1994). The approach provides a mechanism for participants to run a range of "what-if" scenarios, thus allowing the effect of many different assumptions—such as ways in which treatment is delivered, the age of the patients, and the effect of varying the values assigned to the costs and utilities—to be tested.

Key Issues in Building the Simulation Model

To conform to the above objectives the model is split into two distinct parts: the simulation engine and the user interface. The role of the simulation engine is to represent the model's structure and behavior in terms of treatment duration, patient routing, and clock setting. The interface is used as the input/output facility for the health economists (our customer). There are many aspects to the interface. Firstly, the look and feel of the interface to the model are based on the users' requirements—this helps ensure that they are able to "play" with the model. That is, the interface ensures that the model is accessible to them, overcoming the problems that non-experts often find in using and making sense of simulation packages. In this case, the user interface was seen as a medium through which they could communicate with the model.

The interface also had to be built to accommodate all the input factors and output responses that may be important in the ABC RCT. The interface had to offer the users the flexibility to modify the values associated with variables in the model. Since an aim of the model was to identify the important variables prior to data collection, there were no specific values or standard probability distributions for the model's variables at the start of the modeling exercise. We can think of the model parameters as being divided into two main classes: parameters with deterministic (fixed) values such as treatment costs; or those with probabilities such as different types of menopausal symptoms. Dealing with the first type of variables was straightforward; all we had to do was identify the parameters and provide variables in the model to hold their corresponding values. To model the stochastic parameters, we assigned a number of levels to represent each variable and their probabilities based on expert opinions gained through discussion with clinicians and health economists. The probabilities were based on the basic probability axioms (Hines & Montgomery, 1990). That is, each probability value lies between zero and one, with the sum of the values being one. An example in this study is that each patient has a probability associated with falling within one of four age groups (rather than giving a mean and standard deviation for age), where these probabilities are classed as follows:

$$\Pr(Age < 40) = A_1, \Pr(40 <= Age < 50) = A_2, \Pr(50 <= Age < 60) = A_3, \Pr(Age >= 60) = A_4.$$
$$\text{Given that } A_1 + A_2 + A_3 + A_4 = 1.00$$

Note: the interface ensures that all probability values are nonnegative and that they sum to one.

Even though this may seem a simplistic structure for modeling probabilistic behavior, we think it is quite flexible and it is not restricted to standard probability distributions. This method is appropriate for cases where there is no known input data or behavior, and allows users to model input data in any form or shape. The reader is reminded here, that in asking "what-if" questions, users will be able to examine the system's behaviors not just by changing the values of the defining parameters, but also by changing the entire distribution of the input variables, in an easy yet powerful way.

A large sample is required to give the precise estimate of these probabilities in real life. However, the simulation model is used to explore the effect on the results/outputs of the model when these probabilities are distributed in a certain manner. As such, the probability values are presented to users in an accessible manner so that they can be changed to explore their effect on the model. All probability distributions in the model are presented in this way.

The events modeled were the administration of the adjuvant therapies (chemotherapy and ovarian suppression) in addition to Tamoxifen, which is common in all treatments. After the administration of the adjuvant therapy the patient could remain in remission until death or relapse. Relapse was modeled as local/regional recurrence, non-bone metastases and bone metastases. Non-bone and bone metastases are followed by death. The division between non-bone and bone metastases was based on prognosis and intensity of resource use (Hurley et al., 1992; Goldhirsch et al., 1988). Local regional recurrence may be operable and followed by a second remission, or may become an uncontrolled local/regional recurrence, uncontrolled non-bone or uncontrolled bone metastases, which are followed by death. Following a second remission, after operable local/regional recurrence, the patient could be in second remission until death or develop non-bone or bone metastases. The events and their consequent pathways were drawn up using specialist help from clinical oncologists involved in the ABC RCT, and are shown in *Figure 1* and *Figure 2*. In both figures, each node represents a separate event which starts and ends at discrete points in time.

As patients move through the model's pathways, they are assigned attributes. The attributes of patient age, menopausal side effects and toxicity side effects associated with chemotherapy are assigned by sampling from non-specified probability distributions. Utility values (associated with administering the adjuvant therapies, menopausal and/or toxicity side effects, remissions and relapses) and the costs (associated with administering the adjuvant therapies, remissions, relapses, and with the treatments of toxicity side effects and menopausal side effects) were also modeled as attributes. The age at which a patient is diagnosed as having breast cancer is important as it influences the adjuvant therapy that she will receive, the probability of therapy-induced menopausal symptoms and the maximum number of future years of life.

Figure 1. A Simul8 visual representation of the event pathways which make up the Treatment Phase of the model

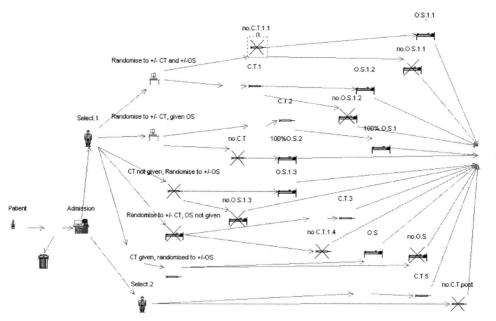

Figure 2. A Simul8 visual representation of the event pathways in the Recurrence Phase of the model

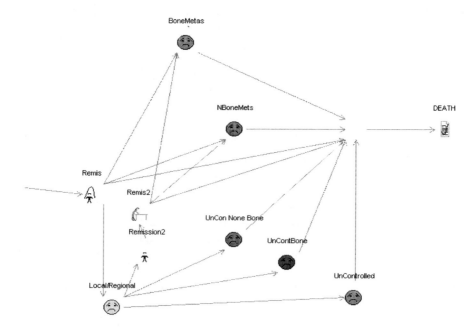

The menopausal side effects were modeled as "none," "mild," "moderate," or "severe." The modeling of menopausal symptoms as a side effect of adjuvant therapy was complicated by the fact that use of Tamoxifen alone can also induce early menopause (Canney & Hatton, 1994). Menopausal effects associated with Tamoxifen alone were thus modeled so that the net effects due to the adjuvant ovarian suppression and/or chemotherapy could be estimated. The model assumed that all women naturally have a permanent cessation of menses once they reached the age of 50. However, before the age of 50 induced menopause resulting from the use of luteinising hormone-releasing hormone (LHRH) agonists is temporary. A proportion of women following the use of chemotherapy or Tamoxifen may also experience temporary menopausal effects. The toxicity side effects were modeled as "none," "mild," "moderate," or "non-fatal major," each assignment depending on the corresponding complications. Chemotherapy itself is based on sets of six-month treatments, each comprised of six monthly treatment cycles. The model allowed for the number of chemotherapy cycles completed to be affected by the toxicity side effects. (This is shown on the sample input screen presented as *Figure 3*).

Cost attributes were assigned to administering the ovarian suppression and each cycle of chemotherapy. A monthly cost was assigned to the mild, moderate and non-fatal major toxicity effects and to the mild, moderate and severe menopausal symptoms. A fixed cost was assigned to local/regional recurrence, regardless of its duration, and monthly costs to the treatment of the non-bone metastases, bone metastases and uncontrolled local/regional recurrence, uncontrolled non-bone and uncontrolled bone metastases.

ABCSim allows patients to carry individual attributes throughout the different states of the model. Each patient is assigned a suitable set from the attributes (detailed in the above discussion) after they are sampled from the corresponding distribution. Routing to any of the post-treatment states (shown in *Figure 2*) is based on the combination of attributes a patient has collected from the treatment states (shown in *Figure 1*). For example, a patient arriving at the first remission ("Remis" in *Figure 2*) from the first arm of treatment (n.o.C.T.1.1 + O.S.1.1 in *Figure 1*) would have a different set of branching probabilities for any of the different recurrences and second remission than those for a patient arriving from the second arm of treatment (C.T.1 + O.S.1.2 in *Figure 1*). A patient's stay in either first or second remission is also influenced by the combination of attributes obtained from the treatment phase. *Figure 3* shows an example of how branching probabilities for the different recurrences can be altered for the different treatment arms. Three points are worth noting here: (i) the duration of a specific treatment is pre-specified and the same for all patients; (ii) the duration of remission is dependent on a patient's treatment history; and (iii) moving between events in the model does not take any time (i.e., the paths between events are purely directional and have zero duration).

It is also worth noting that although there are 14 treatment arms in *Figure 1* and 14 comparison arms in *Table 2*, this is purely coincidental, highlighting the important point that patients from a single treatment arm may belong to more than one comparison arm and vice versa. A patient's treatment history is assigned to her through values for associated attributes, so her information may be accumulated within corresponding comparison groups as she leaves the system. Information about costs, the time a patient spends in the model, and quality adjusted life years (QALYs) are collected from individual patients at the "DEATH" state to be aggregated for economic analysis after the completion of a simulation run. It must be noted that all information is accumulated for each patient throughout her life in the model. It

Figure 3. Branching probabilities for patients arriving from different arms of treatments

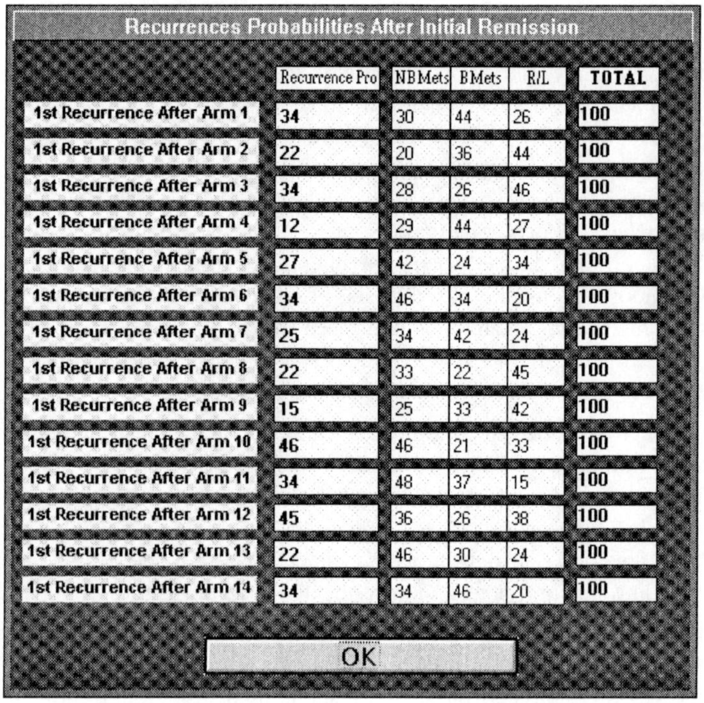

Recurrences Probabilities After Initial Remission	Recurrence Pro	NB Mets	B Mets	R/L	TOTAL
1st Recurrence After Arm 1	34	30	44	26	100
1st Recurrence After Arm 2	22	20	36	44	100
1st Recurrence After Arm 3	34	28	26	46	100
1st Recurrence After Arm 4	12	29	44	27	100
1st Recurrence After Arm 5	27	42	24	34	100
1st Recurrence After Arm 6	34	46	34	20	100
1st Recurrence After Arm 7	25	34	42	24	100
1st Recurrence After Arm 8	22	33	22	45	100
1st Recurrence After Arm 9	15	25	33	42	100
1st Recurrence After Arm 10	46	46	21	33	100
1st Recurrence After Arm 11	34	48	37	15	100
1st Recurrence After Arm 12	45	36	26	38	100
1st Recurrence After Arm 13	22	46	30	24	100
1st Recurrence After Arm 14	34	34	46	20	100

OK

is assumed in the ABC trial that all resources are available to all patients, which means that patient flows through the trial are independent of any external entities. This makes values collected from individual patients independent of each other and identically distributed (i.e., all values are of equal probability in terms of selection through sampling), meaning that classical statistical methods based on normal distribution theory can be applied to the resulting data.

The Computer Model

Simul8 is one of the most widely used discrete event simulation packages. It is perceived as an appropriate simulation tool from both the industrial and the academic community. Although there are various modeling packages such as ARENA, Witness and ProModel, most of these are not as easy to use as Simul8 (Oakshott, 1997). It might be argued that simulation technology has changed so rapidly in the past three years that most packages are intrinsically similar. Simul8 is generally user-friendly and usually no additional skills are needed for programming in order to construct new models. Literature suggests that Simul8 has the ability to animate objects and visually present interaction of objects with each other (Hauge & Paige, 2004). As an additional aid, Simul8's visual logic editor (VLE) allows the ability to organise a system's data in a variety of information stores by way of referencing. Furthermore, these visual logic functions allow the modeler to write simple statements that add dynamic controls to the simulation. Such a performance allows one to record distribution measures of the system under different scenarios. Based on its support features it can be considered as one of the most flexible simulation packages because special plug-ins can be added in order for specific modules to be integrated within the package. One of the main advantages of this Simul8 is the ability to represent large scale and complex systems with less time in contrast with models that have been developed with other simulation packages.

Two pitfalls are identified in the simulation input modeling. These are *the substitution of a distribution by its mean* and *the application of the wrong distribution.* Simul8 can combine some features in order to eliminate the possibilities of existence of these two pitfalls. More

Table 3. Description of Simul8 objects

Simul8 Objects	Activities
Work Entry Point	This object was used to identify patients arriving into the system. Timing plays a large part in this object, therefore in order to specify the arrival timings of the patients an **inter-arrival time** was identified within the **Work Entry Point Properties Dialogue.** **Note: Inter - arrival time** specifies the amount of time between arrivals and is the inverse of the arrival volume. Due to the random nature of arrivals, a **named distribution** was used.
Work Centre	This is an active object and has the ability to perform the work involved in a task. It can pull in work from queues or accept work from other work centres. In addition certain aspects of an item can be changed and on completing processing a **dynamic** decision on what to do with the item can be done. **Note**: Since a work centre is an active object, there is a need to add a resource to go along with the task.
Resources	Resources cannot function unless they are attached to a work centre. For example, resources represent a staff person, specialised piece of equipment and so many others. The purpose of this function is to constrain work centres from operation. **Note:** Without the availability of a resource work centres may not work.
Storage Bin	These passive objects are used as queues or storage areas but in the context of this study there are used to represent waiting areas where patients or items are waiting to be worked on.
Work Exit Point	These passive objects correspond to work items heading out of the system model. In the context of this study, this is a marking point for the patients to leave the system.

precisely Simul8, with the range of appropriate distributions that offer to end-users, can eradicate the application of wrong theoretical distribution. Detailed description of Simul8 objects can be found in *Table 3*. For more detailed description of how to develop models using Simul8 please consult Hauge and Paige (2004).

Another tool which was used for this model was Visual Basic to build an interface (which we called ABCSim) between Simul8 and the users of the model (the health economists). ABCSim provides the facilities for inputting data, displaying the model's outputs, exporting outputs to spreadsheets for further analysis, and saving the different model configurations for comparison and analysis. *Figure 1* and *Figure 2* depict the Simul8 model while *Figure 3*, *Figure 4* and *Figure 5* represent some sample screens from ABCSim.

Inputs to the Model

Input variables include information on treatment duration, costs and utility values; the input variables were identified by expert opinion and the interface to the model allows the user to change their values as required. Probability distributions were required for the permanent and temporarily induced menopause, menopausal and toxicity effects, relapses and second remission. Probability distributions were also required for duration of remissions, the duration of menopausal and toxicity symptoms, and relapses (as the structure in *Figure 2* suggests; see also *Figure 3* for how this is presented in the system). The probability distributions are defined by the user, with the system providing a flexible interface to allow them to be defined and changed as required. Information was also required on the "discount rate,"

Figure 4. A sample input screen from the ABCSim interface program

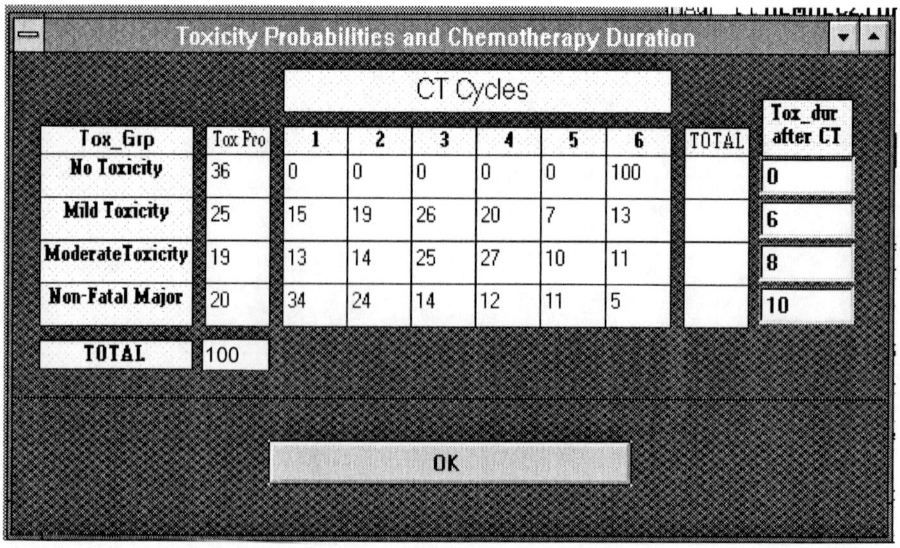

Tox_Grp	Tox Pro	CT Cycles 1	2	3	4	5	6	TOTAL	Tox_dur after CT
No Toxicity	36	0	0	0	0	0	100		0
Mild Toxicity	25	15	19	26	20	7	13		6
Moderate Toxicity	19	13	14	25	27	10	11		8
Non-Fatal Major	20	34	24	14	12	11	5		10
TOTAL	100								

OK

which is another of the system's variables, the value of which is user defined and can be altered as required.

Outputs from the Model

The model's results are, for each of the comparison options, the average cost, average life years, and average QALYs. For each comparison option, the model estimates the average difference in cost, life years and QALYs, undiscounted and discounted. Confidence intervals for these averages are also estimated using normal statistical techniques. Where appropriate the additional cost per life year gained and additional cost per QALY gained, undiscounted and discounted, are presented.

Verification of the Model

The purpose of verification is to make sure that the model is running as it is supposed to be or as is expected (Paul & Balmer, 1993). In this study, the model was verified collectively and individually in terms of the variables. Collective verification was performed by running the model with values for which there are logical expected results, which can easily be estimated without the computerized model. The results produced by the model are then compared to

Figure 5. A sample output screen output (Average Cost/Patient) from the ABCSim interface program

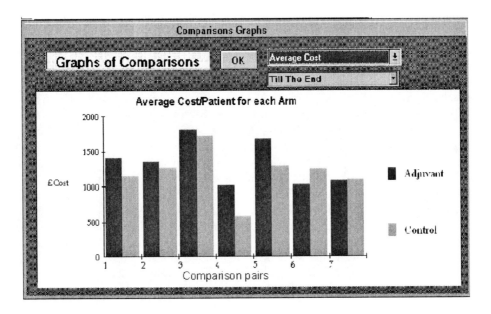

those that are expected. For the individual verification, the technique followed was to set a deterministic value to the variable of interest while resetting other unrelated variables to zero. For example, to verify the average cost for certain patients they were given constant cost values. If the average value was equal to the particular cost value then that cost variable was verified, otherwise the model was revised before repeating the verification.

Validation of the Model

The purpose of validation is to make sure that the developed model is the right model given the objectives of modeling (Paul & Balmer, 1993; Balci, 1997). Given the purpose of this particular modeling exercise and the fact that the modeled system is a non-existent sampling process, the validation process concentrated on the structure and behavior of the model rather than on producing estimated values. For example, it was not feasible to estimate the exact probabilities for which type of recurrence would occur after remission. The validation of the structure is based on making sure that the pathways are the correct representation of the trial. Owing to the wide-ranging discussions and consultations that took place with the stakeholders during the formative stages of the project, the structural validity was considered high. As discussed earlier the model is built based on expert opinion, hence the validation process is mostly based on expert opinions.

Information was supplemented by expert opinion to identify crude estimates of the overall expected effectiveness and costs of the different treatment options. This was used to validate the model's behavior. Given the large number of interactions in the model based on the different factors that may affect one patient at a particular health state, it was necessary to validate these interactions. For example, a patient may experience some side effects from previous treatment while undergoing a new treatment with similar side effects, such as menopausal symptoms. It was important in cases like this to identify the effect of such interactions and their corresponding entities.

It must be noted that interactions in this model are not interactions between different classes of entities competing for different resources. Rather, they are interactions between different attributes of the same entity in general simulation terms. More specifically, these interactions are between the patients' health states and effects from previous or current treatments. Results of such interactions affect the patients' quality of life. Life expectancy is directly affected by the combination of treatment that a patient receives, however the side effects of the treatments affect only the quality of life, rather than life expectancy. This judgement is based on expert opinion elicited during the development of the model. Treatment side effects do not have direct influence on the quantity of life, however life expectancy is affected by the treatment combination. *Table 4* gives a detailed account of the different elements in the interactions integrated in the model based on the validation process. The first section of the table gives the different possible treatment effects experienced by the patients. Effects may occur individually or in combination, and there are 16 possible occurrences (derived from the possible combinations of the levels of toxicity and menopausal symptoms). Health states can be seen in *Figure 1* and *Figure 2*. The "treatment phase" in *Figure 1* is considered as a single health state; each of the seven nodes in *Figure 2* are considered as separate health states. This gives eight health states in total. Having 16 possible effects and eight possible health states gives 128 possible interactions.

Table 4. Elements of the interactions formulated in the model

effects	
Toxicity	Reflects side effects of CT and may affect the course of the treatment. Four levels of toxicity are possible (none, mild, moderate, non-fatal). User defined distribution specifies which level of toxicity a patient will have as a result of chemotherapy treatment. Treatment duration is then determined on basis of severity of toxicity. Post-treatment phase is also affected if effect is existing.
Menopausal Symptoms	Reflects menopausal side effects of CT, OS, and Tamoxifen. Four levels of symptoms are possible (none, mild, moderate, severe). User defined distribution specifies which level of severity a patient will have as a result of each treatment. Post-treatment phase is also affected if effect is existing. Note: "none" effect is only possible for CT and Tamoxifen *Revised (P. 3 editor)*
results	
Utilities	Quality of life utility range is between 0 and 1. Assigned to a patient at each event based on the health state and the effects of toxicity and menopausal symptoms.
Recurrences	Recurrences of cancer possibly occur after adjuvant treatment. Different recurrences exist. The model provides input facilities for each recurrence in terms of duration and recurrence probabilities based on the current health state of the patient.

The second section of *Table 4* gives the expected results arising from such interactions in terms of QALYs. QALYs are calculated by multiplying the utility value of a health state by the duration of that health state (adjusted to years mainly by dividing by 12, as time units in the model are in months). These are summed to give overall QALYs for each patient as she goes through the different health states. The quantity of life—the duration of the post-treatment phase—arising from the probabilities of remission duration and types of recurrences (discussed in the fifth section), is directly affected by the treatment received.

Once the structural and behavioral validation had been established, it was possible to measure the effects and trends of the different variables in order to identify the major factors. For example, it is possible to vary the duration of disease-free periods based on chemotherapy to see the subsequent trend of survival. Validation of some of the estimates in the model had to be deferred, so that the users could alter the estimates as necessary after more data have been collected from the ongoing ABC RCT. This type of validation will depend on the early streams of data generated from the trial. The purpose of such validation is to predict the effects of the different variables on the model's results more accurately.

Use of the Model

As stated earlier in the chapter, the central purpose of the model was to identify the main variables of the ABC RCT and to act as a communication vehicle between the stakeholders.

Accordingly, the model has mainly been used for pilot experimentation without real data, developing an understanding of the interdependencies in the model. Health economists have full control of the model as they are our customer in this exercise. That is, they are able to change all of the variables in the model based on their experimentation plan to establish the significant variables. Health economists are also able to explain the results of their economic evaluation to clinicians using the simulation model.

Health economists follow two steps for identifying the important variables defined in terms of sensitivity to the model's outputs using health economic methods. The first step is to examine whether such variables are important or whether the results are highly sensitive to them. Input sensitivity is measured by the percentage of change in cost-effectiveness ratio (CER), arising from the use of different sets of input variables, compared with initial results arising from initial input values. Initial input values are suggested by experts involved in the trial. Undertaking traditional sensitivity analysis [see, for example, Box & Draper (1987), Kleijnen (1987), or Myer & Montgomery (1995)) would provide a spurious level of accuracy given that the inputs to the model are derived from expert opinion rather than validated input data. Rather, the approach taken in this study was to undertake crude sensitivity analysis by "flexing" the input data and appraising its impact on the model's output. Therefore, the stance is taken that if the change of the CER for a comparison pair (see *Table 2*) is agreed to be significant based on expert opinion, the variable is seen as significantly important with regard to that particular pair.

Usually health economists and clinicians agree on a certain percentage as being significant based on the situation to hand. In our case, the model acts as a communication medium between health economists and clinicians in deciding what would be regarded as a significant change in percentage. This reflects one of the main objectives of this exercise: to facilitate this type of communication between stakeholders.

The second step is to examine the impact of changing the initial value of a variable on the order of the different comparison pairs based on their corresponding CER. This is done by ranking the effects of these variables on the different comparison pairs. The previous step measures the significance of the model's sensitivity to the specific variable(s) within the same pair, whilst the second step measures the significance of the model's sensitivity to the specific variable(s) between the different pairs of comparison. An example is given as part of the following section to demonstrate how this process is conducted.

Results Arising from the Model

An example is given here to demonstrate how the model can be used to identify important variables and explore the impact of their change on the overall results. It should be noted that the aim of this example is to demonstrate relevant aspects of modeling of ABCSim and the data used in this example does not, therefore, necessarily reflect the real data used by health economists for analysis. Rather, the example shows how the model is developed and facilitated to give users full advantage of exploring different situations about the ABC RCT. We are, after all, interested in this chapter in the value of the process of modeling that we adopted, rather than in the specific model developed

Table 5. An example of altering values of the model's variables

	Base value	Change 1	Change 2
OS Cost	£1968	£2952 (50% more)	£3936 (100% more)
CT Cost	£269/cycle	£403.5 (50% more)	£538 (50% more)
CT Side Effects (none, mild, moderate, non-fatal)	0, 69, 30, 1 (probability distribution)	0, 40, 30, 30	0, 20, 20, 60

In this example we will show the results of altering the values of three variables; the cost of ovarian suppression (OS), the cost of a chemotherapy (CT) cycle, and the side effects of having CT. Results from OS cost will be analyzed in depth to give a realistic picture of how these results are analyzed by the users. As mentioned earlier, such changes are often conducted in an arbitrary manner and there is no standard methodology for such changes. This example shows the impact of changing the three variables which are detailed in *Table 5*.

The initial values, like most values in the model so far, are suggested by expert opinion. "Change 1" represents the first change carried out on the variables, and represents an increase on initial costs for OS and CT of 50%. "Change 2" represents a 100% increase. For CT side effects, both changes are concerned with the probability distribution of the different types of effects. For the purpose of this example we populated the model with 2,000 patients arriving in the span of around 13 months, which is one possibility based on the advice of health economists and clinicians involved in the trial. *Tables 5, 6,* and *7* show results for the above example. In each of the tables, model 1 shows results of runs with the initial values, model 2 shows results for "Change 1" and model 3 shows results for "Change 2." Results are given in terms of CERs. CERs are calculated as follows:

$$\frac{\overline{Cost}_A - \overline{Cost}_C}{\overline{Effect}_A - \overline{Effect}_C}$$

while $\overline{Cost}_A$ is the average cost of the adjuvant treatment, $\overline{Cost}_C$ is the average cost for the control treatment, and $\overline{Effect}_A$ and $\overline{Effect}_C$ are average life years gained for adjuvant and control treatments respectively.[1] Since we have a ratio of two averages then the resulting value is a biased estimator for the CER, for a large sample size (Cochran, 1977). However, the existing formula for deriving the standard error (SE) of a ratio (Bratley et al., 1987; Cochran, 1977) is not traceable for CERs. In this case the sum of the differences is not obtainable, as the values are sampled individually rather than as pairs. Also, the numbers of observations from both sets are not necessarily equal (since, for example, patients may die during their time in the model). For this reason we adopted a formula which evaluates the SE of a CER given by O'Brien et al. (1994), and used in healthcare.

Consider

$$R = \frac{\overline{Cost}_A - \overline{Cost}_C}{\overline{Effect}_A - \overline{Effect}_C} \quad , \tag{1}$$

then

$$\text{var}(R) \approx \frac{\frac{\sigma_{CA}^2}{n_A} + \frac{\sigma_{CC}^2}{n_C}}{\left(\mu_{EA} - \mu_{EC}\right)^2} + \left(\mu_{CA} - \mu_{CC}\right)^2 \left[\frac{\frac{\sigma_{EA}^2}{n_A} + \frac{\sigma_{EC}^2}{n_C}}{\left(\mu_{EA} - \mu_{EC}\right)^4}\right] - 2\left(\mu_{CA} - \mu_{CC}\right) \left[\frac{\frac{\rho_A \sigma_{CA} \sigma_{EA}}{n_A} + \frac{\rho_C \sigma_{CC} \sigma_{EC}}{n_C}}{\left(\mu_{EA} - \mu_{EC}\right)^3}\right]. \tag{2}$$

For brevity subscript (*CC*) denotes the cost of treating control patients and (*EC*) is the life years gained by control patients. The same applies for adjuvant patients. *C* and *A* refer to control patients and adjuvant patients respectively. While μ is the average value, σ is the standard error, ρ is the correlation between costs and life years gained, *n* is the number of observations, all for their corresponding subscripts.

It can be seen from the *Tables 5, 6* and *7* that the effects of changing the variables are actually more considerable on some comparison pairs than others. Naturally, in arms where, for example, OS is not prescribed or randomized then the change in CER is irrelevant. A change in such arms is mostly based on sampling fluctuations. *Figure 6* shows that, in terms of changes to OS cost, pair 1 is less sensitive to the 50% increase than to the 100% increase. Yet pair 3 is moderately sensitive for 50% and strongly sensitive for the 100% increase. We also find in *Figure 6* that pair 2 is positively sensitive for 50% increase with a minute negative sensitivity when OS costs are increased by 100%. This is due the fact that both comparators may or may not have women who had OS, and whether the cost will increase or decrease will depend on the number of women who had OS at that particular

Table 6. Cost-effectiveness results of altering OS costs

	Model1	Model2	Model3
Pair 1*	1907.11	3381.34	4778.11
Pair 2	1362.02	2744.88	1205.02
Pair 3	2100.26	5331.62	9123.81
Pair 4	1287.75	2410.69	3553.13
Pair 5	1993.48	2145.78	1382.45
Pair 6	1034.26	1290.88	1010.85
Pair 7	1596.68	1234.19	1297.41

* see Table 2 for identifying comparison pairs

Table 7. Cost-effectiveness results of altering CT costs

	Model1	Model2	Model3
Pair 1	1907.11	1040.41	3126.33
Pair 2	1362.02	3330.07	5829.9
Pair 3	2100.26	922.04	52991.54
Pair 4	1287.75	1043.6	1479.28
Pair 5	1993.48	2333.73	6997.37
Pair 6	1034.26	5727.42	3987.69
Pair 7	1596.68	1411.46	2200.26

Table 8. Cost-effectiveness results of altering CT side effects distribution

	Model1	Model2	Model3
Pair 1	1907.11	2304.07	2581.5
Pair 2	1362.02	2856.9	12982.95
Pair 3	2100.26	2700.44	8703.19
Pair 4	1287.75	1622.74	1226.87
Pair 5	1993.48	2934.4	10385.59
Pair 6	1034.26	2159.65	4692.44
Pair 7	1596.68	1551.47	1963.01

run. For pair 4, results are similar to those of pair 1, with pair 4 being more sensitive to the 100 % increase. For pair 5 we find that there is no change in results as both groups include women who had OS, which makes the proportion of change similar in both groups. The same principle applies for pairs 6 and 7, as there were no OS patients.

The process of variable selection, as mentioned earlier, is based on the percentage of increase or decrease of a factor and its corresponding percentage of change in the CER. This analysis is actually conducted by the health economists. Based on the above example, health economists may decide, in the case where the cost of OS is 100% more than the initial value, that OS data is only important for pair 3. Yet the decision-making process may take other factors into consideration. The purpose of the model is mainly to help the health economists to make better decisions, rather than making the decision for them. This example shows the flexibility of the ABCSim model to accommodate all types of possible change for experimentation, by being able to alter input values (aggregately or individually) and examine their impacts on the model's output. Facilities for such changes are advantageous in this exercise because they are tailor-made with regard to user requirements. Data presented in this example does not represent real data yet we are confident that this model will enable health economists to

Figure 6. Cost-effectiveness results: percentages of change based on model 1 and 2

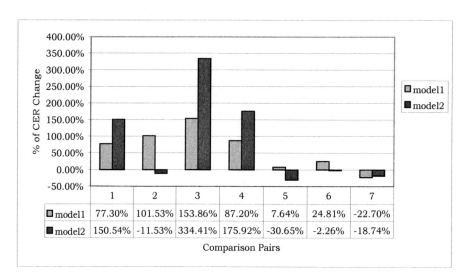

	1	2	3	4	5	6	7
▨ model1	77.30%	101.53%	153.86%	87.20%	7.64%	24.81%	-22.70%
■ model2	150.54%	-11.53%	334.41%	175.92%	-30.65%	-2.26%	-18.74%

Comparison Pairs

gain more insight into their problem and experiment with as many situations as they wish to develop their understanding of the problem space. As mentioned earlier, in addition to so many input variables, there are also output variables other than the CER, such as cost/utility ratios. Exploring the effects of variable changes on outputs enables the users to examine other angles of the problem.

As well as being useful in carrying out detailed experimentation for variable identification, as shown in the previous example, we see the model as a key resource for stimulating and facilitating the development of understanding of the problems associated with adjuvant treatments of breast cancer. These understandings have been facilitated by the model, which has offered a configurable and accessible way for the different stakeholders to develop and negotiate their problem understandings. Involvement of the stakeholders throughout the development of the model also provided them with a sense of ownership and confidence in the model. This is central to the subsequent ABC RCT, which will only be informed by the modeling experience if those responsible for the RCT are prepared to act on the lessons that they learn from the model's use.

Conclusion: Lessons of Transferable Value

Although the focus of this study was solely on clinical trials, the experience of developing the simulation model leads us to suggest simple but extremely valuable lessons, which we feel will be of transferable value across complex and unpredictable healthcare problems. The first relates to the inclusion of stakeholders in the modeling process and the accessibility of the resulting models. The ownership and confidence felt by stakeholders in our case is, we feel, extremely important and may provide an example to others developing models.

This raises an associated area of limitation, though. The chapter clearly reports an example of advocacy modeling, where one takes on board the view of a particular stakeholder for whom one is working. For this particular example, health economists happened to come to us with the problem, asking that we helped them "identify key factors in the ABC trial for economic evaluation." This naturally restricts the domain to their views of the real system, though these clearly change over time with the modeling. The interim and final models may well have been different if the problem owners or the modeling requirements had been different (i.e., if clinicians have been the primary or sole stakeholders). Even though there was a considerable contribution from clinicians with respect to the validity of the model, the final outcomes have mainly been driven by the health economists. As such, we consider this specific exercise as aimed at healthcare managers rather than healthcare professionals, and the model's usefulness to different stakeholder groups should be seen in this light. This highlights the importance of being clear about the purpose of the model and its relation to the stakeholder(s) for whom it has been developed.

The second lesson relates to the perception of models in complex domains such as healthcare. We feel that it is important that they are seen as vehicles for more closely understanding the system being modeled rather than techniques for finding solutions to those problems. Again, we would suggest that stakeholders have to have their expectations managed and see the real value of modeling in developing problem understanding and communication between stakeholders with differing perspectives. If this is not handled well, the stakeholders may well be unhappy when the model either does not provide a workable solution or provides a solution that is unacceptable to stakeholder groups. The closeness of stakeholders to develop negotiated models is important.

A final lesson, which is often underplayed in modeling, is the value of using the model to look at the sensitivity of areas of the model. Since models are only ever approximations of the real system, there are times when we look to simplify the model. Varying the values of parameters in the model and examining the impact on the model's outputs helps us gain a better appreciation of areas on which we should focus our modeling efforts. Given that many healthcare applications are extremely complex and that modeling is both difficult and time-consuming, exploring the sensitivity of the model as it is developed may help us target our efforts more wisely.

In our case, the modeling was further complicated by a lack of data on which to draw in developing and assessing the structure and operation of the model. The approach that was taken in this study to identify key variables was, we would argue, efficient and may form a basis for a methodology to model other systems where input data for modeling is unavailable. The power of such a methodology would lie in its ability to enhance stakeholder understanding of their problem space without the need to undertake large data collection exercises.

References

Balci, O. (1997). Principles of simulation model validation, verification, and testing. *Transaction of the Computer Simulation Society International, 14*(1), 3-12.

Bratley, P., Fox, B.L., & Scharge, L.E. (1987). *A guide to simulation* (2nd ed.). New York: Springer Verlag.

Buxton, M.J., Drummond, M.F., van Hout, B.A., Prince, R.L., Sheldon, T.A., Szucs, T., & Vray, M. (1997). Modeling in economic evaluation: An unavoidable fact of life. *Health Economics, 6*(3), 217-227.

Canney, P.A., & Hatton, M.Q.F. (1994). The prevalence of menopausal symptoms in patients treated for breast cancer. *Clinical Oncology, 6*(5), 297-299.

Cochran, W.G. (1977). *Sampling techniques* (3rd ed.). New York: John Wiley & Sons.

Davies, R., & Davies, H.T.O. (1994). Modeling patient flows and resource provision in health systems. *Omega, International Journal of Management Science, 22*(2), 123-131.

Davies, R., & Flowers, J. (1995). The growing need for renal service. *OR Insight, 8*(2), 6-11.

Davies, R., & Roderick, P. (1998). Planning resources for renal services throughout UK using simulation. *European Journal of Operational Research, 105*(2), 285-295.

Delesie, L. (1998). Bridging the gap between clinicians and health managers. *European Journal of Operational Research, 105*(2), 248-256.

Drummond, M.F., O'Brien, B., Stoddart, G.L., & Torrance, G.W. (1997). *Methods for the economic evaluation of health care programmes.* Oxford: Oxford University Press.

Goldhirsch, A., Gelber, R.D., & Castiglione, M. (1988). Relapse of breast cancer after adjuvant treatment in premenopausal and perimenopausal women: Patterns and prognosis. *Journal of Clinical Oncology, 6*(1), 89-97.

Halpern, M.T., Brown, R.E., Revicki, D.A., & Togias, A.G. (1994). An example of using computer simulation to predict pharmaceutical costs and outcomes. In J.S. Tew, S. Manivannan, D.A. Sadowski, & A.F. Siela (Ed.), *Proceedings of the 1994 Winter Simulation Conference* (pp. 850-855). Lake Buena Vista: ACM.

Hauge, J.W., & Paige, K.N. (2004). *Learning Simul8: The complete guide* (2nd ed.).

Hillner, B.E., Smith T.J., & Desch, C.E. (1993). Assessing the cost-effectiveness of adjuvant therapies in early breast cancer using a decision analysis model. *Breast Cancer Research and Treatment, 25*(2), 97-105.

Hines, W.W., & Montgomery, D.C. (1990). *Probability and statistics in engineering and management science* (3rd ed.). . Singapore: John Wiley & Sons.

HM Treasury. (1997). Appraisal and evaluation in central government. *Treasury Guidance.* London: The Stationary Office.

Hurley, S.F., Huggins R.M., Snyder R.D., & Bishop J.F. (1992). The cost of breast cancer recurrences. *British Journal of Cancer, 65*(3), 449-455.

Klein R.W., Dittus R.S., Roberts S.D., & Wilson J.R. (1993). Simulation modeling and health care decision making, *Medical Decision Making, 13*(4), 347-354.

Lagergren, M. (1998). What is the role and contribution of models to management and research in the health services? A view from Europe. *European Journal of Operational Research, 105*(2), 257-266.

O'Brien, B.J., Drummond, M.F., Labelle, R.J., & Willan, A. (1994). In search of power and significance: Issues in the design and analysis of stochastic cost-effectiveness studies in health care. *Medical Care, 32*(2), 150-163.

Oakshott, L. (1997). *Business modelling and simulation.* London: Pitman Publishing.

Paul, R.J. (1995). Outpatient clinic: The CLINSIM simulation package. *OR Insight, 8*(2), 24-27.

Paul, R.J., & Balmer, D.W. (1993). *Simulation modeling.* Lund: Chartwell-Bratt.

Pidd, M. (1996). *Tools for thinking: Modeling in management science.* Chichester: John Wiley & Sons.

Pitt, M. (1997). A generalised simulation system to support strategic resource planning in healthcare. In S. Andradottir, K.J. Healy, D.H. Withers, & B.L. Nelson (Ed.), *Proceedings of the 1997 Winter Simulation Conference* (pp. 1155-1162). New York: ACM..

Russell, L.B., Gold, M.R., Siegel, J.E., Daniels, N., & Weinstein, M.C. (1996). The role of cost-effectiveness analysis in health and medicine. *Journal of the American Medical Association. 276*(14), 1172-1177.

Schulman, K.A., Lynn, L.A., Glick, H.A., & Eisenberg, J.M. (1991). Cost effectiveness of low-dose zidovudine therapy for asymptomatic patients with Human Immunodeficiency Virus (HIV) infection. *Annals of Internal Medicine, 114*(9), 798-802.

Sonnenberg, F.A., & Beck, J.R. (1993). Markov models in medical decision making. *Medical Decision Making, 13*(4), 322-338.

Weinstein, M.C., Siegel, J.E., Gold, M.R., Kamlet, M.S., & Russell, L.B. (1996). Recommendations of the panel on cost-effectiveness in health and medicine. *Journal of the American Medical Association, 276*(15), 1253-1258.

Endnote

[1]   The model also provides results for discounted QALYs, which means that users can also use discounted QALYs for measuring cost-effectiveness should they prefer to do so.

Chapter IX

Intelligent Simulation Framework for Integrated Production System

Abid Al Ajeeli, University of Bahrain, Bahrain

Abstract

This chapter addresses the problem of modeling finished products and their associated sub-assemblies and/or raw materials. A production system is a set of policies that monitors and controls finished products and raw materials. It determines how much of each item should be manufactured or be kept in warehouses, when low items should be replenished, and how many items should be assembled or be ordered when replenishment is needed. Practical production models rarely optimize or even represent very precise descriptions of realistic situations. The art of model design is to develop commonsense approximations that give enough information to facilitate managerial decision-making. A system with a high degree of intelligence is more robust and able to perform better in terms of lower cost and higher efficiency. The integration concept of manufacturing processing has enjoyed increased popularity among researchers and manufacturers. Integrated production system is a set of mathematical models and policies that monitors and controls raw items, sub-assemblies, and finished products. Although such systems provide a better utilization of resources they have a number of drawbacks due to the complexity of real world problems. Using intelligent simulation techniques will lead to better automated systems.

Introduction

Before starting the discussion of intelligent systems, one needs to introduce a number of terms, including modeling. A model may be viewed as a representation of real life activities while a system can be viewed as a section of reality. A common property of all physical systems is that they are composed of components that interact with one another. The physical laws that govern their behavior determine the nature of the interactions in these systems.

A system, in our case, is an organized group of entities, such as people, equipment, methods, principles, and parts, which come together and work as one unit. A simulation model characterizes a system by mathematically describing the responses that can result from the interactions of a system's entities.

The set of values of variables in a system at any point in time is called the state of the system at that point in time.

System state is the collection of variables, stochastic (can change randomly) and deterministic (not influenced by probability), which contain all the information necessary to describe a system at any point in time.

A discrete event is an instantaneous action that occurs at a unique point in time. A part arriving at a delivery dock, a customer arriving at a bank, and a machine finishing a cycle of production are examples of discrete events. A continuous event continues uninterrupted with respect to time. The temperature of water in a lake raising and lowering during a day, the flowing of oil into a tanker, and chemical conversions are simple examples (Turban, 2001).

Simulation is not strictly a type of model. Models in general represent reality, whereas simulation typically imitates it. Simulation is a technique for conducting experiments. To simulate means to assume the appearance of the characteristics of reality. Simulation involves testing specific values of the decision or uncontrollable variables in the model and observing the impact on the output variables. Simulation is usually used only when a problem is too complex to be treated by numerical or analytical optimization techniques (Bowswell, 1999).

Steady State

A steady state simulation implies that the system state is independent of its initial start-up conditions. Analyses of these models are based on output data generated after the steady

Figure 1. Simulation of die values

```
Main ()
{
    srand(time(0));
    double sum ; long int  C, Iter, freq[arraysize];
    for (  C = 100; C<= 1000000; C*=1.5){
            sum = 0; for(int j=1; j<=6; j++) freq[j] = 0;
            for( long int roll = 1; roll <= C; roll++) ++freq[1+ rand()%6];
            for(int i = 1; i <= 6; i++ ) sum += freq[i]*i;
        cout << C<<'\t'<<sum/C<<endl;
    }
}
```

state conditions are achieved. *Figure 1* shows a simple C++ program simulating die values. *Figures 2, 3,* and *4* show that die values remain at 3.5 approximately.

We run the piece of code in *Figure 1* a number of times as described in the following experiments

Experiment 1: In this experiment, we run the program a number of times with initial value = 100 and final value = 1000000 and increment calculated as C = C * 1.5. Averages of iterations of the outer loop are computed. The x-axis represents iteration numbers. Averages are graphed in *Figure 2*.

Experiment 2: In this experiment, we run the program a number of times with different initial and final values. Averages of iterations of the outer loop are computed. The x-axis represents iteration numbers. Averages are graphed in *Figure 3*.

Experiment 3: This experiment is similar to the above two experiments except that the initial, final, and increment values are different. Output of this experiment is drawn in *Figure 4*.

A terminating simulation runs for a predetermined length of time or until a specific event occurs. Analyses and conclusions are based on output values produced at the stopping point. The results of terminating simulation are usually dependent upon the initial values and quantities used when starting the model. The start-up condition should accurately re-

Figure 2. Output of experiment 1

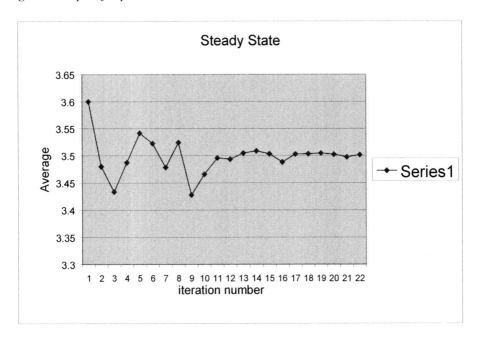

Figure 3. Graph of experiment 2

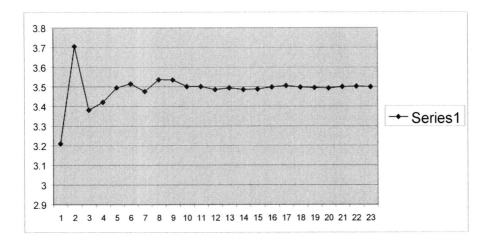

Figure 4. Graph of experiment 3

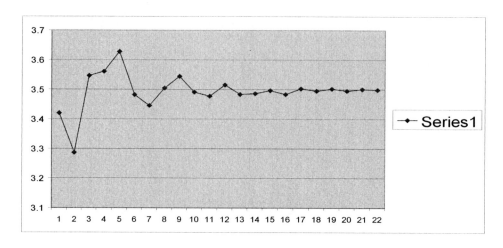

flect start-up circumstance exhibited in the real world system, which is being studied. The decision to employ a steady state or terminating simulation is made during the preliminary planning strategies of a simulation project.

Warm-up period is the amount of time that a model needs to run before statistical data collection begins. Model verification is the operation of the model in its intended manner. Consider a system consisting of a conveyor feeding parts to a machine. A simulation model is developed to analyze parts queuing on the conveyor.

The model can be deemed verified when it reflects the following conditions:

1.   Parts arrive at a desired rate.
2.   Parts serviced at a desired rate.
3.   Parts queued are counted correctly.

Model validation implies that the results generated coincide with the results produced by the system being represented by the model. Sometimes simulation is used to analyze theoretical systems, which do not physically exist. Model validation is not possible prior to hypothesis testing. In these cases, model builders must rely on system experts to establish the reasonableness of the results. Model verification becomes a major element for establishing rational validity.

Random number stream is a sequence of random numbers where each succeeding number is calculated from the previous one. The initial number is referred as the *random number seed.* Random number with values between zero and 1 play an important role in extracting values from probability distributions. A model run involves operating a simulation for a specified period of time with a unique set of random values. An independent model replication entails operating the same model for the same period of time with a different set of random values. Multiple model replications are always required when analyzing results from a stochastic simulation.

Intelligent System

Intelligent techniques can be exploited to model human reasoning in order to model the intangible aspects of a system. Integrated computational intelligence system reduces subjective decisions and increases the potential for real-time automation (Nakagiri, 1994). It has capabilities for assisting, constructing, and maintaining simulation system. This allows human expertise to be coded for future use in inference mechanism. It helps organize production processing. They produce a flexible system based on a set of production rules (Sharma, 1994).

The current available tools and techniques for solving production problems have a number of drawbacks such as large integer problem and an inefficient implementation with some interfacing obstacles (Sharma, 1994). They require sufficient items of information of a specific domain (Wildberger, 1995).

The framework embodies a dynamic system where the raw materials are not known in advance. Hence, the system should provide alternatives for different criteria. That is, different types of raw materials, and different numbers of identical raw materials used in the manufacturing of sub-assemblies and/or finished products.

Processing a large amount of items of information about the system components, control variables, and the interdependency structures creates new challenges on the shoulder of engineers and managers. The integrated system provides a traceability capability for components and their relationships. It provides the manager and engineer with sufficient items of information in order to detect inconsistencies; that is, it has reasoning capabilities on the system objects.

Construction of the Simulation Model

The process of manufacturing of a complex product requires a wide range of knowledge and information updating. It is of great importance to know how the assemblies of the system be integrated, as well as their individual effects on the overall production system. This integrated system would be more efficient if a complete understanding of the behavior of all sub-systems and the relationships among them is automatically available.

The system has two main parts:

a.   The finished products model, and
b.   The raw materials model.

The two models are discussed in turn.

Developing the Finished Product Model

The finished product model has three main components. These components are (Weida, 2001):

- **Set-up cost:** This is the cost of changeover in a production line from making one product to making a different product. Set-up costs favoring large production runs result in larger inventory. The low set-up costs favor smaller runs with fewer inventories.

- **Holding costs:** These are the costs the organization incurs in purchase and storing of the inventory. They include the cost of financing the purchase, storage costs, handling costs, taxes, obsolescence, pilferage, breakage, spoilage, reduced flexibility and opportunity cost. Holding costs are also known as carrying costs. High holding costs favor low inventory levels and frequent orders, while low holding costs favor holding large quantities of inventory.

- **Shortages costs:** This is the cost of not having stock when they are needed. These costs include loss of goodwill, loss of a sale, loss of a customer, loss of profit, and late penalties. Many of these costs are difficult or impossible to measure with any accuracy without using simulation techniques.

The finished product model has the form:

Total Finished Cost = Setup cost + holding cost + shortages cost

From *Figure 5*, the following model, which represents the finished product total cost, is outlined in equation 1.

Rule 1:

$$\text{Total Finished Cost} = \frac{DC_1}{Q} + \frac{(P_1 Q - S)^2}{2P_1 Q} C_2 + \frac{S^2}{2P_1 Q} C_3 \qquad \cdots \quad (1)$$

Figure 5. Finished production cycle

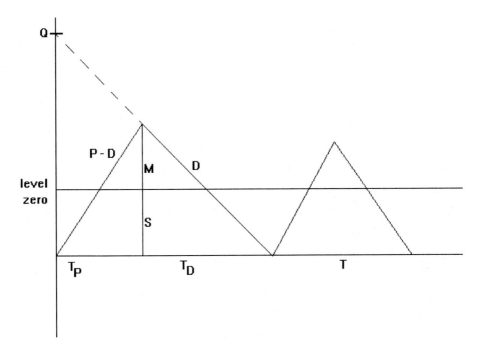

Variables explanation:

D = finished product demand per unit time,

Q = finished production quantity,

C1= set-up cost per item per cycle (or order),

C2= holding cost per item per unit time,

C3= shortages cost per unavailable unit per unit time,

S = shortages quantity,

$P_1 = (1 - D / P)$, where P is production rate per unit time.

In real life situations, all above variable values are not known in advance and they are following a probabilistic distribution. It would be impossible to estimate their values without using simulation.

T is the length of the production cycle, T_P is the actual production time, and T_D is the demand time when there is no production processing (Sharma, 1994).

Raw Material Model

A finished product model is developed through the process of a very complicated interaction of raw materials. Each unit of a finished product may be produced by combining, on average, hundreds of items (raw materials). The number of items from each type required is not evenly distributed. In order to facilitate the formulation and the understandability of the simulated model, we assume that assembled component A_j is made up by raw materials and/or assembled components $R1, R2, ..., Rj$ of kind J_w out of kind w where $J = 1, 2, 3, ..., i_m$ and $1 <= i_m <= m$.

Finished product P_i is made up by assembled components and/or raw materials $A_1, A_2, ..., A_m$ of kind i_m from kind m. This is characterized in *Figure 6*.

The branching of the tree may continue to a finite number of levels. K_j and Z_j are decision variables indicating the time between raw material j releases for level 1 and level 2 respectively. From *Figure 7*, the raw material model is:

Total Raw Material Costs = Ordering Costs + Holding Costs

Figure 6. Structure of raw materials and assembled components

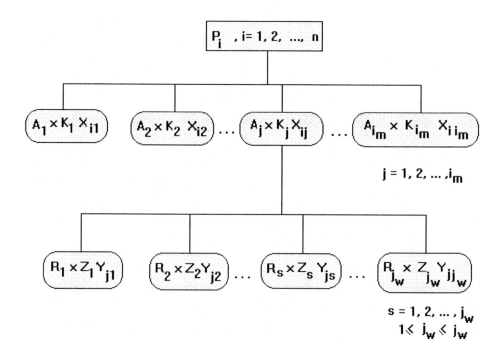

We assume no shortages of raw materials are permitted as situations in real life do not permit shortages of raw materials; otherwise production processes will be stopped. The model would be:

$$\text{Total Raw Material Cost} = \sum_{j=1}^{M} RMC_j(T, K_j)$$

Where **RMCj** is the cost for the raw material number j. The above formula can be described as follows (Nakagiri, 1994):

Rule 2:
$$\text{Total Raw Material Cost} - \sum_{J=1}^{M} \frac{O_j}{TK_j} + \frac{1}{2} \sum_{J=1}^{M} \left[X_J \frac{DT}{P} K_J + 2b_J \right] hc_J \quad , \qquad K_j < 1$$

Or,

Rule 3:
$$\text{Total Raw Material Cost} - \sum_{J=1}^{M} \frac{O_j}{TK_j} + \frac{1}{2} \sum_{J=1}^{M} \left[X_J \frac{DT}{P} + X_J T[K_J - 1] + 2b_J \right] hc_J \quad , \quad K_j >= 1$$

Figure 7. Raw material cycle sketch

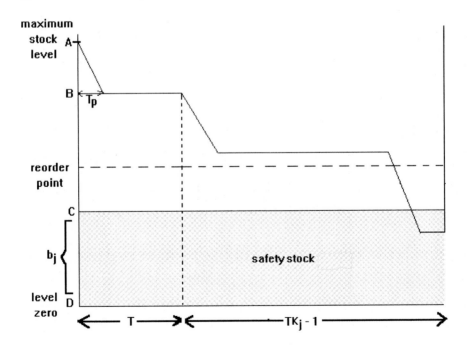

The symbols A, B, C, and D are variables used to help in formulating the raw materials costs model. K_j is a real number that refers to the number of production cycles between any two consecutive replenishments of raw materials.

O_j is the ordering cost for raw material j, which is the cost of placing an order for an item. Ordering costs apply to items the organization purchase. Ordering costs include placing an order, tracking the order, shipping costs, receiving and inspecting the order, and handling the paperwork. b_j is the safety stock for raw material j.

A number of fixed costs were not explicitly mentioned in the formulation of the model. These costs may include labor, machinery, overheads, and so forth. These costs have no effect on the solution of the model. They, the fixed costs, can be added to the total variable costs or they can be added to the set-up costs (administrative costs), holding costs, or shortages costs.

Simulating the Integrated Model

When finished products and raw materials models are combined, the following integrated model is produced.

Total Variable Cost = Integrated Model Cost = Finished Product Total Cost

+ Raw Material Total Cost

In other words,

Rule 4:

$$Tvc\ [\ T,\ Kj,\ S] = \frac{C_1}{T} + \frac{(P_1DT - S)^2}{2P_1DT}C_2 + \frac{S^2}{2P_1DT}C_3 + \sum_{J=1}^{M} RMC_j\ (\ T,\ Kj\ )$$

$$= \frac{C_1}{T} + \frac{1}{2}P_1DTC_2 - SC_2 + \frac{S^2}{2P_1DT}(C2 + C3) +$$

$$\sum_{J=1}^{M} \frac{O_j}{TK_j} + \frac{1}{2}\sum_{J=1}^{M} \left[X_j \frac{DT}{P} K_j + 2b_j \right] hc_j \quad , \quad K_j < 1$$

OR,

Rule 5:

$$Tvc\,[\,T,\,Kj,\,S\,]= \frac{C_1}{T} + \frac{1}{2}\,P_1\,DT\,C_2 \,-\, SC_2 \,+\, \frac{S^2}{2P_1DT}\,(\,C_2 + C_3\,) \,+\, \sum_{J=1}^{M} \frac{O_j}{TK_j}$$

$$+\, \frac{1}{2} \sum_{J=1}^{M} \left[\, X_j\frac{DT}{P} \,+\, X_j T[\,K_j - 1\,] + 2b_j \,\right] hc_j \,, \quad K_j \geq 1 \quad ...(\,3\,)$$

Rule 4 is Rule 1 and Rule 2.

Rule 5 is Rule 1 and Rule 3.

Rule 6: if K_j is less than 1 then Trigger Rule 4.

Rule 7: if K_j is greater or equal to 1 then Trigger Rule 5.

To accomplish the optimal solution from the proposed model, a cost function of state m is computed as follows:

$$\text{let} \quad \underline{f}^{(m)} = \begin{bmatrix} T^{(m)} \\ k_1^{(m)} \\ \vdots \\ k_n^{(m)} \\ S^{(m)} \end{bmatrix}$$

therefore,

$$\underline{f}^{(m+1)} = \underline{f}^{(m)} + \Delta\underline{f}^{(m)}$$

Transformation from one state to the next one is continued until the optimal solution is deduced.

Scientific Validity of the Model

Although analytical solutions are available to many production problems, other real life problems become complicated or impossible to analytically solve. It provides a mental picture of the manufacturing processing. It improves manufacturing capability and flexibility and hence reduces total cost and increases equipment utilization (Benjaafar, 1992).

Items of information about the nature of states, the cost of transforming from one state to another and the characteristics of the objectives can be used to guide the simulation actors more efficiently. These items are expressed in the form of a heuristic evaluation function f (k, g), a function of iteration (nodes) and the objectives. This approach helps pruning fruit-less paths. This is a best-first search that provides guidelines with which to estimate costs (Patterson, 1990).

For each iteration along the path to the objective goal, the heuristic estimation function is:

$$f(k, g) = f_1(k, g_1) + f_2(k, g_2)$$

Where both $f_1(k, g_1)$, $f_2(k, g_2)$ are estimates of cost from the beginning to node k and from node k to last node.

Optimality Criteria

The necessary and sufficient conditions for a function $f(x_1, x_2, ..., x_n)$ to be optimal at a point $x^* = (x_1^*, x_2^*, ..., x_n^*)$, such that its n partial derivatives are zero, the *Hessian matrix* (the second-order partial derivatives) *principal minors* must strictly be positive for a minimum point and negative for a maximum. In our case, the function is Tvc with three variables T^*, S^*, and K_j^* (J=1,2, ..., M). Hessian matrix is:

$$
\begin{pmatrix}
\dfrac{\partial^2}{\partial S^2} & \dfrac{\partial^2}{\partial S \partial T} & \dfrac{\partial^2}{\partial S \partial K_j} \\[2ex]
\dfrac{\partial^2}{\partial T \partial S} & \dfrac{\partial^2}{\partial T^2} & \dfrac{\partial^2}{\partial T \partial K_j} \\[2ex]
\dfrac{\partial^2}{\partial K_j \partial S} & \dfrac{\partial^2}{\partial K_j \partial T} & \dfrac{\partial^2}{\partial K_j^2}
\end{pmatrix}
$$

The principal minors of the Hessian matrix are

$$\Delta_1 = \frac{\partial^2}{\partial S^2}$$

$$\Delta_2 = \begin{vmatrix} \dfrac{\partial^2}{\partial S^2} & \dfrac{\partial^2}{\partial S \partial T} \\[2ex] \dfrac{\partial^2}{\partial T \partial S} & \dfrac{\partial^2}{\partial ST^2} \end{vmatrix}$$

$$\Delta_3 = \begin{vmatrix} \dfrac{\partial^2}{\partial S^2} & \dfrac{\partial^2}{\partial S \partial T} & \dfrac{\partial^2}{\partial S \partial Kj} \\[2ex] \dfrac{\partial^2}{\partial T \partial S} & \dfrac{\partial^2}{\partial ST^2} & \dfrac{\partial^2}{\partial T \partial Kj} \\[2ex] \dfrac{\partial^2}{\partial Kj \partial S} & \dfrac{\partial^2}{\partial Kj \partial T} & \dfrac{\partial^2}{\partial Kj^2} \end{vmatrix}$$

If the conditions $\Delta_1 > 0$, $\Delta_2 > 0$, and $\Delta_3 > 0$ hold then the point (T^*, S^*, Kj^*) is a minimum. These conditions are computed using numerical differentiation of $O(h^4)$; that is, using the central-difference formula:

$$f'' = (-f_2 + 16f_1 - 30f_0 + 16f_{-1} - f_{-2})/(12h^2) + O(h^4)$$

Constraints

Restrictions may be imposed on any variable. For example, K_j may be constrained into:

$$L_j <= K_j <= U_j,$$

Constraints may be imposed on T, S, or the total raw material.

$$\sum_{j=1}^{i_m} X_{ij} < Storage\ Available\ ,\ i=1,2,...,\ m$$
$$1 = < i_m = < m$$

Simulation Results

Experiments have been conducted to demonstrate the viability of the proposed simulation model.

Example 1: In this example, consider the items of information listed below:

m = 4, c1=56, c2=2.59, c3=1.9, p=380, d:=165, p1= 1.0-d/p,
 Raw Material Demand:
x[1]:= 495; x[2]:= 825; x[3]:=165; x[4]:=330;
   Holding Costs:
hc[1]:=0.005; hc[2]:=4.221; hc[3]:=0.401; hc[4]:=10.024;
 Ordering Costs:
 o[1]:=40.87; o[2]:=32.91; o[3]:=14.19; o[4]:=12.23;
 Safety Stocks:
 b[1]:= 495; b[2]:=825; b[3]:=165; b[4]:=330;

After running the simulated model the following output is outlined below:

Minimum costs = $758.442

Optimal Production Cycle = 0.55 months

Shortages Allowed are: 38 units

Reorder Raw Material (1) After 78 Days

Reorder Raw Material (2)   After 12 Days

Reorder Raw Material (3) After 36 Days

Reorder Raw Material (4)   After 12 Days

Example 2: In this experiment, consider the problem listed below:

m= 6; c1:=34;c2:=1.59;c3:=1.2; p:=4400;d:=1155; p1:= 1.0-d/p;

Raw Material Demand:

x[1]:= 1155;x[2]:= 1155; x[3]:=2310;x[4]:=2310;x[5]:=1155;x[6]:=3465;

Holding Costs:

hc[1]:=0.05;hc[2]:=0.021;hc[3]:=0.001;hc[4]:=0.002;

hc[5]:=0.003;hc[6]:=0.01;

Ordering Costs:

o[1]:=4.87;o[2]:=2.91;o[3]:=1.19;o[4]:=3.23;o[5]:=2.74;o[6]:=4.46;

Safety Stocks:

b[1]:=10;b[2]:=20;b[3]:=10;b[4]:=10;b[5]:=10;b[6]:=10;

After running the simulated model the following output is outlined below:

Minimum costs = $248.836

Optimal Production Cycle = 0.37 months

Shortages Allowed are: 182 units

Reorder Raw Material (1)	After 33 Days.
Reorder Raw Material (2)	After 39 Days
Reorder Raw Material (3)	After 81 Days
Reorder Raw Material (4)	After 96 Days
Reorder Raw Material (5)	After 87 Days
Reorder Raw Material (6)	After 42 Days

Example 3: In this experiment, consider the problem outlined below:

m:= 10; c1:=34;c2:=1.59;c3:=1.2; p:=4400;d:=1155; p1:= 1.0-d/p;

x[1]:= 1155;x[2]:= 1155; x[3]:=2310;x[4]:=2310;x[5]:=1155;x[6]:=3465;

x[7]:= 3465; x[8]:=4620; x[9]:= 2310; x[10]:=1155;

hc[1]:=0.05;hc[2]:=0.021;hc[3]:=0.001;hc[4]:=0.002;hc[5]:=0.003;

hc[6]:=0.01;hc[7]:=0.12; hc[8]:= 0.076; hc[9]:=0.0025; hc[10]:=0.11;

o[1]:=4.87;o[2]:=2.91;o[3]:=1.19;o[4]:=3.23;o[5]:=2.74;o[6]:=4.46;

o[7]:=4.43; o[8]:=3.21; o[9]:=8.76; o[10]:=5.01;

b[1]:=10;b[2]:=20;b[3]:=10;b[4]:=10;b[5]:=10;b[6]:=10;
b[7]:=10; b[8]:=20; b[9]:=10;b[10]:=20;

Subject to the following constraints

$T > 0.0$ and $T <= 0.5$,
$S >=   0$ and $S <= 144$,
$1 <= K1 <= 1.6$
$1 <= K2 <= 2.3$
$1 <= K3 <= 3.2$
$1 <= K4 <= 1.9$
$1 <= K5 <= 1.6$
$1 <= K6 <= 2.6$
$1 <= K7 <= 1.4$
$1 <= K8 <= 1.6$
$0 <= K9 <= 2.5$
$1 <= K10 <= 1.6$

After running the simulated model the following output is outlined below:

Minimum costs = $383.70

Optimal Production Cycle = 0.325 months

Shortages Allowed are: 134 units

Reorder Raw Material (1)	After 40 Days
Reorder Raw Material (2)	After 42 Days
Reorder Raw Material (3)	After 42 Days
Reorder Raw Material (4)	After 42 Days
Reorder Raw Material (5)	After 42 Days
Reorder Raw Material (6)	After 42 Days
Reorder Raw Material (7)	After 39 Days
Reorder Raw Material (8)	After 39 Days
Reorder Raw Material (9)	After 60 Days
Reorder Raw Material (10)	After 48 Days

The proposed model provides insight that could not be obtained by separated modules or other methods. The model provides users with a flexible approach of imposing and satisfying constraints.

From example 2, one can conclude that optimal production cycle length (T^*) is 0.37 months and optimal backorders (S^*) permitted is 182 units. K_j^* values are as listed above. The total variable cost is 248.836 Dinars. These numbers are the optimal ones while any other combinations would increase the total variable costs. In order to convince general readers, an evidence of optimality is provided and a number of case studies for sensitivity analysis are conducted.

Example 4: A company is specialized in selling cranes. The sale follows a uniform distribution of values between 1 and 3 cranes per day. After placing an order for a new shipment of cranes, arrival time follows a normal distribution with a mean of 2 weeks and a standard deviation of 0.6 weeks. In the past, the manager has placed an order when the quantity has dropped to 21. Set up a simulation and experiment with various reorder quantities to see which values seems to work best.

Sales Uniform Distribution

Minimum	1
Maximum	3

Normal Arrival Distribution

Mean	2.0
Standard Deviation	0.6

Reorder Quantity	21
Starting Quantity	44

Results

Average Inventory	28.7
Minimum Inventory	0.0
Maximum Inventory	55.0
Number of Stockouts	16

0=No, 1=Yes

Day	Inventory	Sales	Ending Inventory	Orders Pending	Orders	Days Until Arrival	Arrival Date	Arrivals
1	44	1	43	0	0	0	0	0
2	43	1	42	0	0	0	0	0
3	42	3	39	0	1	13	16	0
4	39	1	38	1	0	0	16	0
5	38	3	35	1	0	0	16	0

6	35	1	34	1	0	0	16	0
7	34	3	31	1	0	0	16	0
8	31	2	29	1	0	0	16	0
9	29	2	27	1	0	0	16	0
10	27	1	26	1	0	0	16	0
11	26	2	24	1	0	0	16	0
12	24	2	22	1	0	0	16	0
13	22	2	20	1	0	0	16	0
14	20	1	19	1	0	0	16	0
15	19	1	18	1	0	0	16	0
.	.	.	.	.	.	.	.	.
975	26	2	24	1	0	0	978	0
976	24	3	21	1	0	0	978	0
977	21	1	20	1	0	0	978	0
978	41	1	40	0	1	9	987	21
979	40	1	39	1	0	0	987	0
980	39	1	38	1	0	0	987	0
981	38	2	36	1	0	0	987	0
982	36	1	35	1	0	0	987	0

Extension to Example 4: expand the experiment varying the reorder point as well as the re-order quantity. Experiment with various values. Set up a data table to report results.

Sales Uniform Distribution

Minimum	1
Maximum	3

Normal Arrival Distribution

Mean	2.0
Standard Deviation	0.6

Reorder Quantity	21
Reorder Point	30
Starting Quantity	44

Results

Average Inventory   19.4

Minimum Inventory   0.0

Maximum Inventory   51.0

Number of Stockouts   31

0=No, 1=Yes

Day	Inventory	Sales	Ending Inventory	Orders Pending	Orders	Days Until Arrival	Arrival Date	Arrivals
1	44	1	43	0	0	0	0	0
2	43	2	41	0	0	0	0	0
3	41	3	38	0	0	0	0	0
4	38	1	37	0	0	0	0	0
5	37	2	35	0	0	0	0	0
6	35	3	32	0	0	0	0	0
7	32	3	29	0	1	13	20	0
8	29	3	26	1	0	0	20	0
9	26	1	25	1	0	0	20	0
10	25	1	24	1	0	0	20	0
. . .	.	.	.	.	.	.	.	.
975	11	3	8	1	0	0	981	0
976	8	3	5	1	0	0	981	0
977	5	3	2	1	0	0	981	0
978	2	2	0	1	0	0	981	0
979	0	0	0	1	0	0	981	0
980	0	0	0	1	0	0	981	0
981	21	3	18	0	1	10	991	21
982	18	3	15	1	0	0	991	0

Optimality Testing

When optimality criteria is applied the following numerical results are obtained: Δ_1 = 0.08781833, $\Delta_2 = 52.00098$, and Δ_3 is positive for all Kj, j=1, ..., M. The minor values for j=1,2, ..., 6 are 953.0471, 331.406, 13.88616, 23.11795, 16.06441, 455.7297 respectively. As the principal minors are strictly positive, the point (T^*, S^*, Kj^*) is a minimum of the function $Tvc(T^*, S^*, Kj^*)$. The above statements prove that the proposed model has actually optimal solutions.

Conclusion and Suggestions

The chapter addressed an important application issue: how well an integrated model can describe the real world applications. The idea was to integrate the finished products, sub-assemblies and raw materials in one model. Simulation is used to generate a number of scenarios in what-if analysis approach in order to deal with uncertainty.

Simulation is important for any intelligent system involving uncertainty. Simulation is applicable to complex situations where mathematical techniques do not work or are hard to analytically or numerically optimize. Simulation is used in two general types of situations:

- The probability distributions cannot be expressed in mathematical forms as we have seen in our models.

- The model is too complex. There are too many components, and the model is thus impossible to solve using mathematical methods.

Introducing artificial intelligence in the simulated applications provides a laboratory to generate and examine models and what-if scenarios that involve many uncertainties. The intelligent system can examine not only results but also assumptions, particularly as far as probabilities are concerned.

References

Benjaafar, S. (1992). Intelligent simulation for flexible manufacturing systems: An integrated approach. *Computers and Industrial Engineering, 22*(3), 297-311.

Bowswell, C. (1999, September). Process simulation software offers efficiency and savings. *Chemical Market Reporter, 256*(13).

Graul, M., Boydstun, F., Harris, M., Mayer, R., Bagaturova, O. (2003). *Integrated framework for modeling & simulation of complex production system*s. Knowledge Based Systems Inc. Retrieved from http://www.dtic.mil/ndia/2003systems/graul.pdf

Nakagiri, D., & Kuriyama, S. (1994, March 7-9). A study on production planning for CIM. In *Proceedings of the 16th International Conference on Computers and Industrial Engineering,* Japan (pp.654-657).

Patterson, D.W. (1990). *Introduction to artificial intelligence and expert systems*. Prentice-Hall International.

Sharma, G., Asthana, R.G.S., & Goel, S. (1994). A knowledge-based simulation approach (K-SIM) for train operation and planning. *Simulation, 62*(6), 381-391.

Turban, E., & Aronson, J.E. (2001). *Decision support systems and intelligent systems* (3rd ed.). Prentice Hall.

Weida, N.C., Richardson, R., & Vazsonyi, A. (2001). *Operations analysis using Microsoft Excel*. Duxbury.

Wildberger, A.M. (1995, February). AI & simulation. *Simulation, 64*(2).

Chapter X

Simulation and Modelling of Knowledge-Mining Architectures Using Recurrent Hybrid Nets

David Al-Dabass, Nottingham Trent University, UK

Abstract

Hybrid recurrent nets combine arithmetic and integrator elements to form nodes for modelling the complex behaviour of intelligent systems with dynamics. Given the behaviour pattern of such nodes it is required to determine the values of their causal parameters. The architecture of this knowledge mining process consists of two stages: time derivatives of the trajectory are determined first, followed by the parameters. Hybrid recurrent nets of first order are employed to compute derivatives continuously as the behaviour is monitored. A further layer of arithmetic and hybrid nets is then used to track the values of the causal parameters of the knowledge mining model. Applications to signal processing are used to illustrate the techniques. The theoretical foundations of this knowledge mining process is presented in the first part of the chapter, where the application of dynamical systems theory is extended to abstract systems to illustrate its broad relevance to any system including biological and non physical processes. It models the complexity of systems in terms of observability and controllability.

Chapter Introduction

In this chapter we illustrate the use of mathematical modelling and simulation to discover the reasons for data to behave in certain way. Knowledge mining refers to the process of extracting hidden knowledge from data, where advances in several branches of computer science came together under the title of "data mining" (Klosgen, 2002). Conventional techniques range from simple pattern searching to advanced data visualisation and neural networks. As the aim is to extract comprehensible and communicable scientific knowledge, the approach is rightfully characterised as "knowledge mining." Previous attempts in the field were concerned with machine learning techniques and inductive logic programming (ILP) [see Muggleton (1994) and De Raedt (1992)]. Induction is the process of obtaining general rules from example data. Logic facilitates the explicit encoding of constraints and relevant prior knowledge together with machine generated and testable hypotheses.

However, the aim in this chapter is to gain parametric knowledge about the patterns of behaviour of the data as continuous processes by using systems theory. In the first section, we take an abstract view to apply the theoretical principles to a general class of problem areas that generate the data. These problem areas span science, engineering, medicine and finance and include: i) monitoring the movement of objects and structures to anticipate undesirable effects such as instability of large delicate structures,for example, but not necessarily in space, ii) monitoring patient data to check on reactions to drugs and recovery progress, iii) monitoring the movement of stock values and share indices in financial markets to predict the occurrence and timing of the next peak or trough in their values, iv) observing facial features and body movements in psychological analysis of mental states to predict future behaviour patterns.

Knowledge Models: To understand and control the behaviour of systems in general, models that represent the knowledge embedded within these systems are formulated and used to acquire this knowledge from measurements. In data mining applications, for example, there is a need to determine the causes of particular behaviour patterns. We adopt a systems theory approach by determining the causal system parameters, such as input, natural frequency and damping coefficient. To account for complicated behaviour patterns we illustrate through modelling and simulation the use of a multi-level system of interconnected nodes to generate time varying input and frequencies.

Hybrid Inference Networks: To represent the knowledge embedded within intelligent systems, a multi-level structure is put forward. By its very nature this knowledge is continually changing and needs dynamic paradigms to represent and acquire its parameters from observed data. In a normal inference network the cause and effect relationship is static and the effect can be easily worked out through a deduction process by considering all the causes through a step-by-step procedure, which works through all the levels of the network to arrive at the final effect. However, reasoning in the reverse direction, such as that used in diagnosis, starts with observing the effect and working back through the nodes of the network to determine the causes.

Knowledge Mining: Work in this chapter extends these ideas of recurrent or dynamical systems networks to models where some or all the data within the knowledge base is time varying. The effect is now a time dependent behaviour pattern, which shall be used as an input to a differential process to determine knowledge about the system in terms of time varying causal parameters. As will be argued, these causal parameters will themselves embody knowledge (meta knowledge), which is obtained through a second level process to yield second level causal parameters. The proposed knowledge mining architecture consists of a differential part to estimate the higher time derivative knowledge, followed by a non-linear algebraic part to compute the causal parameters.

Modelling the Complexity of Systems Dynamics

Introduction

In this part we abstract the fundamental principles embedded in the set of application areas referred to above and cast them as abstract systems. The approach we adopt is based on continuous systems theory. The notion of systems dynamics is put forward as an aid to categorise a set of systems according to some behaviour criteria. An abstract system, including the abstract notion of concepts, will change with time under the influence of inputs. Systems will have measurable outputs in a suitable environment. Not all of the system inner (state) variables are directly measurable from their output, or action, sequences, and give rise to the "observability" condition. Furthermore, senses have only limited ability to change the inner state of systems, which in turn leads to the "controllability" condition. The behaviour pattern of systems can be categorised by a set of parameters as well as by inputs. Thus suitable knowledge mining algorithms are needed to achieve categorisation.

Systems Possess Inertia

It takes finite time duration to change a system's variable value and thus it can be said to have inertia. A system variable subjected to an input will not acquire its new value immediately but evolve gradually. Three patterns of behaviour may result: i) the variable will reach its new value without overshooting it, or ii) it will overshoot initially and then undershoot and overshoot several times before eventually reaching it new value, or iii) the overshoot and undershoot become more divergent and the state variable never reaches its intended new value.

System Variables Exhibit Oscillatory Behaviour

This is an extension to the notion of "inertia"—a state variable may embody within its semantics structure multiple "energy storage facilities" that, under the right conditions, such as low damping, can operate sufficiently out of phase to cause semantic energy to flow back and forth in a similar way that a liquid flows between two interconnected reservoirs.

System State Vector: Let X be an n-vector of variables that represent the system.

Input Vector: Let U be an r-vector of variables that represent all inputs to the "system;" for a hybrid intelligent system these input variables will represent humans as well as machines.

Input Matrix: Let the r-vector U influence the system state variable rate of change by a time varying n x r matrix C(t).

Output Vector: Let Z be an m-vector of variables that results in mapping the state variables onto a suitable environment space. No distinction is made here between real physical space and virtual space within a suitable domain (such as software, or as perceived in the mind of a human operative/decision maker) inhabited by a system generating the variables and at least one "other" system monitoring it.

Output Matrix: Let H(t) be an m x n matrix that maps the state variables n-vector X onto the output m-vector Z.

Therefore the mapping between measurable outputs and state variables, that is, "**Outputs as derived from state variables**" **equation**, is:

$$Z(t) = H(t).X(t) \tag{1}$$

State Variable Dynamical Equation: Within a dynamical systems context, the state vector X evolves in time according to:
1. *the values of its present state,*
2. *data from its input, and*
3. *random disturbance.*

For simplicity of the current treatment the random disturbance contribution will be assumed to be negligible. This, however, does by no means imply that disturbances are not important; on the contrary their profound significance is best treated in a separate monograph.

Rate of Change of State: Let X' be the rate of change of state such that it is directly influenced by i) current state value, plus ii) data from its input.

Therefore the following equation defines the dynamics of the state variables:

$$X'(t) = F(t).X(t) + C(t).U(t) \tag{2}$$

Where F(t) is n x n time varying matrix, which determines how the rate of change of each variable is influenced by all the current values of the other state variables; and C(t) is the input matrix defined above.

Thus the system's dynamics are described by a set of first order time-varying linear differential equations.

Observability and Controllability

There are two fundamental ideas involved in modelling the dynamical behaviour of systems, which are related to the way they are changed by data at their input on the one hand and the way their internal (state) variables influence the outputs of the system on the other.

1. Under what conditions can the system state vector be estimated from measurements of outputs?
2. Under what conditions can the state vector be influenced by data received at its input?

Observability

Consider the continuous linear model formulated above:

$$X'(t) = F(t).X(t) + C(t).U(t)$$

$$Z(t) = H(t).X(t)$$

Figure 1. Controllable but unobservable system

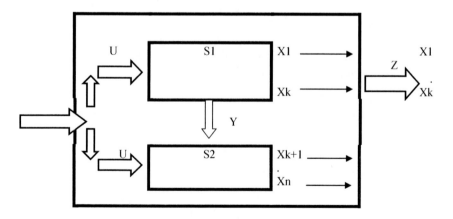

Figure 2. Uncontrollable but observable system

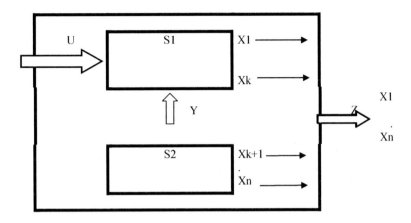

Consider a dynamical system S with state vector X, input vector U and output vector Z, refer to *Figure 1*. Assume that a subset of the states, say Y, forms a subsystem S2 such that they do not influence the other state variables X1 - Xk. Furthermore, the overall systems outputs are derived from X1 - Xk state variables only. It is then clear that no matter how many sequences of Z are measured, it will not be possible to determine Xk+1 - Xn. Such a system is clearly controllable but not observable.

Consider now the system shown in *Figure 2*. The inputs now have no influence on the sub-system S2 whether directly or indirectly through the other subsystem (S1) state variables, despite the fact that all state variables are available for measurement. This is therefore a completely observable but uncontrollable system.

Loosely stated, an observable system is one where measurement of the output is sufficient to determine the value of the state variable that generated the output; that is, when a suitable "observer" or state re-constructer can be formed to estimate the values of the state vector from the output trajectories. The implication, of course, is that the "system" can become unobservable when the values of the system matrices F and Z change to give rise to the configuration shown in *Figure 1*. More rigorously, the following is true.

Definition: a system is observable if the value of its state vector at some time instant t0 X(t0) can be determined from its output trajectories Z(t), t0 <= t <= t1, for some finite t1. If this is true for any t0, the state variables are completely observable.

Observability Matrix: is given in terms of state transition matrix ST and output mapping matrix H as:

$$\text{OM(t0,t1)} = \int_{t0}^{t1} \{ST"(t,t0) . H"(t) . H(t) . ST(t,t0) . dt\} \qquad (3)$$

Where ST" indicates transpose of ST and so forth.

Observability Condition: The monitored system as modelled by the linear system F and
H is completely observable if and only if the symmetric n x n matrix OM is positive
definite for some finite time t1 > t0. PROOF if needed

Constant Parameter System: If the matrices F and H are constant, the observability condition simplifies such that the system is completely observable if and only if the n x
mn matrix

$$H", F"H", \ldots , (F")^{n-1} H" \qquad (4)$$

has rank n.

Controllability

Consider the system given by:

$$X' = F(t).X + C(t).U(t) \qquad (5)$$

For t >= t0, where X(t0) is known but U(t) is not specified. The controllability problem
involves transferring the state vector from X(t0) to some required terminal state X(t1) = X1
where t1 is finite. By suitable change of coordinates, the problem becomes that of transferring from some X(t0) to the origin [now given as 0 or X(t1)] in a finite time.

Definition: The system given above is controllable at time t0 if there exists an input function (as data applied to it's input) U(t) depending on X(t0) and defined over some
finite interval t0 <= t <= t1 for which X(t1) =0. If this is true for all X(t0) and t0, the
system is completely controllable.

Controllability Matrix: Given in terms of the state transition matrix ST and the input
matrix C as:

$$\text{CM}(t0,t1) = \int_{t0}^{t1} \{ \text{ST}(t0,t1) \cdot C(t) \cdot C"(t) \cdot \text{ST}"(t0,t1) \cdot dt \} \tag{6}$$

Where ST" indicates transpose of ST.

Controllability Condition: The monitored system as modelled by the linear system F and C is completely controllable if and only if the symmetric n x n matrix CM is positive definite for some finite time $t1 > t0$.

Constant Parameter System: If the matrices F and C are constant, the controllability condition simplifies such that the system is completely controllable if and only if the n x nr matrix

$$C, FC, \ldots, F^{n-1} C \tag{7}$$

has rank n.

Knowledge Mining for Systems Categorisation

To generalise the treatment we reduce the essential characteristics of these application areas to "abstract systems," which can be modelled mathematically as described earlier in this chapter. System dynamics are defined by differential equations and the matrices F, C and H. The numerical values of these matrices determine the behaviour pattern of the system. Conversely, monitoring the output of systems should enable us to determine not only the values of the state variables but also the values of the parameters that determine the nature of their behaviour.

Consider a system modelled as a second order dynamical system, the state space being its value x and its time derivative x'. To simplify the treatment, and without loss of generality, let the system be driven by a single input variable u and generate a single output represented by the variable value itself, that is, z=x. Furthermore, reduce the resulting 2x2 system matrix F to represent the system dynamics in terms of its natural frequency omega and damping coefficient zeta. This results in the following well-known second order differential equation:

$$\omega^{-2} x" + 2 \cdot \zeta \cdot \omega^{-1} \cdot x' + x = u \tag{8}$$

Categorisation of Systems

Let the above equation be the model to represent any system and its dynamics. Therefore categorisation reduces to finding the three parameters—u, omega and zeta—that completely describe the behaviour pattern of the system.

The process of identifying a particular system thus reduces to monitoring its trajectory and determining these three parameters from it. Comparison between systems reduces to comparing the corresponding values of these parameters derived from the behaviour trajectory of the associated system.

Several algorithms are available for estimating parameters from trajectories depending on i) the number of points on the trajectory, ii) the number of high order time derivatives available at each point on the trajectory an iii) on the assumption whether the parameters are constant or time varying.

For the purpose of this illustration it suffices to consider single point algorithms. These are classified according to the order of the parameter variation used in the derivation, that is, constant, first order polynomial (constant u' but u''=0), 2nd order polynomial (constant u'' but u'''=0), and so forth. They will be given here without proof; the interested reader may refer to Al-Dabass (2002).

Constant Parameters

Consider using the 1st to 4th time derivatives at a single point of the action trajectory generated by the system. Given the second order system:

$$\omega^{-2} x'' + 2. \zeta.\omega^{-1}.x' + x = u \tag{9}$$

We get expressions for estimated ω, estimated ζ, and estimated u:

$$E\omega^2 = [x''. x'''' - x'''^2] / [x'. x''' - x''^2] \tag{10}$$

$$E\zeta = -[E\omega^{-2} x''' + x'] / [2. E\omega^{-1}.x'']$$

$$Eu = E\omega^{-2}. x'' + 2. E\zeta. E\omega^{-1}. x' + x$$

First Order Parameters

Let the first time derivative of u to be non-zero. For simplicity assume that both a and b (the coefficients of x'' and x' to make symbol manipulation easier) to be constant and hence disappear on first differentiation. The extra information needed for u' to be non-zero is extracted from the 5th time derivative of the trajectory. To simplify the expression consider the following "lumped" parameter model:

$$a.x'' + b.x' + x = u \qquad (11)$$

By successive differentiation with regard to t, assuming u' to be non-zero (but u'' and higher time derivatives = 0) and re-arranging for a gives:

$$a = (x'' . x'''' - x'''^2)/(x''''' . x''' - x''''^2) \qquad (12)$$

$$b = (x''.x''''' - x'''.x'''')/(x'''''^2 - x''' . x''''')$$

$$u = a.x'' + b.x' + x$$

Simulation Model

Numerous simulation software packages are available to aid experimentation understanding of the techniques, one such package is Mathcad. The simulation derivative vector is shown in *Figure 3*.

Figure 3. Rows 1 and 2 from the top show a 2nd order system; row 3 is not used; rows 4 and 5 simulate the input parameter as a second order subsystem; rows 6 to 11 produce estimated x (output) and its time derivatives x', x'', x''', x'''', x'''''

$$D(t,x) := \begin{bmatrix} x_2 \\ w \cdot w \cdot x_4 - 2 \cdot z \cdot w \cdot x_2 - w \cdot w \cdot x_1 \\ 0 \\ x_5 \\ (wu \cdot wu \cdot uu) - 2 \cdot zu \cdot wu \cdot x_5 - wu \cdot wu \cdot x_4 \\ G \cdot (x_1 - x_6) \\ G1 \cdot \left[G \cdot (x_1 - x_6) - x_7 \right] \\ G2 \cdot \left[G1 \cdot \left[G \cdot (x_1 - x_6) - x_7 \right] - x_8 \right] \\ G3 \cdot \left[G2 \cdot \left[G1 \cdot \left[G \cdot (x_1 - x_6) - x_7 \right] - x_8 \right] - x_9 \right] \\ G4 \cdot \left[G3 \cdot \left[G2 \cdot \left[G1 \cdot \left[G \cdot (x_1 - x_6) - x_7 \right] - x_8 \right] - x_9 \right] - x_{10} \right] \\ G5 \cdot \left[G4 \cdot \left[G3 \cdot \left[G2 \cdot \left[G1 \cdot \left[G \cdot (x_1 - x_6) - x_7 \right] - x_8 \right] - x_9 \right] - x_{10} \right] - x_{11} \right] \end{bmatrix}$$

Results and Discussion

Mathcad routines were set up to generate the input u as second order system with its own parameters of natural frequency, damping ratio and input. The input subsystem damping ratio was set to 0.05 to generate an oscillatory behaviour for long enough to test the sense tracking algorithm thoroughly. The frequency of the input was set to 16 radians per second, one quarter of the frequency of the system natural frequency. The derivative generation cascade was increased by one to produce the fifth time derivative. The results are shown in *Figure 4*.

The input is the smooth trace, which gives approximately two and a half cycles over a period of one second as expected, that is, 16 radians/s = 2.546 Hertz. The jagged trace is the estimated value of the input using the constant u derivation algorithm, which is failing completely to track the input. The 3rd trace shows the result of the second algorithm, which is managing to track the input much more closely. It is clear that tracking remains stable. It is interesting to note that the first algorithm, while completely failing to track the upper half of the input trajectory, seems to track it well during its lower half but not as well as the second algorithm.

Figure 4. Tracking the input parameter u using a trajectory of the system output

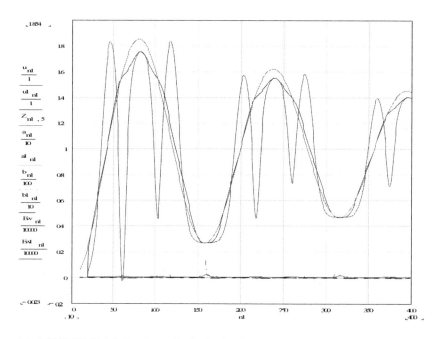

Comments

This part of the chapter attempted to model state as vector variables whose time derivatives are related to current values and input. Systems generate output in their environment; these outputs are measured by other systems to estimate the state values, the inputs and the system parameters that determine the behaviour pattern of the state and outputs. The notions of observable and controllable systems were introduced and their conditions established. Categorisation of systems based on similarity in the parameter values that determine their behaviour was put forward. Suitable estimation algorithms were investigated to illustrate the procedure of extracting these parameters from the output trajectories of the system.

Knowledge Mining Algorithms Using Recurrent Inference Networks

Introduction

Recurrent inference networks are introduced to represent knowledge bases that model dynamic intelligent systems. Through a differential abduction process, the causal parameters of the system behaviour are determined from measurements of its output to extract the knowledge embedded within (Al-Dabass, 2001). The use of dynamical knowledge mining processes ensures that knowledge evolution is tracked continuously (Al-Dabass, Zreiba, Evans, & Sivayoganathan, 2002). Meta-knowledge, defined in terms of the causal parameters of the evolution pattern of this first level knowledge, is further determined by the deployment of second level dynamical processes (Al-Dabass, Evans, Zreiba, & Sivayoganathan, 2002). In data mining applications, for example, there is a need to determine the causes of particular behaviour patterns. Other applications include cyclic tendencies in stock values and sales figures in business and commerce, changes in patient recovery characteristics, and predicting motion instabilities in complex engineering structures (Al-Dabass, Evans, & Sivayoganathan, 2003. Full mathematical derivation is given together with simulations and examples to illustrate the techniques involved

System Modelling and Simulation Using Hybrid Recurrent Networks

Numerous systems in practice exhibit complex behaviour that cannot be easily modelled using simple nets (Al-Dabass, Zreiba, Evans, & Sivayoganathan, 1999). In this part we recast this problem in terms of hybrid recurrent nets, which consist of combinations of static nodes, either logical or arithmetic, and recurrent nodes. The behaviour of a typical recurrent node is modelled as a second order dynamical system. The causal parameters of such a recurrent node may themselves exhibit temporal tendencies that can be modelled in terms of further recurrent nodes. Layers of recurrent nodes are added until a complete account of the behaviour of the system has been achieved (Al-Dabass, Evans, & Sivayoganathan, 2003). Algorithms are given to abduct the values of the parameters of these models from behaviour trajectories of intelligent systems. One novel aspect of the work lies in having a simple hierarchical 6th order linear model to represent a fairly complicated behaviour encountered in numerous real examples in finance, biology and engineering.

Hybrid Recurrent Network Models

Many physical, abstract and biological phenomena exhibit temporal behaviour even when the input "causal" parameters are constant (*Figure 5*).

To model this oscillatory behaviour a second order integral hybrid model is proposed, shown in *Figure 6*. This model is based on the well-known second order dynamical system, which has the following form:

$$\omega^{-2} x'' + 2.\,\zeta.\omega^{-1}.x' + x = u \tag{13}$$

Where x is the output of the node and ω, ζ and u are the natural frequency, damping ratio and input respectively, which represent the three causal parameters that form the input. To configure this differential model as a recurrent network, twin integral elements are used to form a hybrid integral-recurrent net as shown in *Figure 6*.

Figure 5. A Recurrent Node (R-N) exhibits a temporal behaviour at the output despite having constant causal parameters

Figure 6. Hybrid integral-recurrent net to model the temporal behaviour of the node in Figure 1

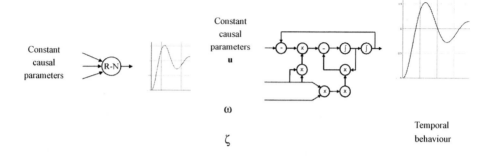

Structure of the Hybrid Integral-Recurrent Net

The net shown in *Figure 6* is a direct representation of Equation 1 and can be derived as follows:

1.   By multiplying both sides by ω^2 we get:

$$x" + 2.\ \zeta.\omega.x' = \omega^2.(u - x) \qquad\qquad (13\text{-}A)$$

OR

$$x" = \omega^2.(u - x) - 2.\ \zeta.\omega.x' \qquad\qquad (13\text{-}B)$$

2.   The output of the net x is fed back to the first subtraction node on the left; as the input from the left of this node is u, the output is $(u - x)$.

3.   The middle input (to the whole net) from the left is ω; it is fed as 2 separate inputs to the multiplication node x to form ω^2 at its output, shown with an up arrow feeding as the lower input of the multiplier node above it, which is the second node from the left in the top chain of nodes.

4.   The output of this multiplier node is therefore $\omega^2.(u - x)$, that is, the RHS of Equation 13-A.

5.   The bottom input from the left (to the whole net) is ζ which is fed as the lower input to the first of the two multipliers in the bottom chain of 2 nodes,- as the top input to this node is ω the output is $\zeta.\omega$. which is multiplied by 2 in the 2nd node in the chain to produce $2.\ \zeta.\omega$.

6.   The last node on the right in the top long node chain is an integrator node that generates x as stated in ii) above. As it is an integrator node, the input to it must therefore be the derivative of x, that is, x'. This is multiplied by the output of the right node in the bottom 2-node chain (which is $2.\ \zeta.\omega$) to produce $2.\ \zeta.\omega.$ x', which is the 2nd term in the LHS ofEquation 13-A or the 2nd term on the RHS ofEquation 13-B.

7.   By subtracting this output from the output of the middle node in the top row, we get the full RHS of Equation 13-B, that is, $\omega^2.(u - x) - 2.\ \zeta.\omega.x'$.

8.   As the output of the second integrator from the right (in the top chain) is the first derivative of x, x', the input to this integrator node must be x", that is, the LHS of Equation 13-B.

9.   Simply connecting the output of the middle node of the top chain (which is $\omega^2.(u - x)$ $- 2.\ \zeta.\omega.x'$) into the input of the 2nd integrator from the right (x") will just complete the equation.

Models of Hierarchical Recurrent Nodes

The output trajectory of the system may be more complex than can be represented by a simple second order differential model. In this case each causal parameter may itself be modelled as having a dynamical behaviour, which may or may not be oscillatory. One such case is where two of the three causal parameters have 2nd order dynamical characteristics, as shown in *Figure 7*.

The 2nd order model of a node in a given layer in the hierarchy is given by Equation 1 above. Starting with the final output node let both u and omega have their own 2nd order dynamics. The input **u** is the output of the following 2nd order system:

$$\omega_u^{-2}\,\mathbf{u''} + 2.\,\zeta_u.\,\omega_u^{-1}.\mathbf{u'} + \mathbf{u} = \mathbf{u}_u \tag{14}$$

The natural frequency ω is the output of the following 2nd order system:

$$\omega_\omega^{-2}\,\omega'' + 2.\,\zeta_\omega.\,\omega_\omega^{-1}.\omega' + \omega = u_\omega \tag{15}$$

Thus the behaviour trajectory is generated by the following 6th order vector differential equation (using Runge Kutta in Mathcad for this example).

First Order Vector Form: To provide a simulation output of the node trajectory, the 2nd order equation is converted to a 2nd order vector differential equation that can be easily computed. This entails assigning a separate state variable to each higher derivative, such that: x1 = x, x2 = x', x3 = x" and so on.

Figure 7. Two of the causal parameters of the final node have temporal behaviour modelled as 2nd order hybrid integral recurrent nets

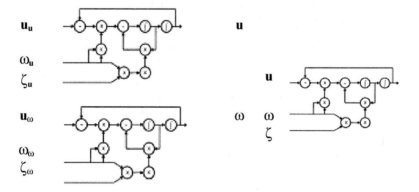

Figure 8. The derivative vector for generating the system output using subsystems for u and omega

$$D(t,x) := \begin{bmatrix} x_2 \\ x_5 \cdot x_5 \cdot x_3 - 2 \cdot z \cdot x_5 \cdot x_2 - x_5 \cdot x_5 \cdot x_1 \\ x_4 \\ (wu \cdot wu \cdot uu) - 2 \cdot zu \cdot wu \cdot x_4 - wu \cdot wu \cdot x_3 \\ x_6 \\ (ww \cdot ww \cdot uw) - 2 \cdot zw \cdot ww \cdot x_6 - ww \cdot ww \cdot x_5 \end{bmatrix}$$

Figure 9. Simulated trajectory of a hierarchical recurrent node (oscillatory trace), with 2 variable inputs: u (upper trace) and omega (lower trace)

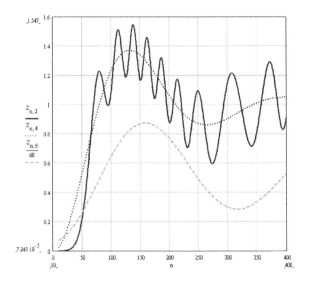

In *Figure 8*, x1 and x2 represent x and x', x3 and x4 represent u and u', and x5 and x6 represent ω and ω' respectively. To generate the trajectory shown in *Figure 9*, the following values were used: for the u subsystem, u started from 0 aiming at $u_u = 1$ at a rate of $\omega_u = 5$ rad/s with $\zeta_u = 0.3$. For the ω subsystem, ω started from 4 rad/s aiming at $u_\omega = 32$ rad/s at a rate of $\omega_\omega = 4$ rad/s with $\zeta_\omega = 0.1$. *Figure 9* shows the resulting compound trajectory of x (oscillatory trace), together with the trajectories for u (upper trace) and ω (lower trace).

Derivative Estimation Using Recurrent Networks

Based on models that describe the behaviour of complex natural and physical systems, a number of explicit static algorithms are developed to estimate the parameters of recurrent second order models that approximate the behaviour of these complex higher order systems (Al-Dabass, Zreiba, Evans, & Sivayoganathan 1999). These algorithms rely on the availability of the time derivatives of the trajectory. In this section, a cascaded recurrent network architecture is proposed to "abduct" these derivatives in successive stages. The technique is tested successfully on parameter tracking algorithms ranging from the constant parameter algorithm that only requires derivatives up to order 4 to an algorithm that tracks two variable parameters and requires up to the 8th time derivatives.

Algorithm for Constant Parameters from Single Point Data

Consider using the 1st to 4th time derivatives at a single point. Given the second order system:

$$\omega^{-2} x'' + 2. \zeta.\omega^{-1}.x' + x = u \qquad (16)$$

Differentiate with respect to t:

$$\omega^{-2} x''' + 2. \zeta.\omega^{-1}.x'' + x' = 0 \qquad (17)$$

divide by x":

$$\omega^{-2} x'''/ x'' + 2. \zeta.\omega^{-1} + x'/ x'' = 0 \qquad (18)$$

and differentiate with respect to t again to give:

$$\omega^{-2}.[(x''. x'''' - x'''^2) / x''^2] + 0 + [(x''^2 - x'. x''') / x''^2] = 0 \qquad (19)$$

Expressions for estimated ω, estimated ζ, and estimated u result as follows:

$$E\omega^2 = [x''. x'''' - x'''^2] / [x'. x''' - x''^2] \qquad (20)$$

$$E\zeta = -[E\omega^{-2} x''' + x'] / [2. E\omega^{-1}.x''] \qquad (21)$$

$$Eu = E\omega^{-2}. x'' + 2. E\zeta. E\omega^{-1} . x' + x \qquad (22)$$

High Order Algorithms

Assume that the first and higher time derivative of u to be non-zero. For simplicity assume that both a and b (the coefficients of x" and x' to make symbol manipulation easier) to be constant and hence disappear on first differentiation. The extra information needed for u', u", u'" and u"" to be non-zero is extracted from the 5th, 6th, 7th and 8th time derivatives of the trajectory. Only the case for the u' is shown here, the others for u" etc are simple extensions of the idea and are left as an exercise for the reader.

$$a.x'' + b.x' + x = u \tag{23}$$

Differentiate with regard to t and assume u' is non-zero to give:

$$a.x''' + b.x'' + x' = u' \tag{24}$$

Differentiate again and set u" = 0 gives:

$$a.x'''' + b.x''' + x'' = 0 \tag{25}$$

Divide Equation 11 by x'" to isolate b:

$$a.x''''/x''' + b + x''/x''' = 0 \tag{26}$$

Differentiate again to eliminate b:

$$a.(x'''''.x''' - x''''^2)/x'''^2 + (x'''^2 - x''.x'''')/x'''^2 = 0 \tag{27}$$

Re-arranging for a gives:

$$E(a) = (x''.x'''' - x'''^2)/(x''''' .x''' - x''''^2) \tag{28}$$

Solve for *b* by substituting a from Equations 3-13 into Equation 26:

$$E(b) = -x''/x''' - a.x''''/x'''$$

which after substituting for a and manipulating gives:

$$E(b) = (x".x''''' - x'''.x'''')/(x'''''^2 - x''' . x''''')$$ (29)

We can now substitute these values for *a* and *b* intoEquation 23 to solve for u,

$$u = a.x" + b.x' + x$$

A Recurrent Architecture to Estimate Time Derivatives

The structure of each cell of the recurrent network is shown in *Figure 10*. The output of each cell feeds the input to the next one to generate the next higher order time derivative, see *Figure 11*. The output of the system and the cascade of 1st order recurrent network filters are simulated using the 4th order Runge-Kutta method in Mathcad. The derivatives vector is

Figure 10. A single stage recurrent sub-net using an integrator in the feedback path to estimate the derivative x' = w(x-E(x)); the net is a low pass filter with a cut off frequency w

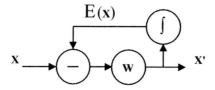

Figure 11. A 2nd order recurrent network to estimate 1st and 2nd time derivatives

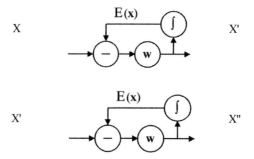

shown in *Figure 12*. *Figure 13* shows a typical set of derivatives estimated from a damped oscillatory trajectory.

Figure 12. A cascade of five recurrent cells plus the 2nd order trajectory model

$$x = \begin{bmatrix} 0 \\ 0 \\ 0 \\ 0 \\ 0 \\ 0 \\ 0 \end{bmatrix} \quad D(t, x) := \begin{bmatrix} x_2 \\ -\omega^2 \cdot x_1 - 2\zeta \cdot \omega \cdot x_2 + \omega^2 \cdot u \\ G \cdot (x_1 - x_3) \\ G \cdot \left[G \cdot (x_1 - x_3) - x_4 \right] \\ G \cdot \left[G \cdot \left[G \cdot (x_1 - x_3) - x_4 \right] - x_5 \right] \\ G \cdot \left[G \cdot \left[G \cdot \left[G \cdot (x_1 - x_3) - x_4 \right] - x_5 \right] - x_6 \right] \\ G \cdot \left[G \cdot \left[G \cdot \left[G \cdot \left[G \cdot (x_1 - x_3) - x_4 \right] - x_5 \right] - x_6 \right] - x_7 \right] \end{bmatrix}$$

$$Z := Rkadapt(x, t0, t1, N, D)$$

Figure 13. A typical set of time derivatives estimated from the trajectory of an oscillatory 2nd order dynamical system

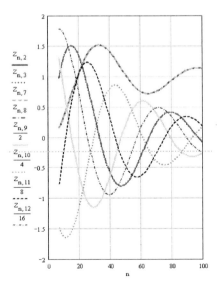

Results And Discussion

First Algorithm using Constant Parameters: This algorithm uses a single time point and four higher order time derivatives. The filter cascade provide a continuous estimate of the 1st to 4th time derivative x', x", x'" and x"". This provides a continuous estimate of all parameters at each point on the trajectory. The results of the estimation are given in *Figure 14*, which shows fast and accurate convergence.

Discussion: Estimated values for constants parameters are very close to the desired set values. The derived algorithms estimate ω, ζ and u for a good range of values: ω from 1 to10, ζ between +/- (0.01 to 1), and u between +/- (0.5-40), and give accurate estimates. Estimation errors decreased as ω increased, particularly for small ζ (less than 0.5): where oscillation provided wide variation in the variables to decrease errors. The differences between the (simulated) system time derivatives (x, x' and x") and their estimates from the filter cascade depended on G (the cut-off frequency): high G provided more accurate estimation of derivatives but made the algorithms prone to noise and vice versa. Another disadvantage of high G from the simulation point of view is that simulation time increases considerably due to the integration routine adapting to ever-smaller steps. The algorithm provides fast convergence.

Results for the Higher Order Algorithm: Mathcad routines are set up to generate the input u as second order system with its own parameters of natural frequency, damping ratio and input. The input subsystem damping ratio was set to 0.05 to generate an oscillatory behaviour for long enough to test the parameter tracking algorithm thoroughly. The frequency of the input is set to 16 radians per second, one quarter of the frequency of the data natural frequency. The derivative generation cascade is increased by one to produce the fifth time derivative. The results are shown in *Figure 15* below.

Figure 14. Estimated constant omega, zeta and u

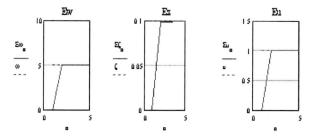

Figure 15. Results of the high order algorithm

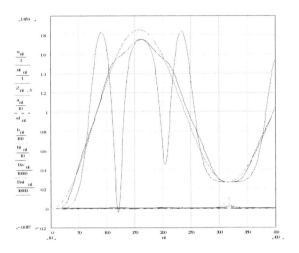

Figure 16. Results of the high order algorithm for one second integration time

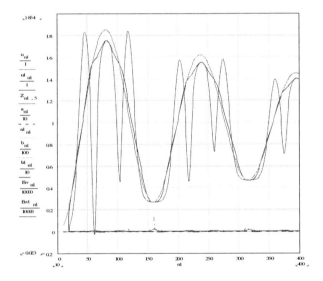

The actual input is shown in pink, which gives approximately one and one quarter cycles over a period of half a second as expected, that is, 16 radians/s − 2.546 Hertz. The red trace shows the results from the previous constant u derivation algorithm, which is failing completely to track the input parameter. The blue trace shows the result of the new algorithm, which is managing to track the input much more closely; however it start to diverge slightly near the peak of the cycle but then returns to track it well right down and round the lower trough of the input trajectory.

To check the quality of tracking as time progresses, a second set of results, *Figure 16*, is obtained with integration time extended to one second to give two and a half cycles. It is clear that tracking remains stable. It is interesting to note that the old algorithm, while completely failing to track the upper half of the input trajectory, seems to track it well during the its lower half but not as well as the new algorithm.

Knowledge Mining in Hybrid Inference Nets

Deduction and Abduction in Inference Networks. To engineer a knowledge base to represent intelligent systems behaviour, a multilevel structure is needed. By its very nature the knowledge embedded within these systems is continually changing and need dynamic paradigms to represent and acquire their parameters from observed data. In a normal inference network the cause and effect relationship is static and the effect can be easily worked out through a deduction process by considering all the causes through a step-by-step procedure, which works through all the levels of the network to arrive at the final effect. On the other hand, reasoning in the reverse direction, such as that used in diagnosis, starts with observing the effect and working back through the nodes of the network to determine the causes: this is termed knowledge mining.

Dynamical Knowledge Mining Processes. These ideas are applied here to recurrent or dynamical systems networks where some or all of the data within the knowledge base is time varying. The effect is now a time dependent behaviour pattern, which is used as an input to a differential abduction process to determine the knowledge about the system in terms of time varying causal parameters. These causal parameters will themselves embody knowledge (meta knowledge), which is obtained through a second level mining process to yield 2nd level causal parameters. These mining processes consist of a differential part to estimate the higher time derivative knowledge, followed by a non-linear algebraic part to compute the causal parameters.

Hierarchical Causal Parameters with Temporal Behaviour. The output trajectory of the system may be more complex than can be represented by a simple second order differential model. In this case each causal parameter is itself modelled as having a dynamical behaviour, which may or may not be oscillatory. One such case is where two of the three causal parameters have 2nd order dynamical characteristics, as was shown in *Figure 9*.

Knowledge Mining Algorithms

The knowledge embedded in such a model is represented by the parameters of the various recurrent nodes. By taking measurements of the system trajectory, tracking algorithms are employed to estimate the values of these parameters on line. Parameter tracking algorithms fall into several categories depending on the manner of accessing relevant information from the trajectory of the data and the order of parameter variation used in the derivation of the relevant models.

Several explicit algorithms for the three usual parameters characterising the behaviour of second order models have been derived (Al-Dabass et al., 1999) based on information available from the system's time trajectory. Leaving the 2nd order model in its 2nd time derivative form and using three points on the trajectory, each providing position, velocity and acceleration, a set of three simultaneous algebraic equations were solved to yield estimates of input, natural frequency and damping ratio. An online dynamical algorithm was then configured to combine estimates of the trajectory time derivatives with these explicit static non-linear functions to provide continuous parameter estimation in real time.

Multipoint Algorithms

For time varying parameters, the time separation between the 3 points on the trajectory has a direct influence on estimation accuracy, where the assumption of constant parameters used in the derivation is no longer valid, and accuracy deteriorates with increasing rate of parameter variation. To reduce the separation effect, a second algorithm is derived that relied on two points only; but, to compensate for this reduction in data, it needs more information from each point in the form of a higher, 3rd time derivative. This results in better estimation of higher rate parameter variation despite the fact that higher derivatives are more sensitive to trajectory measurement noise and estimation errors.

Several algorithms are easily derived to estimate values of causal parameters using as many points from the trajectory as necessary to form a set of simultaneous algebraic equations. The parameters to be estimated form the unknown variables and the trajectory values and their time derivatives form the constant parameters of these equations.

Figure 17. A three layer dynamical inference net

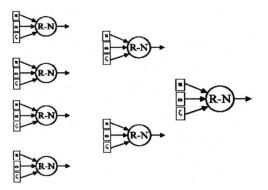

Algorithm 1: Three-Points in x, x' and x". Consider estimating ω, ζ and u using three sets of x, x' and x":

$$\omega^{-2} x_1'' + 2. \zeta.\omega^{-1}.x_1' + x_1 = u \tag{30}$$

$$\omega^{-2} x_2'' + 2. \zeta.\omega^{-1}.x_2' + x_2 = u \tag{31}$$

$$\omega^{-2} x_3'' + 2. \zeta.\omega^{-1}.x_3' + x_3 = u \tag{32}$$

Subtracting 31 from 30 and 32 from 30 to give:

$$\omega^{-2}.(x_1'' - x_2'') + 2. \zeta.\omega^{-1}.(x_1' - x_2') + (x_1 - x_2) = 0 \tag{33}$$

$$\omega^{-2}.(x_1'' - x_3'') + 2. \zeta.\omega^{-1}.(x_1' - x_3') + (x_1 - x_3) = 0 \tag{34}$$

Divide 33 by $(x_1' - x_2')$ and 34 by $(x_1' - x_3')$ to give:

$$\omega^{-2}.(x_1'' - x_2'') / (x_1' - x_2') + 2. \zeta.\omega^{-1} + (x_1 - x_2) /(x_1' - x_2') = 0 \tag{35}$$

$$\omega^{-2}.(x_1'' - x_3'') / (x_1' - x_3') + 2. \zeta.\omega^{-1} + (x_1 - x_3) /(x_1' - x_3') = 0 \tag{36}$$

and subtracting gives:

$$\omega^{-2}.[(x_1'' - x_2'') / (x_1' - x_2') - (x_1'' - x_3'') / (x_1' - x_3')] +$$
$$[(x_1 - x_2) / (x_1' - x_2') - (x_1 - x_3) / (x_1' - x_3')] = 0 \tag{36-A}$$

Using the following notations:

$$\Delta12 = (x_1 - x_2), \quad \Delta'12 = (x_1' - x_2'), \quad \Delta''12 = (x_1'' - x_2'')$$
$$\Delta13 = (x_1 - x_3), \quad \Delta'13 = (x_1' - x_3'), \quad \Delta''13 = (x_1'' - x_3'')$$

we get expressions for estimated ω, estimated ζ, and estimated u:

$$E\omega^2 = [\Delta''13.\Delta'12 - \Delta''12.\Delta'13] / [\Delta12.\Delta'13 - \Delta13.\Delta'12]$$
$$E\zeta = [-E\omega^{-2}.\Delta''12 - \Delta12] / [2. E\omega^{-1}. \Delta'12]$$
$$Eu = E\omega^{-2}. x_1'' + 2. E\zeta.. E\omega^{-1} . x_1' + x_1$$

Algorithm 2: Two-Points and One Extra Derivative. Consider using two sets of x, x', x" and x"'.

$$\omega^{-2} x_1" + 2.\ \zeta.\omega^{-1}.x_1' + x_1 = u \tag{37}$$

$$\omega^{-2} x_2" + 2.\ \zeta.\omega^{-1}.x_2' + x_2 = u \tag{38}$$

Subtracting 38 from 37 and dividing by $(x_1' - x_2')$:

$$\omega^{-2}.(\ x_1" - x_2")\ /\ (\ x_1' - x_2')\ + 2.\ \zeta.\omega^{-1} + (\ x_1 - x_2)\ /(\ x_1' - x_2')\ = 0 \tag{39}$$

Differentiating 39 with respect to t gives:

$$[\omega^{-2}[(\ x_1' - x_2').\ (\ x_1"' - x_2"')]- (\ x_1" - x_2")^2]/(\ x_1' - x_2')^2 + 0 +$$
$$[(\ x_1' - x_2')^2 - (\ x_1 - x_2).\ (\ x_1" - x_2")]\ /(\ x_1' - x_2')\ = 0 \tag{40}$$

Using the following notations:

$$\Delta12 = (x_1 - x_2),\quad \Delta'12 = (\ x_1' - x_2'),$$
$$\Delta"12 = (\ x_1" - x_2")\ \text{and}\quad \Delta"'12 = (\ x_1"' - x_2"')$$

we get expressions for estimated ω, estimated ζ, and estimated u:

$$E\omega^2 = \ [(\Delta'12).\ (\Delta"'12\ ) - (\Delta"12\ )^2]/\ [(\Delta'12\ )^2 - \Delta12\ .\ \Delta"12]$$
$$E\zeta = [-E\omega^{-2}.\Delta"12/\Delta'12 - \Delta12\ /\Delta'12\ ]\ /\ [2.\ E\omega^{-1}\ ]$$
$$Eu = E\omega^{-2}.\ x_1" + 2.\ E\zeta..\ E\omega^{-1}\ .\ x_1' + x_1$$

Single Point Algorithms

Ultimately, the separation effect can only be eliminated if all the information needed for the estimation is derived from a single time point. This is successfully carried out by deriving a third algorithm, termed Algorithm 3, which obtains the additional information from a higher, 4th time derivative. An online algorithm would then consist of two subsystems: i) a cascade of recurrent networks, and ii) static non-linear functions of these derivatives to produce continuous estimates of the three parameters. The first subsystem performs state estimation, as a set of first order observers, to generate continuous trajectories of all time derivatives up to 4th, and simultaneously provides noise filtering. As all the information needed to estimate the parameters are obtained from a single time point on the trajectory, this algorithm proves, as expected, to be the most successful in coping with high rates of parameter variation. However, accurate tracking of parameters when two or more of them were varying simultaneously may still prove to be problematical. The essential assumption of constant parameters in the derivation is one of the fundamental causes of these difficulties.

In this section we relax the constant parameter condition by assuming a linear time variation, that is, constant first derivative but zero second and higher time derivatives of parameters. As may be expected, more information is needed for this new case, which is to be extracted from the system output trajectory by obtaining higher time derivatives. Explicit functions of the parameters are still possible as well as those of their first time derivatives. A set of three equations, one for each parameter, is formulated and numerically computed in real time together with the state estimation vector observer to yield continuous trajectories of the parameters. This is a different technique to that of augmenting the state derivative vector with the parameter derivatives: instead of driving these derivatives with some function of the error between the system and model output, we provide an explicit function that should aid successful and speedy convergence to actual parameter values and provide continuous tracking. This should hold even when the parameters are changing rapidly compared to the system's natural frequency or time constant.

These are classified according to the order of the parameter variation used in the derivation, that is, constant, first order polynomial (constant u' but u''=0), 2nd order polynomial (constant u'' but u'''=0), and so forth.

Algorithm 3: Constant Parameters. Consider using the 1st to 4th time derivatives at a single point. Given the second order system:

$$\omega^{-2} x'' + 2. \zeta.\omega^{-1}.x' + x = u \tag{41}$$

Differentiate with respect to t and divide by x'':

$$\omega^{-2} x'''/ x'' + 2. \zeta.\omega^{-1} + x'/ x'' = 0 \tag{42}$$

and differentiate with respect to t again to give:

$$\omega^{-2}.[(x''. x'''' - x'''^2) / x''^2] + 0 + [(x''^2 - x'. x''') / x''^2] =0 \tag{43}$$

We get expressions for estimated ω, estimated ζ and estimated u:

$$E\omega^2 = [x''. x'''' - x'''^2] / [x'. x''' - x''^2]$$
$$E\zeta = -[E\omega^{-2} x''' + x'] / [2. E\omega^{-1}.x'']$$
$$Eu = E\omega^{-2}. x'' + 2. E\zeta. E\omega^{-1} . x' + x$$

Algorithm 4: First Order Parameters. Let the first time derivative of u to be non-zero. For simplicity assume that both a and b (the coefficients of x" and x' to make symbol manipulation easier) to be constant and hence disappear on first differentiation. The extra information needed for u' to be non-zero is extracted from the 5th time derivative of the trajectory.

$$a.x" + b.x' + x = u \tag{44}$$

Differentiate with regards to t and assume u' is non zero to give:

$$a.x'" + b.x" + x' = u' \tag{45}$$

Differentiate again and set u" = 0 gives:

$$a.x'''' + b.x'" + x" = 0 \tag{46}$$

Divide by x'" to isolate b:

$$a.x''''/x'" + b + x"/x'" = 0 \tag{47}$$

Differentiate again to eliminate b:

$$a.(x'''''.x" - x''''^2)/x'''^2 + (x'''^2 - x" . x'''')/x'''^2 = 0 \tag{48}$$

Re-arranging for a gives:

$$a = (x" . x'''' - x'''^2)/(x''''' . x'" - x''''^2) \tag{49}$$

Solve for *b* by substituting a from Equation 49 into Equation 47:

$$b = -x"/x'" - a.x''''/x'"$$

Substituting for a and manipulating gives:

$$b = (x".x''''' - x'".x'''')/(x'''''^2 - x'" . x''''') \tag{50}$$

We can now substitute these values for *a* and *b* into equation 44 to solve for u,

$$u = a.x" + b.x' + x$$

Simulation: Simulation is carried out using Mathcad (see appendix); the simulation derivative vector is as shown in *Figure 12*. Programs implemented in Mathcad were presented in reference 1 for the 3 categories of parameters: constant and variable with 1st and 2nd order dynamics, and were used to illustrate the generation of time derivatives. In the next section the four algorithms will be applied to test their effectiveness.

Derivative Mining Architecture: The structure of each cell of the recurrent network and the cascade of the derivative architecture is the same as that shown in *Figures 5* and *6*. As before the output of the system and the cascade of 1st order recurrent network filters are simulated using the 4th order Runge-Kutta method in Mathcad (the derivatives vector) and the output trajectory of a system displaying temporal behaviour.

Results and Discussion

Algorithm 3 using Constant Parameters: This algorithm uses a single time point but two further time derivatives compared to Algorithm 1. The filter cascade is increased by one again to provide a continuous estimate of the 4th time derivative x''''. The separation problem disappears altogether now to provide a continuous estimate of all parameters at each point on the trajectory. The program was run, (see appendix) and gave fast and accurate convergence to the set parameter values.

Discussion: Estimated values for constants parameters were close to the desired set values. The derived algorithms estimated ω, ζ and u for a good range of values: ω from 1 to 10, ζ between +/- (0.01 to 1), and u between +/- (0.5-40), and gave accurate estimates. Estimation errors decreased as ω increased, particularly for small ζ (less than 0.5): where oscillation provided wide variation in the variables to decrease errors. The differences between the (simulated) system time derivatives (x, x' and x") and their estimates from the filter cascade depended on G (the cut-off frequency): high G provided more accurate estimation of derivatives but made the algorithms prone to noise and vice versa. Another disadvantage of high G from the simulation point of view is that simulation time increased considerably due to the integration routine adapting to ever-smaller steps. The algorithms provided progressively faster convergence with Algorithm 3 being the fastest to converge.

Results: Further results to those shown earlier are obtained to test the case when one or two of the input causal parameters were changing. *Figure 18* shows the results when u is changing and the result of tracking it using Algorithms 3 and 4: Algorithm 3 results showing large oscillations while those for Algorithm 4 show much smoother tracking. *Figure 19* shows the results of both algorithms tracking the other input ω. Again Algorithm 4 is producing a far smoother estimate than the large oscillatory output of Algorithm 3.

*Figure 18. The **u** causal parameter (pink trace) [a] being tracked using Algorithm 3 (red trace) [b] and Algorithm 4 (blue trace) [c]. The black trace is the node output trajectory [d].*

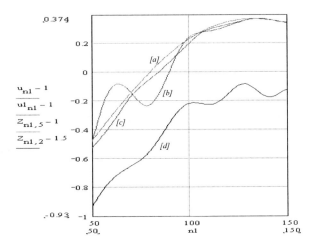

Figure 19. The ω causal parameter (pink trace) [a] being tracked using Algorithm 3 (red trace) [b] and the new algorithm (blue trace) [c]. The black trace is the node output trajectory [d].

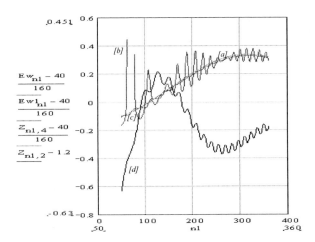

Comments: For a given range of parameters the algorithms worked well, being able to estimate the two causal parameters u and omega with their temporal behaviour; that is, track them while they are changing. The 1st order algorithm worked better than the constant one; *Figure 18* shows a comparison of the two algorithms tracking omega. Algorithms of higher order than 1st showed marginal improvement but in certain cases showed a deteriorating behaviour; *Figure 19* shows a 3rd order algorithm deviating quite markedly from the true trajectory compared to a 1st order algorithm. This is likely to be due to an accumulation of errors in higher derivative values used in the former algorithm.

Knowledge Mining for Signal Processing Applications

A special sixth order dynamical model is proposed to simulate the behaviour of complex signals. The model consists of a two-layer hierarchy of second order dynamics, two of whose parameters are themselves second order. Given the trajectory of the actual complex signal, a recurrent hybrid algorithm is derived to estimate the parameters of the model. Results show good performance of the algorithm in tracking the model parameters online. Suggestions for future directions are given.

Several parameter estimation algorithms were derived earlier in this chapter. These combine estimates of a given trajectory time derivatives, using data from several points on the trajectory, with explicit static non-linear functions to provide continuous parameter estimation in real time. For time varying parameters, the time separation between the points on the trajectory directly influences the estimation accuracy. This due to the assumption of constant parameters used in the derivation is no longer valid, and accuracy deteriorates with increasing rate of parameter variation. This is termed the separation effect.

This effect can only be eliminated if all the data needed for estimation is obtained from a single time point o the trajectory. An algorithm was derived and proved, as expected, to be the most successful in coping with high rates of parameter variation. Accurate tracking of parameters when two of the parameters were varying simultaneously still proved difficult. The constant parameters assumption in the derivation is seen as the fundamental cause here.

Sixth Order Models of Compound Signals

Hierarchical Second Order Models

The signal trajectory is more complicated than can be represented by a simple second order differential model. In this case each parameter may itself be modelled as having dynamics, which may or may not be oscillatory. One such case is where two of the three parameters have 2nd order characteristics, as shown in *Figure 20*.

The 2nd order model of a node in a given layer in the hierarchy is given by:

$$\omega^{-2} x'' + 2. \zeta.\omega^{-1}.x' + x = u$$

To model complicated signals let both u and omega have their own 2nd order dynamics. The input **u** is the output of the following 2nd order system:

$$\omega_u^{-2} \mathbf{u}'' + 2. \zeta_u. \omega_u^{-1}.\mathbf{u}' + \mathbf{u} = u_u$$

Figure 20. Hybrid integral-recurrent net model of a 2ⁿᵈ order system

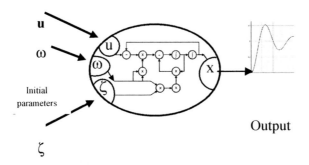

Figure 21. Two of the input parameters of the signal are time varying and modelled with 2ⁿᵈ order dynamics

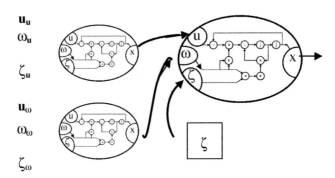

The natural frequency ω is the output of the following 2ⁿᵈ order system:

$$\omega_\omega^{-2}\, \omega'' + 2.\, \zeta_\omega \cdot \omega_\omega^{-1}.\omega' + \omega = u_\omega$$

Thus the behaviour trajectory is generated by the following 6ᵗʰ order vector differential equation (using Runge Kutta in Mathcad for this example), see *Figure 22*.

Where x1 and x2 represent the x and x', x3 and x4 represent u and u', and x5 and x6 represent ω and ω' respectively. To generate the trajectory shown in *Figure 5*, the following values

Figure 22. Simulation vector of a 6ᵗʰ order trajectory (top 2 rows) with u (rows 3 and 4) and omega (rows 5 and 6) of the signal having 2ⁿᵈ order dynamics

$$D(t,x) := \begin{bmatrix} x_2 \\ x_5 \cdot x_5 \cdot x_3 - 2 \cdot z \cdot x_5 \cdot x_2 - x_5 \cdot x_5 \cdot x_1 \\ x_4 \\ (wu \cdot wu \cdot uu) - 2 \cdot zu \cdot wu \cdot x_4 - wu \cdot wu \cdot x_3 \\ x_6 \\ (ww \cdot ww \cdot uw) - 2 \cdot zw \cdot ww \cdot x_6 - ww \cdot ww \cdot x_5 \end{bmatrix}$$

Figure 23. Simulated trajectory of a complex signal of a sixth (red trace) [a], with 2 variable inputs: u (blue) [b] and omega (green) [c]

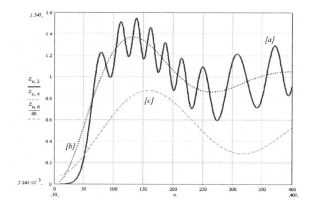

were used: for the u subsystem, u started from 0 aiming at uu=1 at a rate of ω_u = 5 rad/s with ζ_u = 0.3. For the ω subsystem, ω started from 4 rad/s aiming at u_ω = 32 rad/s at a rate of ω_ω = 4 rad/s with ζ_ω = 0.1. The resulting compound trajectory of x (red), together with the trajectories for u (blue dotted) and ω (green dotted) are shown in the graph below.

Results And Discussion

Mathcad routines were set up to generate the input u as second order system with its own parameters of natural frequency, damping ratio and input. The input subsystem damping ratio was set to 0.05 to generate an oscillatory behaviour for long enough to test the parameter tracking algorithm thoroughly. The frequency of the input was set to 10 radians per second; the frequency of the main signal, on the other hand, started from 20 and aimed at 80 with

Figure 24. The signal (red solid) following a damped 2nd order input (blue small dots) with variable natural frequency (green dotted); U: omega=10 rad/s, zeta=0.1, input=1 starting from 0; Omega: omega=5 rad/s, zeta=0.05, input=80 starting from 20 rad/s

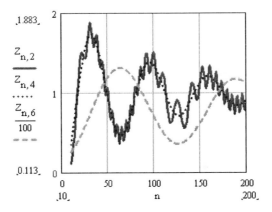

Figure 25. The signal (red) [a] and its estimated high order derivatives: high values during periods of high frequency followed by low valued for low frequency

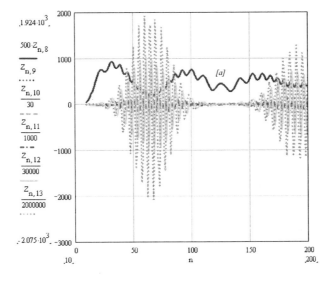

a peak of about 130 radians. The derivative generation cascade was increased by one to produce the fifth time derivative. The results are shown in *Figures 24* to *26* below.

Tracking Two Parameters: For a given range of parameters the algorithm worked well, being able to estimate the two input parameters u and omega with their time varying behaviour; that is, track them while they are changing. The 1st order algorithm worked better than the

Figure 26. Deterioration in the accuracy of estimated input u during periods of low value high derivatives, particularly prominent for the high order estimation algorithm, for example, 3rd order shown in blue [a] against 1st order (red) [b] and actual value (pink) [c]

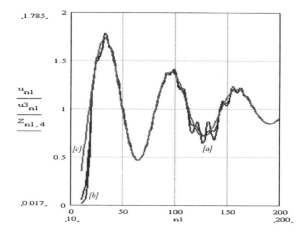

Figure 27. Estimated Omega assuming constant parameters (Ew, very jagged trace) and first order parameters (Ew1, jagged trace) compared to actual (smooth trace)

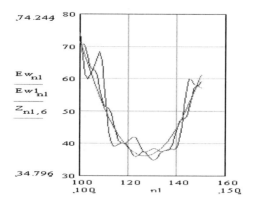

Figure 28. Deterioration of accuracy of estimated Omega with higher order algorithms: 3rd order estimate (Ew3, very jagged trace) is worse than 1st order estimate (Ew1, jagged), actual (smooth)

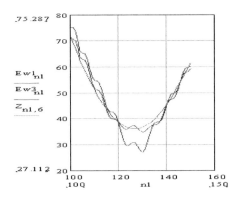

constant one; *Figure 27* shows a comparison of the two algorithms tracking omega. Algorithms of higher order than 1st showed marginal improvement but in certain cases showed a deteriorating behaviour; *Figure 28* shows a 3rd order algorithm deviating quite markedly from the true trajectory compared to a 1st order algorithm. This is likely to be due to an accumulation of errors in higher derivative values used in the former algorithm.

Conclusion

A model for hybrid logic nets was put forward to model the complex behaviour of systems. To estimate the values of the causal parameters a number of parameter knowledge mining algorithms were presented. Two of the algorithms used multiple points from the trajectory, three for the first algorithm and two points for the second. Two single point algorithms were presented: one that assumed constant parameters and used higher time derivatives of the trajectory (up to 4th), and a second algorithm that used additional information from a 5th time derivative of the trajectory to allow one of the parameters, the input parameter u, to have a non-zero first order time derivative.

The fourth algorithm was tested for its ability to track the input parameter for a reduced order model. The test involved the generation of a lightly damped second order recurrent net. The results showed the algorithm maintaining good tracking over an extended period of time. This algorithm proved to be far superior to the third algorithm, which relied on the assumption of constant input in the derivation.

References

Al-Dabass, D. (2002). *Modelling the complexity of concept dynamics.* 47th Meeting of the International Society for the Systems Sciences, Shanghai International Convention Centre, Shanghai, China.

Al-Dabass, D. (2001, July 31). *A Kalman observer computational model for metaphor based creativity.* Panel paper, Workshop on Creative Systems, ICCBR2001, Simon Fraser University, Vancouver. Retrieved from http://ducati.doc.ntu.ac.uk/uksim/dad/web-pagepapers/Paper-1.doc

Al-Dabass, D., Evans, D.J., & Sivayoganathan, K. (2003a). Signal parameter tracking algorithms using hybrid recurrent networks. *International Journal of Computer Mathematics, 80*(10), 1313-1322.

Al-Dabass, D., Evans, D.J., & Sivayoganathan, K. (2003b). Signal processing using hybrid recurrent models for data mining knowledge discovery in financial trajectories. In M.H. Hamza (Ed.), *5th IASTED International Conference on SIGNAL and IMAGE PROCESSING (SIP2003),* August 13-15, 2003, Honolulu, U.S. (pp. 456-460).

Al-Dabass, D., Evans, D.J., & Sivayoganathan, K. (2002a). *A recurrent network architecture for non-linear parameter tracking algorithms.* 2nd International Conference on Neural, Parallel, and Scientific Computations, Morehouse College, Atlanta, GA, U.S.

Al-Dabass, D., Evans, D.J., & Sivayoganathan, K. (2002b). Derivative abduction using a recurrent network architecture for parameter tracking algorithms. In *IEEE Joint International Conference on Neural Networks, World Congress on Computational Intelligence, (IJCNN02),* May 12-17, 2002, Honolulu, U.S. (pp. 1570-74).

Al-Dabass, D., Zreiba, A., Evans, D.J., & Sivayoganathan, K. (2002c). Parameter estimation algorithms for hierarchical distributed systems. *International Journal of Computer Mathematics, 79*(1), 65-88.

Al-Dabass, D., Zreiba, D., Evans, D.J., & Sivayoganathan, K. (1999b, Oct 29). Simulation of noise sensitivity of parameter estimation algorithms. In *Simulation'99 Workshop,* UCL, London (pp.32-35).

Al-Dabass, D., Zreiba, A., Evans, D.J., & Sivayoganathan, K. (1999a, April 7-9). Simulation of three parameter estimation algorithms for pattern recognition architecture. In *UKSIM'99, Conference Proceedings of the UK Simulation Society,* St Catharine's College, Cambridge (pp. 170-176).

Bailey, S., Grossman, R.L., Gu, L., & Hanley, D. (1996). A data intensive approach to path planning and mode management for hybrid systems. In R. Alur, T.A. Henzigner, & E. Sontag (Eds.), *Hybrid Systems III, Proceedings of the DIMACS Workshop on Verification and Control of Hybrid Systems.* LNCS 1066. Springer Verlag.

Baltagi, B. (2005). *Economic analysis of panel data.* Chichester: Wiley.

Berndt, D.J., & Clifford, J. (1996). Finding patterns in time series: A dynamic programming approach. In U.M. Fayyad, G. Piatetsky-Shapiro, P. Smyth, & R. Uthurusamy (Eds.), *Advances in knowledge discovery and data mining* (pp. 229-248). AAAI Press/MIT Press.

Bovet, D.P., & Crescenzi, P. (1994). *Introduction to the theory of complexity.* Prentice Hall.

Cant, R., Churchill, J., & Al-Dabass, D. (2001). Using hard and soft artificial intelligence algorithms to simulate human go playing techniques. *International Journal of Simulation, 2*(1), 31-49.

Cawley, P. (1984). The reduction of bias error in transfer function estimates using FFT-based analysers. *Journal of Vibration, Acoustics, Stress and Reliability in Design,* 29-35.

Close, C.M., & Fredrick, D.K. (1994). *Modelling and analysis of dynamic systems* (2nd ed.). Wiley.

De Raedt, L. (1992). *Interactive theory revision: An inductive logic programming approach.* Academic Press.

Dewolf, D., & Wiberg, D. (1993). An ordinary differential-equation technique for continuous time parameter estimation. *IEEE Transactions on Automatic Control, 38*(4), 514-528.

Doyle, E. (2005). *The economic system.* Chichester: Wiley.

Gersch, W. (1974). Least squares estimates of structural system parameters using covariance function data. *IEEE Transactions on Automatic Control, 19*(6).

Kailath, T. (1978). *Lectures on linear least-squares estimation.* CISM Courses and Lectures No. 140. New York: Springer Verlag.

Kalman, R. (1960). A new approach to linear filtering and prediction problems. *Transactions of SIAM: Journal of Basic Engineering, Series D, 82,* 35-45.

Klosgen, W. (2002). *Handbook of data mining and knowledge discovery.* Oxford University Press.

Koop, G. (2005). *The analysis of economic data.*

Man, Z. (1995). Parameter-estimation of continuous linear systems using functional approximation. *Computers and Electrical Engineering, 21*(3), 183-187.

Mannila, H., Toivonen, H., & Verkamo, I. (1997). *Discovery of frequent episodes in event sequences, data mining and knowledge discovery (Vol. 1)* (pp. 259-289).

Muggleton, S. (1994). *Machine intelligence: Machine intelligence and inductive learning (No.13).* Clarendon Press.

Papazoglou, M., & Ribbers, P. (2005). *E-business: Organizational and technical foundations.*

Ren, M., & Al-Dabass, D. (2001). Simulation of fuzzy possibilistic algorithms for recognising Chinese characters. *International Journal of Simulation, 2*(1), 1-13.

Ren, M., Su, D., & Al-Dabass, D. (1995). An associative memory artificial neural network system. In *ECAC'95-London: Proceedings of European Chinese Automation Conference,* London (pp. 91-96).

Ren, M., Al-Dabass, D., & Su, D. (1996). A three-layer hierarchy for representing Chinese characters. In *Expert Systems '96, 6th Annual Technical Conference of the BCS Specialist Group on Expert Systems,* Cambridge, UK (pp. 137-146).

Samuelson, W. (2005). *Managerial economic.* ISBN: 0471-6636-2X.

Schank, R.C., & Colby, K.M. (Eds.). (1973). *Computer models of thought and language.* Freeman & Co.

Seveance, F.L. (2001). *Systems modelling and simulation: An introduction.* New York: Wiley.

Books

Baltagi, B. (2005). *Economic Analysis of Panel Data.* ISBN: 0470-0145-63.

Doyle, E. (2005). *The Economic System.* ISBN: 0470-8500-19.

Koop, G. (2005). *The Analysis of Economic Data.* ISBN: 0470-0246-82.

Papazoglou, M., & Ribbers, P. (2005). *e-Business: Organizational and Technical Foundations.* ISBN: 0470-8437-64.

Samuelson, W. (2005). *Managerial Economic.* ISBN: 0471-6636-2X.

Appendix

Using Mathcad for Simulation

Refer to Figure 1. The figure is a screen print of a typical Mathcad code to simulate a 2nd order (main) system whose: i) input u is the output of a 2nd order sub-system, and ii) its natural frequency is the output of a 1st order sub-system.

Appendix Figure 1.

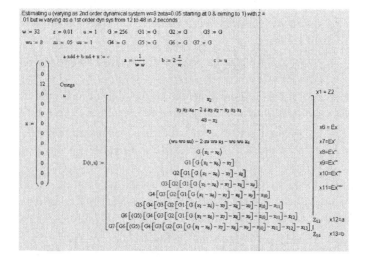

1. The input sub-system: u is changing as the output of a 2nd order sub-system with an initial value of 0, natural frequency wu = 8 c/s, and damping coefficient zu = 0.05, see the 2nd row of parameters after the 2 line text at the top. u is given the variable name x4 in the simulation derivative vector and has an initial value 0, shown as the 4th value down in the initial value vector on the left. u aims at a final value, that is, its own sub-system input, uu of 1.

2. The main system: The main system is given the variable names x1 and x2 and is placed in the 1st and 2nd rows of the derivative vector. Its input u is variable x4 and its natural frequency is variable x3.

3. The natural frequency sub-system: the natural frequency is given the variable x3 and placed in the 3rd row of the derivative vector. It is a 1st order system with an input, final value, of 48 and hence the derivative expression is 48 - x3 in row 3 of the derivative vector. It has an initial value of 12, shown as the 3rd entry down the initial value vector on the left.

4. Derivative estimation: row 6 upwards in the derivative vector consist of numerical solutions of a sequence of 1st order systems driven by the difference (error) between their output and input. The dynamics of such systems will generate an estimate of the derivative of the input as a by-product of integration as the error (between input and output) reduces in the time solution of the process. Row 6 will generate the 1st derivative, which is fed into the next row to generate the 2nd derivative and so on. However, as the Mathcad program is "declarative" (and not sequential as a traditional computer program) the whole derivative generation process has to be replicated in each row to determine the higher derivatives.

Figure 2 is a screen print of the arithmetic part of the knowledge mining "architecture." The calculated values of the integration process shown in Figure 1 are stored in a time matrix Z and elements are accessed by a suitable time index to give the values of the derivates used in the expressions to calculate the parameters.

Appendix Figure 2.

$$Z := \text{Rkadapt}(x,0,2,400,D) \qquad n := 20..400$$

$Z_{n,1}$ time

$Z_{n,2}$ $\quad= \blacksquare \qquad x_1$

$Z_{n,3}$ $\quad= \blacksquare$ xdot $\quad x_2$

$Z_{n,7}$ $\quad= \blacksquare$ Ex $\quad x_6$

$Z_{n,8}$ $\qquad$ Edx $\quad x_7$

$Z_{n,9}$ $\qquad$ Ex2dot $\quad x_8$

$Z_{n,10}$ $\qquad$ Ex3dot $\quad x_9$

$Z_{n,11}$ $\qquad$ Ex4dot $\quad x_{10}$

$Z_{n,12}$ $\qquad$ Ex5dot $\quad x_{11}$

$Z_{n,13}$ $\qquad$ Ex6dot $\quad x_{12}$

$$a_n := \frac{Z_{n,8} \cdot Z_{n,10} - \left(Z_{n,9}\right)^2}{Z_{n,9} \cdot Z_{n,11} - \left(Z_{n,10}\right)^2} \qquad Ew_n := \sqrt{\frac{1}{a_n}}$$

$$b_n := \frac{\left(-Z_{n,8} - a_n \cdot Z_{n,10}\right)}{Z_{n,9}}$$

$$u_n := a_n \cdot Z_{n,9} + b_n \cdot Z_{n,8} + Z_{n,7}$$

$$a1_n := \frac{Z_{n,9} \cdot Z_{n,11} - \left(Z_{n,10}\right)^2}{Z_{n,10} \cdot Z_{n,12} - \left(Z_{n,11}\right)^2} \qquad Ew1_n := \sqrt{\frac{1}{a1_n}}$$

Chapter XI

Simulating Theory-of-Constraint Problem with a Novel Fuzzy Compromise Linear Programming Model

Arijit Bhattacharya, The Patent Office, Bouddhik Sampada Bhawan, India

Pandian Vasant, Universiti Teknologi Petronas, Malaysia

Sani Susanto, Parahyangan Catholic University, Indonesia

Abstract

This chapter demonstrates development of a novel compromise linear programming having fuzzy resources (CLPFR) model as well as its simulation for a theory-of-constraints' (TOC) product mix problem using MATLAB® v. 7.04 R.14 SP.2 software. The product-mix problem considers multiple constraint resources. The developed CLPFR model helps in finding a robust solution with better profit and product mix solution in a non-bottleneck situation. The authors simulate the level of satisfaction of the decision maker (DM) as well as the degree of fuzziness of the solution found using the CLPFR model. Simulations have been carried out with MATLAB® v. 7.04 R.14 SP.2 software. In reality, the capacities available for some resources are not always precise. Some tolerances should be allowed on some constraints. This situation reflects the fuzziness in the availability of resources of the TOC product mix problem.

Introduction

Simulation is a method that allows the analysis of complex systems through mathematically valid means. Through a software interface, the user creates a computerized version of a model (Peterman, 2006). Among other things, "model abstraction is a method for reducing the complexity of a simulation model while maintaining the validity of the simulation results with respect to the question that the simulation is being used to address" (Frantz, 2006). Model abstraction is the intelligent capture of the essence of the behaviour of a model without all the details of how that behaviour is implemented in code (Frantz, 1996). Researchers in the field of artificial intelligence (AI) have also been developing techniques for simplifying models, determining whether model results are valid and developing tools for automatic model selection and manipulation (Frantz, 2006).

Vast literature exists in the field of modelling and simulation. Fishwick and Zeigler (1992) reported substantial parallels between their work and the researches in qualitative simulation. Miller et al. (1992), Fishwick (1992) and Fishwick et al. (1994) provide general rationale and approaches for synergizing traditional simulation and AI modelling & simulation techniques. Weld (1992) reported model sensitivity analysis for qualitative models to formalize an approximation approach. Nayak (1992) described an alternative approximation approach based on the causal relationships of model parameters.

It is to be noted that simulation and modelling has a wide applicational range in military sciences. Sisti and Farr (2005) dealt with the wide variety of research issues in simulation science addressed by government, academia and industry, and their application to the military domain, specifically to the problems of the intelligent analyst.

Advancement in model abstraction research deals with the application and adaptation of the concept of "qualitative reasoning" which is borrowed from the field of AI (Sisti & Farr, 2005). Qualitative simulation concerns itself with getting away from the idea of "exactness" (Sisti & Farr, 2005). Some of the ancillary topics of research in qualitative simulation, as suggested by Sisti & Farr (2005), are: fuzzy modelling, random set theory, possibility theory, rough sets and Dempster-Shafer theory (DS theory) and ordinal optimisation. The common factor among all of these fields is that all of these strive to represent "intermediate degrees of truth" (uncertainty) in such a way as to attain optimal answers, or ranges of answers, as opposed to an optimum answer to 10-decimal place precision (Sisti & Farr, 2005).

In this chapter the authors present, first, a novel fuzzy compromise linear programming (CLPFR) model to solve a product mix problem under theory of constraint (TOC). The problem contains multiple constraint resources. The developed CLPFR model helps in finding a robust solution with better profit and product mix solution. Later, the authors simulate the level of satisfaction of the decision maker (DM) as well as the degree of fuzziness of the solution found using the CLPFR model. Simulations have been carried out with MATLAB® v. 7.04 R.14 SP.2 software. A thorough interpretation and discussion of the outcome of the product mix decision using the CLPFR model has also been presented in this chapter.

Earlier Susanto, Bhattacharya, Vasant, and Suryadi (2006) introduced the CLPFR model to optimize product mix of a chocolate manufacturing firm. Susanto, Vasant, Bhattacharya, and Kahraman (2006) reported a "compromise linear programming having fuzzy objective function coefficients" (CLPFOFC) with fuzzy sensitivity. Their work was also applied to solve a chocolate manufacturing firm's product mix decision using the CLPFOFC model.

In reality, the capacity available for some resources are not always precise, since, for example the company manager can ask workers to work overtime or add more materials from suppliers. Therefore, some tolerances should be allowed on some constraints. This situation reflects the fuzziness in the availability of resources. This problem is called as fuzzy compromise linear programming (CLPFR) having fuzzy resources.

The TOC Problem

Enormous volume of works exists in the arena of product mix decision under TOC heuristic using linear as well as integer programming models. Luebbe & Finch (1992) compared the TOC and linear programming using the five-step improvement process in TOC. They categorized the TOC as a manufacturing philosophy and linear programming (LP) as a specific mathematical optimization technique. It was stated that the algorithm could optimize the product mix as integer LP (ILP) (Luebbe & Finch, 1992).

Balakrishnan and Cheng (2000) reported that LP was a useful tool in the TOC analysis. They (Balakrishnan & Cheng, 2000) showed that some of Luebbe and Finch's (1992) conclusions were not generalizable. Finch and Luebbe's (2000) argued that Balakrishnan and Cheng (2000) did not compare LP with TOC. Finch and Luebbe (2000) commented that Balakrishnan & Cheng's (2000) work was a comparison of LP with one of many techniques sometimes incorporated in TOC.

Hsu and Chung (1998) presented an algorithm using dominance rule classifying non-critically constrained resources into three levels for solving the TOC product mix problem.

Plenert (1993) discussed an example having multiple constrained resources in order to delineate that the TOC heuristic didn't provide an optimal feasible solution. Lee and Plenert (1993) demonstrated that TOC was inefficient when new product was introduced. Lee and Plenert (1993) observed that the solution from TOC during introduction of new product produced a non-optimal product mix. They (Lee & Plenert, 1993; Plenert, 1993) used an ILP formulation that identified a product mix. The product mix fully utilized the bottleneck. Their conclusion was that ILP solution was more efficient than the TOC heuristic. Mayday (1994) and Posnack (1994) criticized Lee and Plenert (1993) and Plenert (1993).

Coman and Ronen (2000) formulated a production outsourcing problem as a LP problem and identified an analytical solution. Coman and Ronen (2000) argued that the TOC solution was inferior to the LP-enhanced solution since it computed the throughput relative to a no-production alternative while the LP solution computed the throughput based on the contractor's mark-up.

Onwubolu (2001) compared the performance of the Tabu search-based approach to both the original TOC heuristic, the ILP solution and the revised TOC algorithm. Further, large-scale difficult problems were generated randomly and solved by Onwubolu (2001). The research work of Boyd and Cox (2002) was focused to compare the TOC solution to the product mix problem with an optimal solution given by LP or ILP.

Bhattacharya and Vasant (2006) developed and subsequently used fuzzy-LP with smooth S-curve membership function (MF) in making product mix decisions under TOC heuristic more explicit. Bhattacharya, Vasant, Sarkar, and Mukherjee (2006) introduced a fully fuzzi-fied and intelligent TOC product mix decision. In order to avoid linearity in the real life application problems, especially in product mix decision problems, a non-linear function such as modified MF was used in the above works. This MF was used when the problems and its solutions were independent (Vasant, 2003; Bhattacharya & Vasant, 2006; Bhattacharya, Vasant, Sarkar, & Mukherjee, 2006).

Now, let us discuss the product mix problem when multiple constrained resources exist. The effectiveness of the CLPFR model will be delineated in solving the said product mix problem under TOC. Hsu and Chung (1998) and Onwubolu and Mutingi (2001) illustrated the said product mix problem as shown in Figure 1. The same problem of Hsu and Chung (1998) is solved in this chapter in order to compare the developed CLPFR model with that of the earlier proposed methodologies.

The problem of Hsu and Chung (1998) can be modelled as a dual simplex LP problem with a view to maximize the throughput when multiple resource constraints exist. In their (Hsu & Chung 1998) problem, four different types of products, namely, R, S, T & U, are to be produced wherein seven different resources, A to G, exists. Each resource has a capacity of 2,400 minutes. Table 1 illustrates loads required for producing one unit of each of the products R, S, T and U.

Table 2 shows loads on each of the resources. It is seen from Table 2 that only resource G is underutilized and resource E runs in its full capacity, while resources A-D and F are over-loaded. Resource B is the capacity constraint resource (CCR) as it is the most overloaded and the said CCR is indicated in Table 2 using a vertical upward arrow. Now, throughput per constraint resource minute needs to be calculated for finding out the required number of products to be produced within the available capacity of each resource per week.

Table 1. Loads required for producing four products

Products	Weekly market potential (units)	Unit selling price (US$ / unit)	Processing time per unit (min)							Raw material cost per unit (US$ / unit)	Throughput per unit (US$ /unit)
			A	B	C	D	E	F	G		
R	70	90	20	5	10	--	5	5	20	10	80
S	60	80	10	10	5	30	5	5	5	20	60
T	50	70	10	5	10	15	20	5	10	20	50
U	150	50	5	15	10	5	5	15	--	30	30

Figure 1. Modified product mix problem of Hsu and Chung (1998)

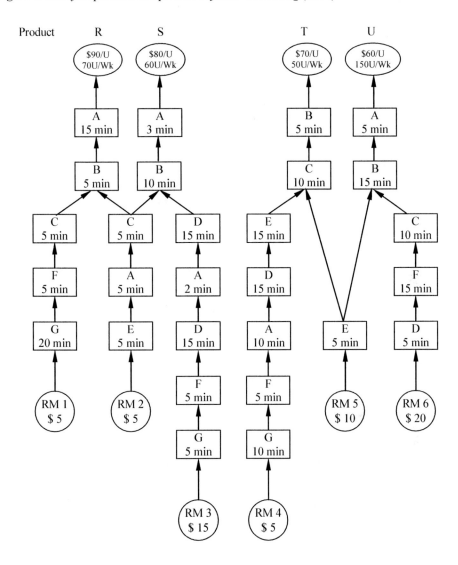

The total throughput is $(70 \times 80 + 60 \times 60 + 50 \times 50 + 80 \times 30) = 14100$. *Table 3* formalizes capacity utilization for each of the resources. From *Table 3* it is identified that the CCR is still there with resources A and D, as these two resources exceed the available maximum capacity of 2,400 minutes. Thus, it appears that product mix solution under TOC heuristic [particularly the problem of Hsu & Chung (1998)] is infeasible when multiple constraint resources exist.

In the spirit to maintain the Hsu and Chung's (1998) mathematical formulation, some inconsistencies have been found in their paper. The inconsistencies are as follows:

Table 2. Load calculations and constrained resources

Products	Weekly market potential (units)	Capacity per unit for resources (min)						
		A	B	C	D	E	F	G
R	70	1400	350	700	--	350	350	1400
S	60	600	600	300	1800	300	300	300
T	50	500	250	500	750	1000	250	500
U	150	750	2250	1500	750	750	2250	--
Total load (min)		3250	3450	3000	3300	2400	3150	2200
Available capacity (min)		2400	2400	2400	2400	2400	2400	2400
Overload (min)		850	1050	600	900	0	750	-200
Capacity utilization (%)		131.42	143.75	125	137.50	100	131.25	91.67

↑

Table 3. Load calculation after removing CCR at B

Products	Weekly market potential (units)	Capacity per unit for resources (min)						
		A	B	C	D	E	F	G
R	70	1400	350	700	--	350	350	1400
S	60	600	600	300	1800	300	300	300
T	50	500	250	500	750	1000	250	500
U	80	400	1200	800	400	400	1200	--
Total load (min)		2900	2400	2300	2950	2050	2100	2200
Available capacity (min)		2400	2400	2400	2400	2400	2400	2400
Overload (min)		500	0	-100	550	-350	-300	-200
Capacity utilization (%)		120.83	0	95.83	122.92	85.42	87.5	91.67

↑   ↑

1.   There should be an arrow connecting node E to C for product T, since it more complies with the resource E usage per unit of product T;

2.   The resource C usage per unit of product U, should be changed from 15 min to 10 min, so it complies with resource C constraint [see Figure 1 of Hsu & Chung (1998), more specifically the resource C usage for product U (15 minutes), with the corresponding values in Table 1 and Table 2 of their paper, which is 10 minutes]; and

3.   The price per unit of RM6, should be changed from US$10 to US$20, so that it complies with the objective function coefficient of U in the objective function (equation 1 of Hsu & Chung, 1998),that is, US$30. To maintain the Hsu and Chung's (1998)

mathematical formulation, a modification to the price of per unit RM6 from the existing US$10 to US$20, is suggested in *Figure 1*.

Extensive published literatures depict that TOC heuristic is implicit for solving product mix decision problem when multiple constrained resources exist. The same was also reported by Onwubolu and Mutingi (2001). Moreover, TOC-based product mix decisions can never be better than a correctly formulated LP approach (Souren et al., 2005).

In the next sections a detailed computational analysis with the developed CLPFR model will be illustrated. The CLPFR model makes the TOC heuristic more explicit in making product mix decision when multiple constrained resources exist.

"CLPFR" Algorithm

The development of compromise linear programming having fuzzy resources (CLPFR) algorithm passes through the following steps:

Step 1: Formulating the crisp linear programming problem

Step 2: Determining the resources whose availability are to be fuzzified and subsequent determination of their tolerances

Step 3: Defining the membership functions representing the fuzziness of the ith resource, i = 1, 2, ..., m, m being the number of resource whose availability are to be fuzzified

Step 4: Solving the following crisp LP:

max **cx**
subject to

$(\mathbf{Ax})_i \leq b_i$
$\mathbf{x} \geq \mathbf{0}$
i = 1, 2, ..., m

Step 5: Solving the following LP with the ith constraint tolerances:

max **cx**
subject to

$$(\mathbf{Ax})_i \le b_i + t_i$$

$$t_i \ge 0$$

$$\mathbf{x} \ge \mathbf{0}$$

$$i = 1, 2, ..., m$$

Step 6: Defining the membership function representing the degree of the optimality of the solution

Step 7: Defining the following linear programming problem:

$$\max_{\mathbf{x} \ge 0} \min\{\mu_0(\mathbf{x}), \mu_1(\mathbf{x}), ..., \mu_m(\mathbf{x})\}$$

Step 8: Converting the LP of Step 7 into the following equivalent compromise linear programming problem:

max α

subject to

$$\mu_0(\mathbf{x}) \ge \alpha$$

$$\mu_i(\mathbf{x}) \ge \alpha$$

$$\alpha \in [0,1]$$

$$\mathbf{x} \ge \mathbf{0}$$

$$i = 1, 2, ..., m$$

Step 9: Obtaining an equivalent compromise solution to Step 8 by using the following equivalent compromise linear programming problem:

max α

subject to

$$\mathbf{cx} \ge z^1 - (1 - \alpha)(z^1 - z^0)$$

$$(\mathbf{Ax})_i \le b_i + (1 - \alpha)t_i,$$

$$\alpha \in [0,1]$$

$$\mathbf{x} \ge \mathbf{0}$$

$$i = 1, 2, ..., m$$

TOC Product Mix Problem Formulation Using "CLPFR"

This chapter improves the solution of the problem reported earlier by Hsu and Chung (1998) fuzzifying the availability of resources A to G. For illustration purpose, the fuzzification is carried out introducing following tolerances t_1 = 120 minutes, t_2 = 240 minutes, t_3 = 180 minutes, t_4 = 120 minutes, t_5 = 320 minutes, t_6 = 240 minutes and t_7 = 180 minutes to resource A, B, C, D, E, F and G, respectively.

As a first step of the proposed CLPFR model, the crisp LP model is to be converted into an equivalent CLPFR using the following identified decision variables:

x_1 = the number of product R to be produced (units)

x_2 = the number of product S to be produced (units)

x_3 = the number of product T to be produced (units)

x_4 = the number of product U to be produced (units)

According to the developed CLPFR algorithm, the following steps have been computed for the product mix problem under TOC adopted from Hsu and Chung (1998):

Step 1: The crisp linear programming problem is formulated as follows:

The objective function for the crisp LP is:

maximize profit $z = 80x_1 + 60x_2 + 50x_3 + 30x_4$

subject to the following constraints

Constraint-1 (for resource A): $20x_1 + 10x_2 + 10x_3 + 5x_4 \leq 2400$

Constraint-2 (for resource B): $5x_1 + 10x_2 + 5x_3 + 15x_4 \leq 2400$

Constraint-3 (for resource C): $10x_1 + 5x_2 + 10x_3 + 10x_4 \leq 2400$

Constraint-4 (for resource D): $0x_1 + 30x_2 + 15x_3 + 5x_4 \leq 2400$

Constraint-5 (for resource E): $5x_1 + 5x_2 + 20x_3 + 5x_4 \leq 2400$

Constraint-6 (for resource F): $5x_1 + 5x_2 + 5x_3 + 15x_4 \le 2400$

Constraint-7 (for resource G): $20x_1 + 5x_2 + 10x_3 + 0x_4 \le 2400$

Constraint-8 (demand for product R): $x_1 \le 70$

Constraint-9 (demand for product S): $x_2 \le 70$

Constraint-10 (demand for product T): $x_3 \le 70$

Constraint-11 (demand for product U): $x_4 \le 70$, and

non-negativity constraints: $x_1, x_2, x_3, x_4 \ge 0$

Step 2: The constraints to be fuzzified are constraint numbers (1) to (7). The tolerances for each of the resources are as follows:

- $t_1 = 120$ minutes for the availability of resource A (resource-1),
- $t_2 = 240$ minutes for the availability of resource B (resource-2),
- $t_3 = 180$ minutes for the availability of resource C (resource-3),
- $t_4 = 120$ minutes for the availability of resource D (resource-4),
- $t_5 = 320$ minutes for the availability of resource E (resource-5),
- $t_6 = 240$ minutes for the availability of resource F (resource-6), and
- $t_7 = 180$ minutes for the availability of resource G (resource-7).

Step 3: The membership functions representing the fuzziness of the ith resource (constraint) are defined in the following fashion as illustrated.

Let t_i be the tolerance of the availability of the ith resource. The fuzziness of this resource is defined by the fuzzification of the ith constraint, $(\mathbf{Ax})_i \le b_i$, through the fuzzy set i with membership function having triangular fuzzy number:

$$\mu_i(x) = \begin{cases} 1, & \text{if } (\mathbf{Ax})_i \le b_i \\ 1 - \dfrac{(\mathbf{Ax})_i - b_i}{t_i}, & \text{if } b_i \le (\mathbf{Ax})_i \le b_i + t_i \\ 0, & \text{if } (\mathbf{Ax})_i > b_i + t_i \end{cases}$$

This triangular membership function represents the degree of satisfaction for the ith constraint.

In the case discussed, the following membership functions represent the fuzziness of constraints (1) to (7) respectively:

$$\mu_1(x) = \begin{cases} 1, & \text{if } (20x_1 + 10x_2 + 10x_3 + 5x_4) \leq 2400 \\ 1 - \dfrac{(20x_1 + 10x_2 + 10x_3 + 5x_4) - 2400}{120}, & \text{if } 2400 \leq (20x_1 + 10x_2 + 10x_3 + 5x_4) \leq 2520 \\ 0, & \text{if } (20x_1 + 10x_2 + 10x_3 + 5x_4) > 2520 \end{cases}$$

$$\mu_2(x) = \begin{cases} 1, & \text{if } (5x_1 + 10x_2 + 5x_3 + 15x_4) \leq 2400 \\ 1 - \dfrac{(5x_1 + 10x_2 + 5x_3 + 15x_4) - 2400}{240}, & \text{if } 2400 \leq (5x_1 + 10x_2 + 5x_3 + 15x_4) \leq 2640 \\ 0, & \text{if } (5x_1 + 10x_2 + 5x_3 + 15x_4) > 2640 \end{cases}$$

$$\mu_3(x) = \begin{cases} 1, & \text{if } (10x_1 + 5x_2 + 10x_3 + 10x_4) \leq 2400 \\ 1 - \dfrac{(10x_1 + 5x_2 + 10x_3 + 10x_4) - 2400}{180}, & \text{if } 2400 \leq (10x_1 + 5x_2 + 10x_3 + 10x_4) \leq 2580 \\ 0, & \text{if } (10x_1 + 5x_2 + 10x_3 + 10x_4) > 2580 \end{cases}$$

$$\mu_4(x) = \begin{cases} 1, & \text{if } (30x_2 + 15x_3 + 5x_4) \leq 2400 \\ 1 - \dfrac{(30x_2 + 15x_3 + 5x_4) - 2400}{120}, & \text{if } 2400 \leq (30x_2 + 15x_3 + 5x_4) \leq 2520 \\ 0, & \text{if } (30x_2 + 15x_3 + 5x_4) > 2520 \end{cases}$$

$$\mu_5(x) = \begin{cases} 1, & \text{if } (5x_1 + 5x_2 + 20x_3 + 5x_4) \leq 2400 \\ 1 - \dfrac{(5x_1 + 5x_2 + 20x_3 + 5x_4) - 2400}{320}, & \text{if } 2400 \leq (5x_1 + 5x_2 + 20x_3 + 5x_4) \leq 2720 \\ 0, & \text{if } (5x_1 + 5x_2 + 20x_3 + 5x_4) > 2720 \end{cases}$$

$$\mu_6(x) = \begin{cases} 1, & \text{if } (5x_1 + 5x_2 + 5x_3 + 15x_4) \leq 2400 \\ 1 - \dfrac{(5x_1 + 5x_2 + 5x_3 + 15x_4) - 2400}{240}, & \text{if } 2400 \leq (5x_1 + 5x_2 + 5x_3 + 15x_4) \leq 2640 \\ 0, & \text{if } (5x_1 + 5x_2 + 5x_3 + 15x_4) > 2640 \end{cases}$$

$$\mu_7(x) = \begin{cases} 1, & \text{if } (20x_1 + 5x_2 + 10x_3) \leq 2400 \\ 1 - \dfrac{(20x_1 + 5x_2 + 10x_3) - 2400}{180}, & \text{if } 2400 \leq (20x_1 + 5x_2 + 10x_3) \leq 2580 \\ 0, & \text{if } (20x_1 + 5x_2 + 10x_3) > 2580 \end{cases}$$

Step 4: The solution to the following crisp LP is carried out:

max **cx**

subject to

$(A\mathbf{x})_i \le b_i$

$\mathbf{x} \ge 0$

$i = 1, 2, ..., 7$

Let $\mathbf{x}^0$ be the solution and $z^0 = \mathbf{c}\mathbf{x}^0$ be the optimal value.

Therefore, the solution to the crisp LP problem of *Step 1* is as follows:

$\mathbf{x}^0 = (50.667 \quad 38.1667 \quad 50 \quad 101)$,

the optimal value is $z^0 = 11\ 873.33$

Step 5: The following crisp LP is solved with the constraint tolerances

max **cx**

subject to

$(A\mathbf{x})_i \le b_i + t_i$

$t_i \ge 0$

$\mathbf{x} \ge 0$

$i = 1, 2, ..., 7$

Let $\mathbf{x}^1$ be the solution and $z^1 = \mathbf{c}\mathbf{x}^1$ be the optimal value.

The formulation of the crisp LP with the constraint tolerances will be as follows:

Objective function: maximize profit $z = 80x_1 + 60x_2 + 50x_3 + 30x_4$

subject to the constraints

Constraint-1" (for resource A): $20x_1 + 10x_2 + 10x_3 + 5x_4 \le 2520$

Constraint-2" (for resource B): $5x_1 + 10x_2 + 5x_3 + 15x_4 \leq 2640$

Constraint-3" (for resource C): $10x_1 + 5x_2 + 10x_3 + 10x_4 \leq 2580$

Constraint-4" (for resource D): $0x_1 + 30x_2 + 15x_3 + 5x_4 \leq 2520$

Constraint-5" (for resource E): $5x_1 + 5x_2 + 20x_3 + 5x_4 \leq 2720$

Constraint-6" (for resource F): $5x_1 + 5x_2 + 5x_3 + 15x_4 \leq 2640$

Constraint-7" (for resource G): $20x_1 + 5x_2 + 10x_3 + 0x_4 \leq 2580$

Constraint-8 (demand for product R): $x_1 \leq 70$

Constraint-9 (demand for product S): $x_2 \leq 70$

Constraint-10 (demand for product T): $x_3 \leq 70$

Constraint-11 (demand for product U): $x_4 \leq 70$, and

Non-negativity constraints: $x_1, x_2, x_3, x_4 \geq 0$

The solution is:

$$\mathbf{x}^1 = \begin{pmatrix} 52.2667 & 39.7667 & 50 & 115.4 \end{pmatrix}$$

and the optimal value is: $z^1 = 12\,529.33$

Note that the LP constraints in *Step 4* are contained in the LP constraints in *Step 5*, thus it is clear that the following relation is trivial:

$$z^1 = \mathbf{cx}^1 \geq z^0 = \mathbf{cx}^0$$

Step 6: The following membership function is defined to represent the degree of optimality of the solution:

$$\mu_0(\mathbf{x}) = \begin{cases} 1, & \text{if } \mathbf{cx} > z^1 \\ 1 - \dfrac{z^1 - \mathbf{cx}}{z^1 - z^0}, & \text{if } z^0 \leq \mathbf{cx} \leq z^1 \\ 0, & \text{if } \mathbf{cx} < z^0 \end{cases}$$

As discussed above, the membership function representing the degree of the optimality of the solution is as follows:

$$\mu_0(\mathbf{x}) =$$

$$\begin{cases} 1, & \text{if } (80x_1 + 60x_2 + 50x_3 + 30x_4) > 12\,529.33 \\ 1 - \dfrac{12\,529.33 - (80x_1 + 60x_2 + 50x_3 + 30x_4)}{656}, & \text{if } 11873.33 \leq (80x_1 + 60x_2 + 50x_3 + 30x_4) \leq 12\,529.33 \\ 0, & \text{if } (80x_1 + 60x_2 + 50x_3 + 30x_4) < 11\,873.33 \end{cases}$$

Steps 7 & 8: So far we have introduced 8 (eight) membership functions as follows:

- $\mu_0(\mathbf{x})$ represents the degree of optimality of the solution
- $\mu_1(\mathbf{x}) \dots \mu_7(\mathbf{x})$, each represents the degree of satisfaction for constraints (1) to (7)

The main aim is to maximize the value of all of these membership functions. In reality, since such aim is never possible to be achieved, a compromise is required. Since all of these membership functions are non-dimensional, one can apply the *max-min* method for the compromise.

Thus the problem is formulated as follows:

$$\max_{x \geq 0} \ \min \ \{\mu_0(\mathbf{x}), \mu_1(\mathbf{x}), \dots, \mu_7(\mathbf{x})\}$$

or, equivalently the *compromise linear programming problem*

max α

subject to

$\mu_0(\mathbf{x}) \geq \alpha$

$\mu_i(\mathbf{x}) \geq \alpha,\ i = 1, 2, ..., 7$

$\alpha \in [0,1]$

$\mathbf{x} \geq \mathbf{0}$

Step 9: The solution to the equivalent compromise linear programming problem results in some algebraic manipulations.

max α

subject to

$\mathbf{cx} \geq z^1 - (1 - \alpha)(z^1 - z^0)$

$(\mathbf{Ax})_i \geq b_i + (1 - \alpha)t_i,\ i = 1, 2, ..., 7$

$\alpha \in [0,1]$

$\mathbf{x} \geq \mathbf{0}$

After some algebraic manipulation, we get the following linear programming problem:

max α

subject to

$80x_1 + 60x_2 + 50x_3 + 30x_4 \geq 12\,529.33 - 656(1 - \alpha)$ or $80x_1 + 60x_2 + 50x_3 + 30x_4 - 656\alpha \geq 11873.33$

$20x_1 + 10x_2 + 10x_3 + 5x_4 \leq 2400 + (1 - \alpha)t_1$ or $20x_1 + 10x_2 + 10x_3 + 5x_4 + 120\alpha \leq 2520$

$5x_1 + 10x_2 + 5x_3 + 15x_4 \leq 2400 + (1 - \alpha)t_2$ or $5x_1 + 10x_2 + 5x_3 + 15x_4 + 240\alpha \leq 2640$

$10x_1 + 5x_2 + 10x_3 + 10x_4 \leq 2400 + (1 - \alpha)t_3$ or $10x_1 + 5x_2 + 10x_3 + 10x_4 + 180\alpha \leq 2580$

$0x_1 + 30x_2 + 15x_3 + 5x_4 \leq 2400 + (1 - \alpha)t_4$ or $0x_1 + 30x_2 + 15x_3 + 5x_4 + 120\alpha \leq 2520$

$5x_1 + 5x_2 + 20x_3 + 5x_4 \leq 2400 + (1 - \alpha)t_5$ or $5x_1 + 5x_2 + 20x_3 + 5x_4 + 320\alpha \leq 2720$

$5x_1 + 5x_2 + 5x_3 + 15x_4 \leq 2400 + (1 - \alpha)t_6$ or $5x_1 + 5x_2 + 5x_3 + 15x_4 + 240\alpha \leq 2640$

$20x_1 + 5x_2 + 10x_3 + 0x_4 \leq 2400 + (1 - \alpha)t_7$ or $20x_1 + 5x_2 + 10x_3 + 0x_4 + 180\alpha \leq 2580$

$x_1 \leq 70$

$x_2 \leq 70$

$x_3 \leq 70$

$x_4 \leq 70$, and

the non-negativity constraints: $x_1, x_2, x_3, x_4 \geq 0$

The Product Mix Solution and Discussions

The results of TOC problem using CLPFR model, obtained with the aid of the WinQSB® software, are tabulated in *Table 4*.

Table 4 is converted into *Table 5* showing optimal combination of products to be produced.

From the definitions of μ_0, μ_1, ..., μ_7 for the optimal solution, the following values are obtained:

$$\mu_0(x) = \mu_0(51.4667, 38.9667, 50, 108.2000) = 0.5000$$
$$\mu_1(x) = \mu_1(51.4667, 38.9667, 50, 108.2000) = 0.5000$$
$$\mu_2(x) = \mu_2(51.4667, 38.9667, 50, 108.2000) = 0.5000$$
$$\mu_3(x) = \mu_3(51.4667, 38.9667, 50, 108.2000) = 1$$
$$\mu_4(x) = \mu_4(51.4667, 38.9667, 50, 108.2000) = 0.5000$$
$$\mu_5(x) = \mu_5(51.4667, 38.9667, 50, 108.2000) = 1$$
$$\mu_6(x) = \mu_6(51.4667, 38.9667, 50, 108.2000) = 1$$
$$\mu_7(x) = \mu_7(51.4667, 38.9667, 50, 108.2000) = 1$$

Let us now examine and discuss on each of the value for $\mu_0, \mu_1, ..., \mu_7$ and α.

Discussions on μ_0

The value of the optimal solution with no tolerance in constraints is $z^0 = 11873.33$. From the definition of μ_0, the value of $z^0 = 11873.33$ corresponds to the value 0. The value of the optimal solution using maximum tolerance in each of the first seven constraints is $z^1 = 12529.33$. From the definition of μ_0, the value of $z^1 = 12529.33$ corresponds to the value 1. Thus, the optimal value of the TOC objective function is:

$$80(51.4667) + 60(38.9667) + 50(50) + 30(108.2000) = 12201.34$$

By linear interpolation, this optimal value corresponds to the degree of optimality of the solution, μ_0, which is equal to 0.5000.

Discussions on μ_1

When the usage of resource A is less than, or equal to the current capacity, that is, 2,400 hours, no tolerance for resource A is required. This situation corresponds to $\mu_1 = 1$. This

Table 4. Solutions from WinQSB® software

	Decision variables	Solution values
1	x_1	51.4667
2	x_2	38.9667
3	x_3	50.0000
4	x_4	108.2000
5	α	0.5000

Table 5. The optimal combination of products

Products	Quantity to be produced
R	51.4667
S	38.9667
T	50
U	108.2000

implies that there is no violation of the boundary situations for the original constraint of availability of resource A.

Let us discuss another case when the usage of resource A is greater than, or equal to the capacity having maximum tolerance, that is, 2,520 hours. This situation corresponds to μ_1 = 0 indicating maximum violation of the boundary situations for the original constraint of availability of resource A.

The optimal solution for the usage level of resource A is:

$$20x_1 + 10x_2 + 10x_3 + 5x_4 = 2460.00.$$

By linear interpolation this value corresponds to the degree of satisfaction of the constraint for resource A.

Discussions on μ_2

When the usage of resource B is less than, or equal to the current capacity, that is, 2,400 hours, no tolerance for this resource is warranted. This situation corresponds to $\mu_2 = 1$. This indicates no violation of the boundary conditions for the original constraint of availability of resource B.

When the usage of resource B is greater than, or equal to the capacity having maximum tolerance, that is, 2,640 hours, the situation corresponds to $\mu_2 = 0$. This situation indicates maximum violation of the boundary conditions for the original constraint of availability of resource B.

Since the optimal solution for the usage level of resource B is:

$$5x_1 + 10x_2 + 5x_3 + 15x_4 = 2520.00,$$

then, by linear interpolation, it is found that this value corresponds to the degree of satisfaction of the constraint for resource B, $\mu_2 = 0.5000$.

Discussions on μ_3

Let us discuss on the first boundary condition. The usage of resource C is less than, or equal to the current capacity, that is, 2,400 hours. It indicates that no tolerance for resource C is warranted and the situation corresponds to $\mu_3 = 1$. It is to be noted that for this condition no violation of the boundary conditions for the original constraint of availability of resource C is present.

For the second boundary condition, the usage of resource C is greater than, or equal to the capacity having maximum tolerance, that is, 2,580 hours. This situation corresponds to $\mu_3 = 0$. This is an indication of maximum violation of the boundary conditions for the original constraint of availability of resource C.

The usage level of resource C in the optimal solution is:

$$10x_1 + 5x_2 + 10x_3 + 10x_4 = 2291.50.$$

Therefore, from the definition of μ_3, we get $\mu_3 = 1$, that is, no violation in the usage level of resource C.

Discussions on μ_4

The first boundary condition for the usage of resource D is less than, or equal to the current capacity, that is, 2,400 hours. This condition indicates that no tolerance for resource D is warranted. This situation corresponds to $\mu_4 = 1$, which implies no violation of the boundary conditions for the original constraint of availability of resource D.

The second boundary condition teaches that the usage of resource D is greater than, or equal to the capacity having maximum tolerance, that is, 2,520 hours. Simulating within this boundary values, $\mu_4 = 0$ is obtained. The resulted value indicates maximum violation of the boundary conditions for the original constraint of availability of resource D.

The usage level of resource D in the optimal solution is:

$$0x_1 + 30x_2 + 15x_3 + 5x_4 = 2460.00$$.

Using linear interpolation technique it is found that this value corresponds to the degree of satisfaction of the constraint for resource D as $\mu_4 = 0.5000$

Discussions on μ_5

If the usage of resource E is less than, or equal to the current capacity, that is, 2,400 hours, then no tolerance for resource E is required. This situation corresponds to $\mu_5 = 1$. The value of μ_5 indicates no violation of the boundary conditions for the original constraint of availability of resource E.

If the usage of resource E is greater than, or equal to the capacity having maximum tolerance, that is, 2,720 hours, then this situation corresponds to $\mu_5 = 0$. This value of μ_5 implies maximum violation of the boundary conditions for the original constraint of availability of resource E.

The usage level of resource E in the optimal solution is:

$$5x_1 + 5x_2 + 20x_3 + 5x_4 = 1993.167$$.

From the definition of μ_5, we get $\mu_5 = 1$, that is, no violation of the boundary conditions in the usage constraint of resource E.

Discussions on μ_6

For the usage constraint of resource F, if the same is less than, or equal to the current capacity, that is, 2,400 hours, then no tolerance for this resource warranted. The corresponding value for μ_6 under this situation is equal to 1. This value restricts any violation of the boundary conditions for the original constraint of availability of resource F.

If the usage constraint of resource F is greater than, or equal to the capacity having maximum tolerance, that is, 2,720 hours, then this situation corresponds to the value $\mu_6 = 0$. This validates the situation of maximum violation of the boundary conditions for the original constraint of availability of resource F.

The usage level of resource F in the optimal solution is:

$$5x_1 + 5x_2 + 5x_3 + 15x_4 = 2325.167$$.

Therefore, from the definition of μ_6, we get $\mu_6 = 1$, that is, no violation of the boundary conditions in the usage constraint of resource F.

Discussions on μ_7

The first boundary condition binds the usage of resource G within less than, or equal to the current capacity, that is, 2,400 hours. This implies that no tolerance for resource G is required. This situation corresponds to $\mu_7 = 1$, indicating no violation of the boundary conditions for the original constraint of availability of resource G.

The second boundary condition for the usage of resource G is greater than, or equal to the capacity with maximum tolerance, that is, 2,580 hours. This condition corresponds to $\mu_7 = 0$. This value of μ_7 indicates maximum violation of the boundary conditions for the original constraint of availability of resource G.

It is to be noted that the usage level of resource G in the optimal solution is:

$$20x_1 + 5x_2 + 10x_3 + 0x_4 = 1724.168.$$

Using the definition of μ_7, one gets $\mu_7 = 1$, that is, no violation of the boundary conditions in the usage constraint of resource G.

Discussions on α

For the TOC problem having fuzzy constraints, the chief aim is to achieve the following two goals:

- to achieve the maximum throughput, if necessary, by exploiting the use of all the constraints tolerance available; and at the same time
- to maintain to the level of resource usage such that no single constraint is violated.

Such an aim is not achievable often, and therefore, a compromise is required to achieve these two goals. This compromise is made with the application of *max-min* principle to the following two parameters:

- the degree of optimality of the solution, represented by the membership function μ_0, and
- the degree of satisfaction of constraints for resources A to G, represented by the membership functions $\mu_1, \mu_2, ..., \mu_7$ respectively.

In *Step 7* of the algorithm, the problem formulated is to maximize the value of $\alpha = \min \{\mu_0,$ $\mu_1, ..., \mu_7\}$. From the previous discussions on the values of $\mu_0, \mu_1, \mu_2, ..., \mu_7$, it is clear why one obtains $\alpha = 0.5000$ in the optimal solution. The value of α indicates that the values of the degree of optimality of the solution, and the degree of satisfaction of the constraints for resources A to G will not be less than 0.5000.

Simulations Using MATLAB®

Another simulation phase of the algorithm with MATLAB® software is described under this heading. This simulation is carried out with an aim to sense the degree of fuzziness of the solution and the level-of-satisfaction of the decision maker (DM). Degree of fuzziness gets induced in the set of solutions due to imprecision of the tolerance values of the CLPRF algorithm. Induction of fuzziness in the solution will affect the level-of-satisfaction of the DM. This level-of-satisfaction is one kind of human "emotion" of the decision makers, which is guided by many factors while making a decision. Moreover, the degree of fuzziness and the level-of-satisfaction of the DM are not tangible quantities. Therefore, sensitivity simulation using a suitable and flexible membership function is the only solution to grab the emotion of DM as well as the degree of fuzziness present in the solution of CLPFR algorithm. Let us now begin with formulating a suitable membership function so as to simulate the sensitiveness of the solution found using the CLPFR algorithm.

In order to solve the issue of degeneration, in fuzzy problems, Leberling (1981) employed a non-linear logistic function, for example a tangent hyperbola that has asymptotes at 1 and 0. The logistic membership function has similar shape as that of tangent hyperbolic function employed by Leberling (1981) but it is more flexible than that of the tangent hyperbola of Leberling (1981). It should be emphasized that some non-linear MFs such as *S*-curve MFs are desirable for use in real life product mix decision problems than that of linear MFs (Vasant, 2004).

The generalised logistic function (Leberling, 1981) is given by:

$$f(x) = \frac{B}{1 + Ce^{gx}},$$

where B and C are scalar constants and γ, $0 < \gamma < \infty$, is a fuzzy parameter measuring the degree of vagueness, wherein $\gamma = 0$ indicates crisp.

The generalized logistic MF (Bhattacharya & Vasant, 2006; Vasant, 2004) is defined as:

$$f(x) = \begin{cases} 1 & x < x_L \\ \dfrac{B}{1 + Ce^{gx}} & x_L < x < x_U \\ 0 & x > x_U \end{cases}$$

The *S*-curve MF is a particular case of the logistic function. The said *S*-curve MF has got specific values of *B*, *C* and γ. The logistic MF is re-defined as $0.001 \leq \mu(x) \leq 0.999$. In real-life problems, the physical capacity requirement cannot be 100%. Thus, the range $0.001 \leq \mu(x) \leq 0.999$ is selected. At the same time, the capacity requirement cannot be 0% (Bhattacharya & Vasant, 2006; Vasant, 2004; Bitran, 1980).

$$\mu(x) = \begin{cases} 1 & x < x^a \\ 0.999 & x = x^a \\ \dfrac{B}{1 + Ce^{gx}} & x^a < x < x^b \\ 0.001 & x = x^b \\ 0 & x > x^b \end{cases}$$

In this simulation procedure the relationship between the degree of possibility, μ, and the level-of-satisfaction, φ, is $\mu = (1 - \varphi)$. Rule-based codes have been generated using MATLAB®'s M-file for simulating the sensitiveness of the solution found using the CLPFR algorithm. These codes help a decision maker to vary the values of the coefficient α in the interval (0,1). Thus, the DM is able to have the optimal throughput (Z) for a particular value of level-of-satisfaction (φ) and degree of fuzziness (γ). *Table 6* illustrates throughput (Z) simulation data at disparate degree of fuzziness (γ) and level-of-satisfaction (φ) of the DM. It is observed from *Table 6* that the characteristics plot simulating a relationship among all these three parameters will behave as a monotonically increasing function.

In the first row, second column of *Table 6*, there are two inputs and one output data. The inputs are γ = 3 and μ = 0.10 and the corresponding output is Z = US$11908. As the μ increases the Z values decrease for any particular γ value. This indicates that decrease in level-of-satisfaction (φ) results in decrease in the profits. The first row indicates that the fuzziness dominates for a very poor level-of-satisfaction of the decision maker because at poor level of satisfaction, higher degree of fuzziness gets associated with the output itself.

Figure 2 is a surface and contour simulation illustrating the behavioural patterns of Z-values with respect to the degree of possibility (μ) at disparate degree of fussiness (γ). It is to be noted that the higher the level of satisfaction values (φ), the lesser will be the dominance of the degree of vagueness (γ). Thus higher level of outcome of decision variable for a particular level-of-satisfaction point results in a lesser degree of fuzziness inherent in the said decision variable.

Figure 3 depicts a 2-D contour simulation illustrating relationship between level-of-satisfaction (μ) and degree of fuzziness (γ). Lower μ values indicate higher level-of-satisfaction (φ) of the decision made and the corresponding degree of fuzziness (γ) will be low. This is because of the relationship $\mu = (1 - \varphi)$.

Figure 4 illustrates a 2-D contour simulation depicting characteristics showing the relationship between the throughput (Z) and the degree of possibility (μ). *Figure 5* simulates relationship between the throughput (Z) and the degree of fuzziness (γ). From all these simulations it is evident that the decision with the proposed CLPFR methodology is to be made with higher level-of-satisfaction with lesser degree of fuzziness. The characteristic simulations guide a DM in deciding his/her level-of-satisfaction with an allowable degree of fuzziness of the decision made.

Conclusion

It has been found from the CLPFR model that the throughput of the product mix problem under TOC is US$12,201.34. Hsu & Chung (1998) used dominance rule technique and their throughput was US$11,873 whereas Onwubolu and Mutingi (2001) tackled the same problem with a throughput of US$11,860. The TOC heuristic results in a throughput of US$14,100. TOC solution is not free from bottleneck and multiple constraint resources exist. Therefore, the previous solutions to the product mix problem of Hsu and Chung (1998) were not optimal. Bhattacharya and Vasant (2006), and Bhattacharya, Vasant, Sarkar, and Mukherjee (2006) tried to solve Hsu and Chung's (1998) product mix problem using a modified S-curve MF, which resulted in a robust solution. But their throughput was comparatively less than the solution presented in this chapter. *Table 7* elucidates a thorough comparison among all the solutions of the Hsu and Chung's (1998) product mix problem.

The proposed CLPFR model finds out a robust optimal solution to the Hsu and Chung's (1998) product mix problem. The fuzzy plots simulate DM's preferences in selecting his/her choice of level-of-satisfaction as per a predetermined degree of fuzziness while making the product mix decision. Further extension of the CLPFR model simulating with a suitably designed smooth logistic membership function (which is of course a more realistic assumption) may increase throughput trading off suitably among decision variables and other constraints.

Table 6. Throughput (Z) simulation at disparate fuzziness (γ) and level-of-satisfaction (φ) of the DM

Z	μ = 1 − φ								
γ	0.1	0.2	0.3	0.4	0.5	0.6	0.7	0.8	0.9
3.0	11908	11889	11883	11880	11877	11876	11875	11874	11874
3.4	11918	11894	11886	11882	11879	11877	11876	11875	11874
3.8	11931	11901	11890	11884	11881	11878	11877	11875	11874
4.2	11946	1.1910	1.1912	0.0596	0.0403	0.0271	0.0176	0.0103	0.0046
4.6	0.3675	0.1923	0.1198	0.0798	0.0544	0.0368	0.0239	0.0140	0.0062
5.0	0.4498	0.2469	0.1575	0.1064	0.0732	0.0499	0.0326	0.0192	0.0086
5.4	0.5391	0.3118	0.2046	0.1408	0.0982	0.0675	0.0444	0.0264	0.0118
5.8	0.6323	0.3859	0.2617	0.1843	0.1306	0.0910	0.0605	0.0362	0.0164
6.2	0.7264	0.4671	0.3283	0.2373	0.1717	0.1217	0.0820	0.0496	0.0227
6.6	0.8186	0.5529	0.4027	0.2995	0.2220	0.1606	0.1102	0.0678	0.0314
7.0	0.9073	0.6403	0.4826	0.3697	0.2813	0.2084	0.1462	0.0918	0.0434
7.4	0.9912	0.7269	0.5655	0.4458	0.3485	0.2651	0.1908	0.1230	0.0596
7.8	1.0697	0.8110	0.6489	0.5253	0.4216	0.3294	0.2440	0.1621	0.0813
8.2	1.1427	0.8912	0.7307	0.6057	0.4982	0.3997	0.3049	0.2096	0.1095
8.6	1.2103	0.9670	0.8096	0.6851	0.5760	0.4736	0.3718	0.2649	0.1452
9.0	1.2729	1.0380	0.8848	0.7622	0.6532	0.5490	0.4427	0.3268	0.1887
9.4	1.3308	1.1043	0.9557	0.8358	0.7282	0.6239	0.5154	0.3935	0.2398
9.8	1.3843	1.1660	1.0222	0.9056	0.8002	0.6970	0.5882	0.4628	0.2973
10.2	1.4339	1.2235	1.0845	0.9714	0.8687	0.7674	0.6594	0.5329	0.3595
10.6	1.4799	1.2770	1.1428	1.0332	0.9334	0.8345	0.7283	0.6022	0.4246
11.0	1.5226	1.3269	1.1972	1.0911	0.9943	0.8980	0.7942	0.6698	0.4908
11.4	1.5625	1.3734	1.2480	1.1454	1.0515	0.9580	0.8568	0.7348	0.5566
11.8	1.5996	1.4169	1.2956	1.1963	1.1053	1.0145	0.9161	0.7969	0.6210
12.2	1.6344	1.4576	1.3401	1.2439	1.1558	1.0677	0.9721	0.8559	0.6833
12.6	1.6670	1.4957	1.3820	1.2887	1.2032	1.1178	1.0249	0.9119	0.7430
13.0	1.6976	1.5316	1.4212	1.3308	1.2479	1.1650	1.0747	0.9648	0.8000
13.4	1.7263	1.5653	1.4582	1.3705	1.2900	1.2095	1.1218	1.0149	0.8543
13.8	1.7534	1.5970	1.4931	1.4078	1.3296	1.2514	1.1662	1.0622	0.9058
14.2	1.7790	1.6270	1.5259	1.4431	1.3671	1.2910	1.2082	1.1070	0.9547
14.6	1.8032	1.6553	1.5571	1.4765	1.4025	1.3285	1.2479	1.1495	1.0011
15.0	1.8261	1.6822	1.5865	1.5081	1.4361	1.3641	1.2855	1.1897	1.0452
15.4	1.8478	1.7076	1.6144	1.5380	1.4679	1.3977	1.3213	1.2279	1.0870
15.8	1.8684	1.7318	1.6409	1.5665	1.4981	1.4297	1.3552	1.2641	1.1268
16.2	1.8880	1.7547	1.6661	1.5935	1.5268	1.4601	1.3874	1.2986	1.1646
16.6	1.9066	1.7766	1.6901	1.6192	1.5542	1.4891	1.4181	1.3314	1.2006
17.0	1.9244	1.7974	1.7130	1.6438	1.5802	1.5167	1.4474	1.3627	1.2350
17.4	1.9414	1.8173	1.7348	1.6672	1.6051	1.5430	1.4753	1.3926	1.2678
17.8	1.9576	1.8363	1.7556	1.6895	1.6288	1.5681	1.5019	1.4211	1.2991
18.2	1.9730	1.8544	1.7756	1.7109	1.6515	1.5922	1.5274	1.4484	1.3290
18.6	1.9878	1.8718	1.7946	1.7313	1.6733	1.6152	1.5518	1.4745	1.3577
19.0	2.0020	1.8884	1.8129	1.7509	1.6941	1.6372	1.5752	1.4995	1.3851
19.4	2.0156	1.9044	1.8304	1.7697	1.7140	1.6583	1.5976	1.5234	1.4115

Table 6. continued

Z	$\mu = 1 - \phi$								
γ	**0.1**	**0.2**	**0.3**	**0.4**	**0.5**	**0.6**	**0.7**	**0.8**	**0.9**
19.8	2.0287	1.9197	1.8472	1.7877	1.7332	1.6786	1.6191	1.5464	1.4367
20.2	2.0412	1.9344	1.8633	1.8050	1.7516	1.6981	1.6397	1.5685	1.4610
20.6	2.0533	1.9485	1.8788	1.8217	1.7693	1.7168	1.6596	1.5897	1.4843
21.0	2.0649	1.9621	1.8937	1.8377	1.7863	1.7348	1.6787	1.6102	1.5067
21.4	2.0760	1.9751	1.9081	1.8531	1.8026	1.7521	1.6971	1.6298	1.5283
21.8	2.0868	1.9877	1.9219	1.8679	1.8184	1.7688	1.7148	1.6487	1.5491
22.2	2.0971	1.9999	1.9352	1.8822	1.8336	1.7849	1.7318	1.6670	1.5692
22.6	2.1071	2.0116	1.9481	1.8960	1.8482	1.8004	1.7483	1.6846	1.5885
23.0	2.1168	2.0229	1.9605	1.9094	1.8624	1.8154	1.7642	1.7016	1.6072
23.4	2.1261	2.0338	1.9725	1.9222	1.8761	1.8299	1.7795	1.7180	1.6252
23.8	2.1351	2.0444	1.9841	1.9346	1.8893	1.8439	1.7943	1.7339	1.6426
24.2	2.1438	2.0546	1.9953	1.9467	1.9020	1.8574	1.8087	1.7492	1.6595
24.6	2.1522	2.0645	2.0061	1.9583	1.9144	1.8705	1.8226	1.7641	1.6758
25.0	2.1604	2.0740	2.0166	1.9696	1.9263	1.8831	1.8360	1.7784	1.6915
25.4	2.1683	2.0833	2.0268	1.9805	1.9379	1.8954	1.8490	1.7923	1.7068
25.8	2.1759	2.0923	2.0366	1.9910	1.9492	1.9073	1.8616	1.8058	1.7216
26.2	2.1834	2.1010	2.0462	2.0013	1.9600	1.9188	1.8738	1.8189	1.7360
26.6	2.1906	2.1094	2.0554	2.0112	1.9706	1.9300	1.8857	1.8316	1.7499
27.0	2.1975	2.1176	2.0644	2.0208	1.9808	1.9408	1.8972	1.8439	1.7634
27.4	2.2043	2.1255	2.0731	2.0302	1.9908	1.9513	1.9083	1.8558	1.7765
27.8	2.2109	2.1332	2.0816	2.0393	2.0004	1.9616	1.9192	1.8674	1.7893
28.2	2.2173	2.1407	2.0898	2.0481	2.0098	1.9715	1.9297	1.8787	1.8016
28.6	2.2235	2.1480	2.0978	2.0567	2.0189	1.9811	1.9399	1.8896	1.8137
29.0	2.2296	2.1551	2.1056	2.0650	2.0278	1.9905	1.9499	1.9003	1.8254
29.4	2.2354	2.1620	2.1132	2.0732	2.0364	1.9997	1.9596	1.9106	1.8368
29.8	2.2412	2.1687	2.1205	2.0811	2.0448	2.0086	1.9690	1.9207	1.8478
30.2	2.2467	2.1752	2.1277	2.0888	2.0530	2.0172	1.9782	1.9305	1.8586
30.6	2.2522	2.1816	2.1347	2.0962	2.0609	2.0256	1.9871	1.9401	1.8691
31.0	2.2574	2.1878	2.1415	2.1035	2.0687	2.0338	1.9958	1.9494	1.8793
31.4	2.2626	2.1938	2.1481	2.1107	2.0763	2.0418	2.0043	1.9585	1.8893
31.8	2.2676	2.1997	2.1546	2.1176	2.0836	2.0496	2.0126	1.9673	1.8990
32.2	2.2725	2.2055	2.1609	2.1244	2.0908	2.0572	2.0207	1.9760	1.9085
32.6	2.2773	2.2111	2.1670	2.1309	2.0978	2.0647	2.0285	1.9844	1.9177
33.0	2.2820	2.2165	2.1730	2.1374	2.1046	2.0719	2.0362	1.9926	1.9268
33.4	2.2865	2.2219	2.1789	2.1437	2.1113	2.0790	2.0437	2.0006	1.9356
33.8	2.2909	2.2271	2.1846	2.1498	2.1178	2.0859	2.0510	2.0084	1.9442
34.2	2.2953	2.2322	2.1902	2.1558	2.1242	2.0926	2.0581	2.0161	1.9526
34.6	2.2995	2.2371	2.1956	2.1616	2.1304	2.0992	2.0651	2.0235	1.9608
35.0	2.3037	2.2420	2.2010	2.1673	2.1365	2.1056	2.0719	2.0308	1.9688

Figure 2. A surface and contour plot simulating the behavioural patterns of Z-values (using S-curve MF) with respect to the degree of possibility (μ) at disparate fussiness (γ)

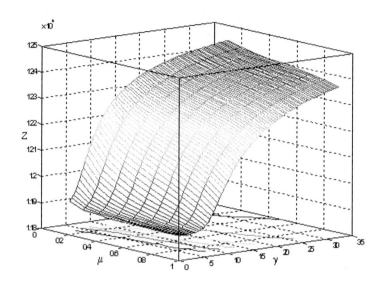

Figure 3. A 2-D contour simulation depicting relationship between the level-of-satisfaction (μ) and the degree of fuzziness (γ) of the decision made

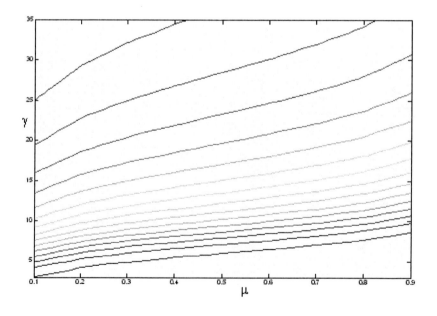

Figure 4. A 2-D contour characteristics simulation depicting relationship between the throughput (Z) and the degree of possibility (μ)

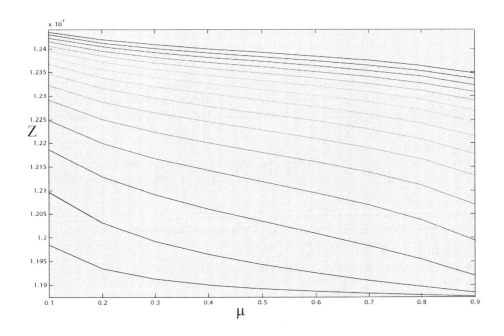

Figure 5. A 2-D contour characteristics simulation showing relationship between the throughput (Z) and the degree of fuzziness (γ)

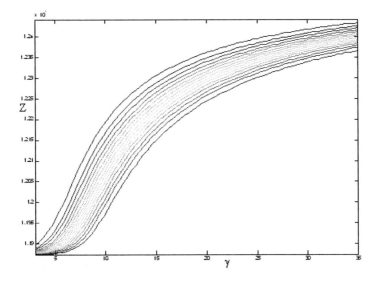

Table 7. Throughput comparison for the product-mix problem

Problem	No. of Resources	TOC solution	LP solution	Dominance rule solution (Hsu & Chung, 1998)	GA solution (Onwubolu & Mutingi, 2001)	FLP solution (Bhattacharya & Vasant, 2006)	CLPFR solution
Hsu & Chung (1998)	7	14100	----	11873	11860	11873	12201.34

References

Balakrishnan, J., & Cheng, C.H. (2000). Theory of constraints and linear programming: A re-examination. *International Journal of Production Research, 38*(6), 1459-1463.

Bhattacharya, A., & Vasant, P. (2006). Soft-sensing of level of satisfaction in TOC product-mix decision heuristic using robust fuzzy-LP. *European Journal of Operational Research, 177*(1), 55-70.

Bhattacharya, A., Vasant, P., Sarkar, B., & Mukherjee, S.K. (2006). A fully fuzzified, intelligent theory-of-constraints product-mix decision. *International Journal of Production Research.* Available online on November 17.

Bitran, G.R. (1980). Linear multiple objective problems with interval coefficients. *Management Science, 26*, 694-706.

Coman, A., & Ronen, B. (2000). Production outsourcing: A linear programming model for the theory-of-constraints. *International Journal of Production Research, 38*(7), 1631-1639.

Finch, B.J., & Luebbe, R.L. (2000). Response to 'Theory of constraints and linear programming: A re-examination.' *International Journal of Production Research, 38*(6), 1465-1466.

Fishwick, P.A. (1992). An integrated approach to system modeling using a synthesis of artificial intelligence, software engineering, and simulation methodologies. *ACM Transactions on Modeling and Computer Simulation, 2*(4), 1-27.

Fishwick, P.A., & Zeigler, B.P. (1992). A multimodel methodology for qualitative model engineering. *ACM Transactions on Modeling and Computer Simulation, 2*(1), 52-81.

Fishwick, P.A., Narayana, N., Sticklen, J., & Bonarini, A. (1994). A multimodel approach to reasoning and simulation. *IEEE Transactions on Systems, Man, and Cybernetics, 24*(10), 1433-1449.

Frantz, F.K. (1996). *Model abstraction techniques* (Rome Laboratory Tech. Rep.). Rome, New York: IFSB, Air Force Research Laboratory.

Frantz, F.K. (2006). *A taxonomy of model abstraction techniques.* Retrieved January 30, 2006, from http://www.rl.af.mil/tech/papers/ModSim/ModAb.html

Hsu, T.-C., & Chung, S.-H. (1998). The TOC-based algorithm for solving product mix problems. *Production Planning and Control, 9,* 36-46.

Leberling, H. (1981). On finding compromise solutions in multi-criteria problems using the fuzzy min operator. *Fuzzy Sets and Systems, 6,* 105-118.

Lee, T.N., & Plenert, G. (1993). Optimizing theory of constraints when new product alternatives exist. *Production and Inventory Management Journal, 34*(3), 51-57.

Luebbe, R., & Finch, B. (1992). Theory of constraints and linear programming: A comparison. *International Journal of Production Research, 30*(6), 1471-1478.

Maday, C.J. (1994). Proper use of constraint management. *Production and Inventory Management Journal, 35*(1), 84.

Miller, D., Firby, J., Fishwick, P., Franke, D., & Rothenberg, J. (1992). AI: What simulationists need to know. *ACM Transactions on Modeling and Computer Simulation, 2*(4), 269-284.

Nayak, P. (1992). Causal approximations. In *Proceedings of the Tenth National Conference on Artificial Intelligence* (pp. 703-709). Cambridge, MA: AAAUMIT Press.

Onwubolu, G.C. (2001). Tabu search-based algorithm for the TOC product mix decision. *International Journal of Production Research, 39*(10), 2065-2076.

Onwubolu, G.C., & Mutingi, M. (2001). A genetic algorithm approach to the theory of constraints product mix problems. *Production Planning and Control, 12*(1), 21-27.

Peterman, M. (2006). *Simulation Nation: Process simulation is key in a lean manufacturing company hungering for big results.* Retrieved January 30, 2006, from http://www.qualitydigest.com/may01/html/simulation.html

Plenert, G. (1993). Optimizing theory of constraints when multiple constrained resources exist. *European Journal of Operational Research, 70,* 126-133

Posnack, A.J. (1994). Theory of constraints: Improper applications yield improper conclusions. *Production and Inventory Management Journal, 35*(1), 85-86.

Sisti, A.F., & Farr, S.D. (2006). *Modeling and simulation enabling technologies for military applications.* Retrieved January 30, 2006, from http://www.rl.af.mil/tech/papers/ModSim/mil001.html

Souren, R., Ahn, H., & Schmitz, C. (2005). Optimal product mix decisions based on the theory of constraints? Exposing rarely emphasized premises of throughput accounting. *International Journal of Production Research, 43*(2), 361-374.

Susanto, S., Vasant, P., Bhattacharya, A., & Kahraman, C. (2006). Product-mix decision with compromise LP having fuzzy objective function coefficients (CLPFOFC). In D. Ruan, P. D'hondt, P.F. Fantoni, M.D. Cock, M. Nachtegael, & E.E. Kerre (Eds.), *The 7ᵗʰ International FLINS Conference on Applied Artificial Intelligence* (pp. 315-320). Genova: World Scientific Publishing Company, Imperial College Press.

Susanto, S., Bhattacharya, A., Vasant, P., & Suryadi, D., (2006). Optimising product-mix with compromise linear programming having fuzzy resources (CLPFR). In Y.-C. Liao, & C.-T. Wu (Eds.), *The 36ᵗʰ International Conference on Computers and Industrial Engineering* (pp. 1544-1555). Taipei: National Tsing Hua University.

Vasant, P. (2003). Application of fuzzy linear programming in production planning. *Fuzzy Optimization and Decision Making, 3*, 229-241.

Vasant, P. (2004). Industrial production planning using interactive fuzzy linear programming. *International Journal of Computational Intelligence and Applications, 4*(1), 13-26.

Weld, D.S. (1992). Reasoning about model accuracy. *Artificial Intelligence, 56*(2-3), 255-300.

Chapter XII

Business Process Reengineering in the Automotive Area by Simulator-Based Design

Torbjörn Alm, Linköping University, Sweden

Jens Alfredson, Saab AB, Aerosystems, Sweden

Kjell Ohlsson, Linköping University, Sweden

Abstract

The automotive industry is facing economic and technical challenges. The economic situation calls for more efficient processes, not only production processes but also renewals in the development process. Accelerating design work and simultaneously securing safe process outcome leads to products in good correspondence with market demands and institutional goals on safe traffic environments. The technique challenge is going from almost pure mechanical constructions to mechatronic systems, where computer-based solutions may affect core vehicle functionality. Since subcontractors often develop this new technology, system integration is increasingly important for the car manufacturers. To meet these challenges we suggest the simulator-based design approach. This chapter focuses on human-in-the-

loop simulation, which ought to be used for design and integration of all car functionality affecting the driver. This approach has been proved successful by the aerospace industry, which in the late 1960s recognized a corresponding technology shift.

Introduction

For the automotive industry, the recent years have been characterized by huge economic losses among some major companies. This occurs from time to time and usually initializes efforts, which can be described as business process reengineering. We have seen much of this concerning the production parts of the companies and in the flow of components and sub-systems from subcontractors. Just-in-time deliveries and lean production are buzzwords we all have heard. But, has anyone noticed something similar from the R&D side? Of course, most people interested in cars have read about shared platforms and so forth over a number of models. But this is not business process reengineering, this is just technique rationalization. Since a long time ago the design process is computer supported in many ways, but what steps could the automotive industry take now in order to improve the design process? We believe that a more extensive use of virtual prototyping and simulation could be that answer. This statement is supported by the ongoing technology shift for all kinds of ground vehicles; from purely mechanical artifacts to more complex systems with computerized functions, more convenient to implement in a simulated environment than the old mechanical solutions ever were.

Figure 1. Concept car from GM (How GM's Hy-wire works, http://auto.howstuffworks. com/hy-wire.htm, January 10, 2006)

The main purpose of this chapter is to give an overview of the ongoing technology shift inside the vehicles (see *Figure 1* as a symbol of this shift) and to couple this to simulation possibilities and thereby introduce the business process simulator-based design (SBD). Our perspective is human-machine interaction (HMI) and therefore we address human-in-the-loop simulators, but we are quite aware on the fact that simulation could and even must be used on other levels in order to optimize and verify more technical functions. This is also a part of the SBD approach, but not specifically addressed in this chapter.

The authors of this chapter have more or less life-long experience from this way to proceed in R&D activities in aircraft design projects. Since the beginning of the 1970s, the SBD approach has been extensively used in the aerospace industry with the initial purpose to get safe design answers at early stages of development in order to avoid late changes at high cost levels. Later on, with more mature simulation tools and expertise, simulator evaluation was introduced also for final system certification, with only a minor part of functionality left for flight tests. The main reason behind this introduction was cost-effectiveness. This way to work has not yet become state-of-the-art in the automotive industry, so the purpose of this chapter is to contribute to this change by sharing our experience. During the last five years we have worked on this theme in our laboratory resources at Linköping University and today our simulator facility for ground vehicle system design has reached a level of effectiveness close to what we have used for aircraft design in the past. Thus, this chapter is more based on our own experience than on other sources.

Concluding this introduction, we would like to give some thoughts on the competence profile in the design departments of the car manufacturers. It is our impression that the technology shift in cars has a rather poor correspondence with the competence profile. Car manufacturing has its roots in mechanical engineering and this still affects the competence profile irrespective of the fact that the new technology represents around 50% of the built-in technique in today's cars. This has opened the market for subcontractors with more IT-based competence and put new demands on the car manufacturer—as system integrator. Simulation is a powerful tool also in system integration work, not the least from the HMI perspective, where there is a strong need for more holistic approaches than we have seen so far. The human driver is a very important part of the joint system and thus capabilities and limitations of the human must be key issues in the development of the future car systems.

Human Capabilities and Limitations

Humans have gradually, but slowly changed during the course of the evolution. Still, people today are in principle not better off than previous generations when it comes to car driving or similar operator work that requires perceptual acuity, fast reactions, simultaneous processing of large amounts of information and decision-making under uncertainty, boring -stimuli deprived surveillance conditions, and so forth. The demands on ordinary cars today are that they should be able to drive and manage safely with basic skills and a minimum of skilled training. However, the traffic environments have become more complex and the traffic volumes have increased tremendously during the last few decades in most countries. Despite compulsory basic training of driving skills and vehicle licensing in most countries, the accident rates are not decreasing as much as expected for a number of reasons. One important contributor to accidents is still poorly designed systems that allow drivers to override safety systems, act inappropriately, misperceive the situation, misunderstand fellow drivers' intentions, and make wrong decisions. Accordingly, much traffic safety gains are to be found at a system level.

The human being has a great potential to learn and adapt to the environment and perform excellently in different contexts. However, the human being also suffers from severe limitations concerning, for instance, vigilance (the ability to detect changed stimuli conditions), selective attention, working memory, modality separation [referring to the difficulties to separate information from different sensory modalities like the visual, auditory and haptic, see, for instance, Spence & Driver (2004)], and decision-making under stress. On top of these general limitations, many drivers have slow reaction times and limited motoric skills, which are to be regarded as major system constraints, since they have an impact on total system performance. Additionally, many drivers suffer from different kinds of impairments of permanent or temporary character that will influence driving behavior negatively.

All the above-mentioned advantages and disadvantages connected to human drivers have to be considered in vehicle design in general and in interface/interaction design in particular. In order to meet all these requirements an SBD approach will be of uttermost importance, since a vehicle simulator will enable researchers and car developers to maintain control of a large number of variables and make efficient manipulations of crucial variables under severe or even lethal driving conditions. Accordingly, the vehicle simulator is used to study causal relationships behind driver behavior during specified circumstances, which seldom is possible in a real life environment for a number of reasons. For instance, it is not possible to expose drivers to real accidents from an ethical point of view. Studies of peoples' reactions to different cockpit layouts, decision support systems, warning systems, and communication devices and their impact on driving behavior are preferably conducted in a simulator environment. Studies of the effects of drugs, medicine, alcohol, and so forth are not recommended in real traffic environments, but could safely be conducted in a simulator environment. A simulator constitutes a forgiving environment with possibilities to log all kinds of data (concerning environment, vehicle, traffic situation, individual status, etc.), which would be too expensive, too cumbersome, and too invasive or obtrusive to collect in a real traffic environment. The simulator also allows control of the environment and crucial parts of the general context. Hence, the simulator will take care of simulated features of the vehicle, the general context, the traffic environment, the driver (and sometimes even passengers), other drivers and pedestrians.

One area where SBD is almost necessary to deploy is for development of adaptive multimodal interaction in vehicles (c.f., Belotti, De Gloria, Montanari, Dosio, & Morreale, 2005), since knowledge about human capabilities and meta-cognitive skills with bearing on driving behavior is grossly missing within the designer community. In a simulated environment with partly simulated driving functions, it is completely safe to test alternative ways of mitigating bad traffic behavior by utilizing spare capacity in other modalities. Driving is primarily based on vision (visual-motoric loops), but the auditory modality has been underused until a few years ago, when suddenly a number of systems utilized the auditory modality as a major feedback channel to the driver. Haptic feedback is still rare in cars, although it has always been present since the first car emerged (the natural haptic feedback due to the movement of the car in relation to the road). However, new ways of handling information and detecting changes of different kinds by means of innovative tactile or kinesthetic feedback may be efficiently explored in a simulator environment.

A large number of performance measures can be used together with subjective ratings of a large number of parameters and a vast number of logged vehicle and system parameters in order to evaluate efficiency, effectiveness and "feeling" of an HMI related design solution, where safety and comfort are key issues. Important questions relate to the ranges of peoples' capabilities and capacity limitations in different respects. Without this knowledge, it is easy to design systems that either work well during normal operations, but crack down during extraordinary driving or information conditions, or do not work at all because system performance goes far beyond the potential human performance envelope. Accordingly, there are a number of new systems implemented in modern cars that have never been subjected to appropriate HMI testing, where optimal mode, modality and media selection is lacking. Hence, we have systems installed in modern cars that are jeopardizing traffic safety by providing non-optimal design solutions and sometimes even devastating cluttering and information disaster. How to display messages, warnings and automatic control functions in operation is not a question of what is possible with modern information technology. It is still a question about what is optimal from a total system perspective, including the temporary capability of the driver with a certain driving task at hand in a particular context, in the actual traffic situation, with a special vehicle with its known and unknown features.

How human capabilities and limitations are defined, operationalized and measured is also of crucial importance. An ideal situation, where the driver is in every respect in full control, is never at hand in modern cars and traffic environments, which has become too complicated for the average driver to understand completely. Thus, a large number of functions have to be automated in modern cars. Which functions should be automated and to which degree are pertinent questions that are ample objects for SBD.

Vehicle Technology Trends

For almost a century, automotive technology had its roots in mechanical engineering with minor supplements of electrical engineering. Today the car has become a mechatronic system where computer and communication technology dominate the system design. More than 50% of the total system is computerized or computer controlled today. In the good old days a car function was possible to understand by just looking at a blueprint. Today, however, these possibilities are radically reduced, since most new functionality is expressed in software code. This apparently deteriorated situation could be compensated for by the use of simulation, since a software-based function is easier to implement in a simulation at all levels (e.g., mathematic oriented desktop simulation or full-scale human-in-the-loop simulation) compared to a mechanical construction. This possibility is also the basic foundation for the SBD approach outlined in this chapter. Therefore, it might be valuable to go inside this new area of in-vehicle systems (IVS).

In-Vehicle Systems

It is not possible and not even useful in this context to completely walk through a whole car in order to describe various systems and the way they all are developing over time. Hence, the ambition here is to give an overview of system classes with focus on what is close to the driver, that is, systems where human-machine interaction (HMI) has relevance.

Such HMI related systems could be classified in the following four groups:

- Primary control systems (PCS: primarily steering, throttle control, brakes and related automation)
- Advanced driving assistance systems (ADAS: for example, visual enhancement systems, obstacle warning, other alarm systems and related automation)
- In-vehicle information systems (IVIS: including communication functions and other information related functions that require driver-vehicle interaction not directly related to the driving control loop)
- Non-integrated systems (any system that the driver might bring into the vehicle, for example, Palms, separate cell phones, GPS), also called nomad systems

Systems in the latter group are not further discussed, since they by definition are not vehicle systems. However, the use of these products has an obvious impact on traffic safety and therefore it is most beneficial that this kind of products is moving into the IVIS group. Hopefully this tendency may result in safer solutions. There is a vast number of new communication and entertainment systems that will soon be integrated in cars or in the driving environment.

The increased number of new ADAS and IVIS together with more intelligent primary control systems in modern cars calls for holistic design approaches from many perspectives. One aspect is about system architecture and another is about HMI.

Most ADAS and IVIS are implemented in the vehicles as "isolated" systems. This means that they have their own sensors, separate software-based functionality, and separate devices for driver interaction. In the holistic approach we suggest, the strategy will be to share resources (e.g., sensors, displays, and computers) as much as possible and to coordinate the driver interface over all systems (Alm, Ohlsson, & Kovordanyi, 2005). This goal is possible to achieve within the SBD concept, while the currently dominating design processes are less supportive.

Drive-by-wire system is an expression, which the automotive industry has adapted from the aviation field, where fly-by-wire systems have existed for more than a decade. In both areas, the expression refers to primary control systems, which are converted from mechanical solutions to computer-based functions. The wire part in the expression refers to the shift from mechanical couplings between the fly stick and the rudders to electric wires, which mediate a signal from the stick to some actuator close to the rudders. Somewhere along this wire, there is a computer involved in the process, understanding the goal of the pilot's input and translating this information to an optimal rudder response.

The reason behind this technology shift in aviation was to reduce aircraft weight and open up for more sophisticated (intelligent) principles for flight control. Similar demands will also be the driving force in the vehicular area. We know from aviation that this paradigm shift will put a strong pressure on the HMI aspects. Since there is no natural (physical) feedback in such a system, this has to be simulated in order to give the pilot information on how the system understands the pilot input and operates. Moreover, this feedback has to be designed so that the tactile signals correspond with the pilot's mental model of the ongoing activity and its goals instead of the actual positions or movements of the rudder planes (as in the mechanical systems). The parallel for cars is the combination of steer-by-wire and four-wheel operation.

In our view on HMI related IVS design we have defined three principal levels. The first is the measurement level, where data from various sources is collected, for example, driver data, vehicle systems data, and environmental data. The second level includes the intelligent part for data analysis and automated "decision–making." On the third level we have the presentation and control resources. In *Figure 2* we present an IVS sketch based on this concept and complemented with a "box" for actuators, which are mandatory in all drive-by-wire and automated systems. In a holistic design approach, sharing principles should minimize the number of technical resources. This means that more than one system could use data from the same sensor or use the same display for presentation.

Figure 2. A principle IVS design approach for HMI related systems. The arrows represent various in-vehicle systems and their basic data-flow. Systems symbolized by continuous lines do not use any second level resources, while the other systems depend deeply on these intelligent functions (Alm, Ohlsson, & Kovordanyi, 2005).

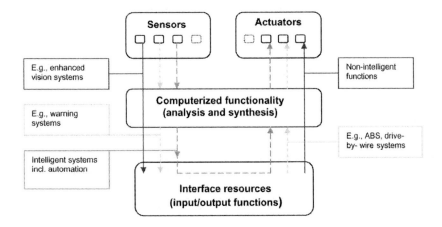

Design Trends for Primary Control Systems

In the PCS area most systems are still dominated by driver controlled mechanical solutions; that is, the driver has the complete control of all actions with no interference of computers and automation. The first system where this concept changed was braking when ABS (Automatic Braking Systems) was introduced. As a result of this change the driver was no longer in charge of all actions. Hard braking input by the driver results in an automated braking procedure, where the system itself controls the action in order to have an optimal outcome. The system also gives the driver tactile feedback through the brake pedal.

Traction control was the next area for automation, in this case with no influence of the driver. In most traction control systems a warning lamp indicates that the system is in action and thereby also informs the driver on the impaired road conditions.

Steering was mentioned earlier and we believe that the change to steer-by-wire will appear in the future. Today we have seen experimental cars with this new technology, but with limited efforts on the HMI side. In order to further develop the interaction with the driver, there is a strong need for a simulator-based approach, which has been the case in the aircraft industry since many years for all kinds of fly-by-wire systems and all other HMI related functions.

ADAS and IVIS Trends

These areas are expanding rapidly. Some aspects on these systems are already mentioned. Another view on this development is that subcontractors to the car industry produce most systems. This fact obstructs holistic efforts both technically with respect to the principle of shared resources and from the HMI perspective. One system in isolation may, for instance, show great positive impact on safety, while in combination with other systems the outcome may be strongly reduced or even negative. Drivers might also try to override distracting and annoying systems [see, for instance, Goodrich & Boer (2000)]. Here the car manufacturers have a great responsibility as system integrators and the need for tools to fill up this role is obvious. Again, a simulator is a perfect test bench for such efforts.

In addition to systems entirely located in the vehicle, there will be a growth of infrastructural systems, which will deliver information to the cars in an integrated way. Today we can see examples of non-integrated traffic information, for example, a congestion warning, coming through the car radio. Tomorrow this information may be integrated in the car navigation system in a similar way as it uses GPS (global positioning system) signals today. In this example, the internal navigation system will calculate the best way to get around the obstacles. Other systems appearing in the future could be coupled to speed restrictions and intelligent traffic signs.

Again, the purpose of each system may be lawful, but it certainly will make the driver situation more complicated due to the range of systems and the total information content in the cockpit environment. We argue the way to overcome some of these problems is to take a holistic approach to the design issues using the SBD methodology.

Glass Cockpits

To conclude this part on vehicle technology trends there also is a need to say something about the development of the driver's workplace from a hardware perspective. In the traditional car cockpit each input function has its own device, for example, steering wheel, pedals and buttons and each output function, for example, speedometer, rev counter, thermometer, has its iron made needle and scale. For different warning messages lamps in various colors and shapes are used. Since a number of years, information on LCD displays is used for more detailed and supplementing information. Two examples are radio panel and temperature display.

Again, it could be useful to look upon the corresponding development in the aerospace area. In the end of the 1960s, the hardware solutions were quite similar to the above description. For many years, however, information on electronic screens has replaced the so-called iron instruments in military and commercial aircraft. Many input functions are located on touch screens. This development opens for more flexible and mode-dependant interface and interaction design. This is the background of the glass cockpit notion.

A similar development could be expected also for ground vehicles and we already see some steps in this direction. Such a development trend will also open for realizing the shared resource principle suggested above.

Simulator-Based Design and Evaluation

The obvious first step in a simulator-based design process concerning HMI related issues is to define the system itself and the features of the specific system – PCS, ADAS or IVIS. Departing from this functional description it should be possible to use the structure shown in *Figure 2* for defining the main parts of the system, which normally consists of some sensor (or other source for information), functions for analysis and synthesis and driver interface. In some cases, mostly for PCS, actuators must also be added. It may be emphasized that the purpose is to make this definition from the SBD perspective and not entirely with respect to the final commercial product. This specific design philosophy will be addressed as a background for the more detailed definition guideline.

SBD Philosophy

The basic philosophy in SBD is to make as many shortcuts as possible in order to evaluate or just demonstrate a solution or to look at different concepts for the system design. This way of working is crucial for extending the number of alternative solutions, that is, open up for an iterative approach (*Figure 3*) within a limited period of time. The more iterations you can have the safer solution. An important precondition for shortcuts is that the simulator and the complete simulation can give all necessary data from the vehicle including driver inputs and the surrounding scenario with its participants (other vehicles, pedestrians, etc.). In other words, all input data to a system is already known in the simulator and thus makes

Figure 3. The main steps in the SBD process. The dotted arrows indicate examples of iterations, which is a crucial part of the SBD approach.

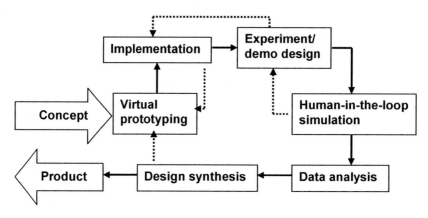

it unnecessary to include real sensors in the simulation. In many cases this is even impossible. For instance, an infrared (IR) sensor (camera) does not work, since the environmental presentation in the simulator facility gives no temperature differences between different objects as it does in the real world. A video camera for rear information can work, but is not necessary, since also rear objects are known in the simulation.

Another part of the basic philosophy is to have a pragmatic view on reality. It is not necessary to build a copy of a specific environment if the purpose is to design a vehicle system. The only ambition should be that the environment model supports the purpose of the simulation. If you, for example, want to design a system for traffic detection in urban street crossings, it is necessary to have buildings at certain distances from the runway. If all houses look the same it makes no difference in the test, but will shorten the environment development time radically, and this is the core ambition in the SBD concept. Another departure from reality is the number of provoking events. In real life many such events luckily appear very seldom in a driver's life, but in the simulation these events could appear frequently, for example, a moose crossing in dark conditions or sudden icy conditions on the road. This possibility to raise the event frequency gives the designer much more data for system analysis than from a corresponding study in real driving. Again, this is an advantage for simulation and thus the SBD concept.

System Definitions

Sensors. As already mentioned, there is little or no need for using real sensors in the simulator. Depending on what kind of sensor one wants to include in the simulation, the solutions for achieving its data may differ. For the vehicle, on one hand, data connected to the primary control system (e.g., steering wheel angles and velocity, accelerator and brake pedal control) are available, since they are key parameters in the simulator. On the other hand, oil pressure, engine temperature, and similar functions are not normally included in the data arsenal. In the case of having such information as input to a prototype system other solutions might be considered. If, for instance, the data is needed in a warning system, it could be triggered by a virtual actuator in the simulation scenario and then sent as a signal from the simulation to the warning system application.

All kind of sensors for information acquisition concerning external objects can be replaced by the existing object information in the simulated scenario. However, this must be done according to the capacity of the intended real sensor. If, for instance, the real sensor is a video camera located at some place on the vehicle exterior, the operational envelope for this camera and its position must be used as borderlines in the simulation. Objects outside these borders should not result in inputs to the prototype system, while objects inside the envelope should.

Infrastructure-based sensors or other external information sources should be realized in the simulation in an equal way as other events in the driving scenario. These events are usually initialized by so-called triggers (trigger points) along the road. This means that when the vehicle passes a certain point an event is initialized. How this is realized in detail depends on the type of simulator system.

Computerized functionality. This, naturally, is the core part of the whole system prototype. Most certainly, this set of algorithms and coded functions will be very close to the corresponding solution in the final commercial product. However, depending on in which stage of system development the prototype is used (from an early concept to close to realization) the sophistication of the functionality could vary. Simple rules instead of more intelligent functions could be good enough in the concept phase in order to have some experience of the new complete system including its interface functions. Functions for data handling and analysis could include classification, data fusion, filtering, rule structures, human performance models, interface agents, and so forth, while the synthesis part is more application dependant. If a certain value is fulfilled then something will happen in the interface or will affect some actuator. Except for the degree of sophistication mentioned above, there are no obvious shortcuts available in this part of the system prototype.

Driver Interface. The driver interface corresponding to the chosen system is (as for the intelligent functions behind) the part of the system we really want to manipulate in order to finally have an optimal HMI solution. It is also obvious that these solutions are very application dependant, but the definition work could follow the scheme below:

In the first step one has to decide on which modality for interaction is most effective to use for the system in question. When interaction is involved you have to think of both input and output functions and it is not necessary that you will have the same modality choice in both directions. Visual output and manual input is, for example, an often-occurring combination.

In the second step, there is a media choice that is what artifact (displays, buttons, etc.) in the cockpit environment you could use for the intended action. Sometimes the solution for this question is to add some new facility to the cockpit. But, following the principle of shared resources this should be done very carefully. The last step is the representation design. How should this visual alarm signal look like in detail? Or, how could this speech-based driver input be realized, in a natural language way or by commands? In a glass cockpit environment, you will have all opportunities to vary this representation solution until the final best design is reached.

Scenario Production

Scenario production is a key element in the SBD concept. The purpose of the scenario is to challenge the prototyped system. The first step is to construct the environment model, the virtual world. In the long run, there will be a number of different worlds available in the simulator facility, which could be reused. The ideal situation is to have a set of environments with different features; city models with different street sizes, crossings with and without traffic lights together with appropriate buildings, gas stations, and so forth, and also different rural roads and highways with typical properties like exits, rest areas, and so forth. If possible and depending on the kind of simulator brand, these models could be combined with different conditions like darkness, daylight, rain, and so forth. Otherwise, there is a need for a number of condition related versions of each original model. To facilitate the environment production there are a number of commercial software tools at the market. The freedom of choice could, however, be restricted depending on the simulator brand, and in some cases the only possibility is to ask the simulator company to deliver the additional virtual world you want.

The next step in the scenario production is to put life into the virtual world. This is not only restricted to cars and people, but also how they perform and when they do it related to where in the world the own car is at a certain moment. These interventions of other players in the scenario could be more or less intelligent. In high-end solutions the other players interact with the own car in quite a normal way. It is even possible to have manned cars, which means other simulator nodes, included in distributed simulations. Again, the possibilities to produce these interactive scenarios are strictly coupled to each simulator brand.

Tools

Hardware. A simulator cockpit meant to fully support the SBD methodology must follow the above mentioned glass cockpit concept irrespective of the corresponding solutions in the real car. This will open for investigations of alternative solutions in the car system design with no need for hardware reconstruction. These solutions will be virtual prototypes completely based on software. A good strategy is to start with an ordinary car without engine, gearbox, wheels, and so forth, and maybe cut the cabin behind the front seats. The main parts of the reconstruction work are to implement a programmable electrical force feedback motor for steering, replace the iron instruments with electronic displays and connect steering wheel, pedals and other control devices to the simulator software. It is also beneficial to install equipment for voice/audio interaction, and maybe also devices for tactile information or feedback. A good principle is to have the simulator cockpit as generic as possible in order to avoid project-related workshop reconstruction.

Another important issue is to have a platform for the simulator consisting of standard computers in a cluster solution. This will make the simulator scalable and thus able to meet increasing demands for new functions, both simulator and vehicle related functions. If the simulator user wants to add a supplementary view to the environment presentation (e.g., rear presentation), this will be done by adding one additional projector, one screen, and a computer to handle this new part of the simulation.

A last issue in the simulator set-up is whether or not to invest in a movement platform. This decision has a major impact on the economic side. According to our opinion, in most cases this is overkill for a simulator aimed for design issues. As mentioned above, the view on reality must be pragmatic. This means that this question has to be thoroughly analyzed in relation to the purposes of the future simulations. This fidelity question will be further addressed later in this chapter.

Software. We strongly recommend the use of commercial software tools wherever this is possible. Such an approach will make it convenient for engineers outside the software field to build environmental models, to set up simulation scenarios, and implement new functions in the simulated car. This means that most of the work could be carried out without software consultants or support from the simulator supplier. However, the real situation in the specific simulator facility will depend on the chosen simulator brand. The fact is that there are too few simulator suppliers who really support this way of work. One obvious reason for this is that the SBD approach is far from established, which makes it troublesome for simulator suppliers to invest in this niche. However, some signs on a development in this direction have recently appeared (e.g., Dangelmaier, Marberger, Wenzel, & Widlroither, 2004; Strobl & Huesmann, 2004) with a corresponding view on the main requirements of such simulators:

- Open and dynamic architecture
- In principle real-time events
- Integrated and synchronized data logging
- Software based modular functionality
- Easy access to scenario production

In general, when looking for development tools, there are tools used in other business areas that could be used also in the suggested SBD work. For instance, in the field of graphical user interfaces—more specifically in the interactive Web page design -- there are tools available that could be reused in automotive applications. Another, perhaps more close possibility, is coupled to CAD and the 3-D extensions for modeling. Also in the market for computer games production there are existing tools, which could be transferred to automotive SBD.

Situation Management and Validity Issues

It is important to provide a context of valid situations all the way through the development process in order for the design process to be effective through early simulations to the final design. Understanding how a situation is managed in valid situations helps guide the design process into a rational path. Through a simulator-enhanced, iterative design process the important features of the design could be detected and optimized.

With an iterative development approach, it is possible to find a way forward, as the understanding of the use context and driver behavior is increasing. Gained knowledge in early development cycles can be used to form an understanding that helps direct the efforts in later development cycles. To do this in a structured approach a description of the process in terms of its components, including the human aspects, is needed. There is an increase in complexity of cars of today. Methods and techniques for designing and developing are increasingly important to produce efficient and usable solutions. Recently attempts have been made to address this matter through a variety of approaches, such as user-centered design, usability engineering, ecological interface design and many others. Depending on how a situation is represented, different means for interpretations and solutions emerge. Severe consequences in safety-critical human activities often arise from problematic interaction between humans and technology. The ongoing development within the car industry towards integration of complex information systems into the car is a typical example of a process dealing with complex socio-technical problems, where misunderstandings and mishaps can lead to severe consequences. Computer supported systems must, in general, provide good representations of reality and the inherent tasks to offer the efficiency improvement, both regarding performance and life cycle cost that they are intended for. For the driver of a car it is also important that the information system is not distracting, so that the main task, driving the car safely, is satisfactory performed. New possibilities created by technological achievements give insights into the need of developing design methods suitable for coping with the grand demands of modern contexts. It is impossible to foresee all consequences of a suggested design of a future system. This calls for an iterative development process. Only if ideas and implementations are tested in an iterative process, it is possible to avoid the pitfalls of bad human-system interaction. What is needed are state-of-the-art methods, simulations and scenarios explicitly suited for iterative development of human-system interaction.

It is important to have a strategy for choosing evaluation methodology, since it will be difficult to use the results from an unstructured approach in an iterative evaluation process. The purpose of the evaluation has to be clear. It can be to demonstrate new features and concepts, or to critically test some hypotheses to see if they could be proven false. To make good demonstrations, all parts of the technical implementation should run smoothly so that all the aspects that are intended to be demonstrated are shown. However, if the purpose of the evaluation is to critically test some hypothesis, the technical implementation might be pushed to, or beyond, its limitations so that we know what will go wrong, when it does. Of course, other purposes may exist, but the point is that the purpose should be explicitly determined before the choice of evaluation methodology.

Not all aspects of a real world situation are possible to mimic in a simulation environment. However, all aspects of a real world situation are not equally important for the purpose of the simulation. Therefore, a selective fidelity approach to simulators used for car development is motivated through a cost-benefit perspective. Some aspects are important to include some are not. Therefore, it is important to analyze the task and the situation relevant for the simulation. The fidelity level has to be assessed according to the objectives of the study.

When this is done you have to choose what to implement in your simulation. An often-underestimated approach is the use of horizontal prototypes as a complement to vertical prototypes. Horizontal prototypes include wide functionality rather than deep. We have to have horizontal prototypes to be able to test human aspects of driving in early phases of an iterative design process. If we wait too long before we test those aspects we may find that it is too late to change our initial postulates at a reasonable cost. The benefit with a horizontal prototype is that we have access to a test environment that is "broad," so that holistic approaches could be tested. What we intentionally reject are the details of a vertical prototype, with all of its functionality. We have to make a choice between the two, since it is too costly to have high fidelity in every sense. Better is to use selective fidelity and to model aspects actually relevant for the purpose of the evaluation. Here, emulated future systems and Wizard of Oz simulation (see, Dahlbäck, Jönsson, & Ahrenberg, 1993; Frazer & Gilbert, 1991), with mimicked functionality is well-suited.

Validity is of course very important. Assuring that the simulation is based on a theoretical foundation will make it possible to show construct validity (Mischel, 1981). To have concurrent validity the simulation has to be related to other simulations. By relating the simulation to empirical experiences, it is possible to show predictive validity. To have face validity, it should be agreed that the simulation addresses what is intended as this is generally understood and/or as it is defined especially for the regarded purposes. A methodological approach to overcome certain fundamental validity problems is to conduct comparative experimental studies. Central to this approach is, according to Alm, Ohlsson, and Kovordanyi (2005), "that *different system designs are evaluated under exactly the same conditions and the simulator validity is the same for all design alternatives and thereby become counter-balanced.*" On one hand, the desire to have plausible scenarios, real primary driving functions, colloquial driving tasks and realistic secondary tasks reflect the ambition of ecological validity. On the other hand, it is demonstrated elsewhere that drivers have a poor recognition memory for ordinary driving situations, which could be interpreted that validity is not as important as claimed, for instance, by car developers (e.g., Chapman & Groeger, 2004). However, reliability in data collection is also important, both internal reliability and reliability over time, in order to be able to draw scientific conclusions.

Another important aspect to regard is that the situation that we are designing for is not a situation of today, but a future situation, with a future car system. Assessment of future systems is, of course, more difficult than assessment of existing systems. An iterative evaluation process including future systems therefore has to regard specific problems related to the assessment of such systems. When assessing a prototype of a future system we will encounter the envisioned world problem, where we as developers have to envision a world that does not yet exist, and try to figure out what is important there (Dekker, 1996; Dekker & Woods, 1999). Often developers tend to focus on the technological advances and do not put as much effort into understanding the new conditions that humans in the envisioned world will experience. When some technology is changing, people's understanding of the domain and their tasks change as well.

A problem with assessing future systems is that the prototypes are not as complete as an existing system. If they are, the cost of refusing the ideas implemented in the prototypes could be unjustified. Comparing a prototype with an existing system, or comparing several prototypes with each other therefore demands effort in thinking through which hypotheses are possible and relevant to test. A related problem is that users, who often have to be included in an evaluation to make it valid, are biased in their behavior and preferences by the fact that they have experience of existing systems, but not of future systems.

The best we can do is to describe the future situation. Only if we know the characteristics of situations that the car is going to be used in, we know to develop presentation and maneuvering support for the driver. When we have realized that this is situation dependant, the next step is to enhance the driver's situation management through the technical design. One of the major reasons for using the concept of situation management is pragmatic: applying the concept to the design might help in creating good design. Since situation management is used to evaluate human-machine interfaces, the concept should enhance the design process, and increase the benefits of the result of the improved design process. The concept of situation management can be used for at least three purposes in the design process:

1. Enhancing the design process by providing rational sub-goals created from an understanding of situation management.
2. Enhancing the actual design by helping the designer to understand the design criteria, and their relative importance.
3. Ensuring that the design is satisfying by testing the design in relation to situation management.

Situation management includes both presentation and maneuvering support, and is therefore a wider and more applicable term to interface design than situation awareness (SA), which is focused on the detection, understanding and prediction of surrounding elements. However, design recommendations, based on Endsley's three-level model of SA, have been presented to help designers to make a better design (Endsley, 1995). It could very well enhance both the design process as well as the design itself. However, by broaden the use of the SA concept within the field of design it might be possible to find even better applications to enhance the design and the design process, for example, by making the design more sensitive to the characteristics of a specific context.

Two Project Examples

To conclude this chapter we briefly present two examples of simulator use for design issues. The first example is about a car application with a conceptual question on system design, where two major approaches for night vision systems were tested against each other, while the second project addressed a more specific display design question. This study was an aircraft project and the results were used later on guiding the real design in a fighter aircraft.

Night Vision Systems

Night vision systems are already introduced on the market. Most products could be considered as vision enhancement systems, which mean that all sensor information will be passed to the driver and presented continuously. There has been some criticism on night vision systems and other ADAS as well based on the risk for negative behavioral adaptation (e.g., Kovordanyi, Ohlsson, & Alm, 2005; Smiley, 2000). Night vision systems were introduced in order to increase the sight distance and visibility in dark conditions and thereby give a contribution to safety. One obvious effect of negative behavioral adaptation in this context is that the driver would increase speed.

The purpose was to study this problem and some other questions using two different concepts. One was the approach with continuous presentation on the windshield and the other was a warning system based on the same kind of sensor presentation. This was activated only if an upcoming object was measured and evaluated as a dangerous object. This system would be designed as the warning system example shown in *Figure 2*. In the simulator, however, the two alternative systems were realized in quite a different way. The display presentation was simply a black and white copy of a central view of the environmental presentation, over-laid this presentation through a separate projector. The analysis procedure (level 2 in *Figure 2*) was replaced by predetermined triggers in the scenario setting at realistic distances (= real sensor range) from dangerous objects.

This way to take shortcuts in order two get answers to research/design questions is normal procedure in our approach to simulator-based research/design. In this particular case we wanted to know if negative behavioral adaptation could be avoided or at least minimized by using a warning system concept for night vision instead of the traditional continuous IR-display implemented, for instance, in Cadillac and BMW high-end models. The result of this research question was that average driving speed was 5 km/h higher for the original system. However, six of the 23 subjects paid no attention to the prescribed speed limit (90 km/h) in any experimental condition. This group drove at an average speed that was 12 km/h higher with the continuous system than with the re-designed system.

Figure 4. Participants' subjectively experienced mental workload measured on the NASA-TLX scale. White bars show average workload for the original system. Light grey bars show average workload for the re-designed system. Vertical error bars indicate standard deviation.

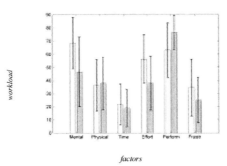

factors

The most evident result in this experiment was that all subjects (100% of 23 subjects) preferred the new system, that is, the situation-dependent display. The main reason for this preference, as reported in interviews, was that the subjects found it too demanding to observe the presentation more or less continuously in the original display. This result is partly supported by the NASA-TLX data (Hart & Staveland, 1988, *Figure 4*) where mental demands and effort showed the same trends, being higher for the original system. Eye-tracking data indicated that when using the continuous display, more than 50% of the driving time was spent fixating the NV display.

Measurement of drivers' reaction to obstacles was based on distance to the obstacle at the time for subjects signaling target detection. We have data only from 13 subjects in both conditions due to loss of data for the first 10 subjects. However, the balance between groups was still intact, since the order of trials was a, b; b, a; a, b; and so forth.

As can be seen in the diagram above (*Figure 5*) the measures showed an almost haphazard pattern between no detection (collision) and detection at a distance of 450 m. It is also evident that target size had little impact on the detection results. The results are probably mainly dependent on the subjects' way of sharing their time between the display and the road environment.

Figure 5. Drivers' reaction to obstacles, moose and pedestrians, when using the original system with continuous presentation

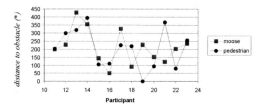

Figure 6. Drivers' reaction to the obstacles, when using the re-designed system with unlit display during uneventful periods

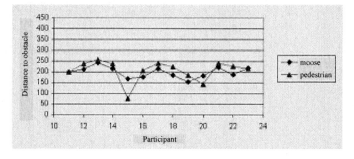

Figure 7. The distribution of distances to obstacles between the two experimental conditions

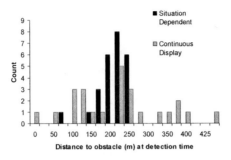

When using the situation-dependent display the detection pattern was more consistent (*Figure 6*), which was anticipated, since the display appearance had a warning effect. However, there were no significant differences in average measured distance to targets between the groups so the most evident difference was the inconsistent reaction performance when using the original system compared with the new system (*Figure 7*).

An F-test for the equality of sample variance indicated that the spans in the two data sets are significantly different, $F(24, 24) = 8.83$, $p < .001$, with the range in the distance to targets at the time for detection being significantly greater in the continuous display condition.

The experiment was carried out by a group of second year master students at Linköping University (Alin Nilsson, Alm, Arabloui, Gemoll, Heyden, Törngren, & Wetterström, 2005) and is also reported in two conference papers (Kovordanyi, Alm, & Ohlsson, 2006; Alm, Kovordanyi, & Ohlsson, 2006).

These results are not further discussed, since the purpose of this result presentation was to exemplify typical measurements used in the SBD approach. As can be noticed we used a whole battery of methods in order to capture interesting data; performance measures, a workload assessment method, eye-tracking data, and also more open interviews.

Another issue we like to emphasize is that we put two separate design solutions into competition. Both system alternatives were evaluated under exactly the same conditions, which make it possible to conclude that the re-designed system showed better performance and was more appreciated by the participants. We do not pretend to say anything on how important such a system is for traffic safety in absolute terms. Such questions will fall outside the scope of simulator-based design.

Color-Coding as Means for Target Discrimination

Some years ago Saab Aerosystems and The Swedish Defense Research Agency conducted research on the effects of, and the design of, color displays for fighter aircraft. In a study on visual search and situation awareness the Swedish multi-role combat aircraft Gripen's monochrome color scheme was compared to two chromatic (dichrome, polychrome) color schemes (Derefeldt, Skinnars, Alfredson, Eriksson, & Andersson, 1999).

A real-time simulation of an air-to-air mission with seven male fighter pilots was carried out in Saab's fixed-base simulator for display and control design (PMSIM), which is an advanced generic flight simulator for a number of tasks. The pilot had to perform dual tasks: 1) to track a maneuvering aircraft within specified limits by using the head-up display, and 2) to detect the appearance of a target on the head-down display. Reaction times were measured, and the pilots made rankings of situation awareness and performance under different phases of the mission. The results showed that color displays were more beneficial than monochrome displays, for example, the ranks on situation and performance were higher, and with complex background on the tactical map, the reaction time for the alternative with polychrome color was significantly shorter.

Since the color schemes were specified in a standardized color appearance system, and because of the adequate fidelity in the simulator, according to current requirements, the design of the applied color schemes in this study was possible to use as a ground for further design efforts towards the use of color in the display in today's version of the fighter aircraft. To let the design efforts made in an early study phase be transformed into the design of a final product is a cost-effective way of using the benefits of simulators.

Conclusion

The purpose of this chapter was, coupled to the ongoing technology shift in cars and trucks, to propose a renewed design process where simulator-based design is the core activity. SBD is possible to perform on each stage of a design project, from the early conceptual phases to the final evaluation. The main activities in SBD are virtual prototyping and simulation and for all HMI related systems, this should be carried out with the human driver in the loop.

The outcome of SBD is twofold. It will accelerate the design work or alternatively give opportunities to an extended number of iterations within a given time frame and it will bring safer product solutions to the market.

Using fixed-base simulators and standard computer platforms could minimize the economic investment in simulator resources. This is possible since the validity problems must be related to the specific purpose of each simulation and not to a general level where the validity is coupled to the experienced reality in the simulator. That kind of reality is only possible to achieve with a moving base and that will bring the simulator investment to much higher levels. It should be emphasized that this fixed-base strategy also is prevalent in the aerospace industry as shown in the case study from Saab Aerosystems above. In fact, the whole SBD success story has its roots in the aerospace industry, which ought to be a strong incentive also for the automotive industry.

References

Alin Nilsson, I., Alm, H., Arabloui, S., Gemoll, H., Heyden, A., Törngren, K-J., & Wetterström, R. (2005). *To see in darkness: A comparison between two night vision systems.* (In Swedish.) Course report HKGBB5, Linköpings Universitet.

Alm, T., Kovordanyi, R., & Ohlsson, K. (2006). Continuous or situation-dependent night vision presentation in automotive applications. In *Proceedings of the Human Factors and Ergonomics Society 50th Annual Meeting.*

Alm, T., Ohlsson, K., & Kovordanyi, R. (2005). Glass cockpit simulators: Tools for IT-based car systems design and evaluation. In *Proceedings of Driving Simulator Conference - North America 2005,* Orlando, FL.

Belotti, F., De Gloria, A., Montanari, R., Dosio, N., & Morreale, D. (2005). COMMUNICAR: Designing a multimedia, context aware human-machine interface for cars. *Cognition, Technology and Work, 7,* 36-45.

Chapman, P., & Groeger, J. (2004). Risk and the recognition of driving situations. *Applied Cognitive Psychology, 18,* 1231-1249.

Dahlbäck, N., Jönsson, A., & Ahrenberg, L. (1993). Wizard of Oz studies: Why and how. In *Proceedings of Workshop of Intelligent User Interfaces,* Orlando, FL.

Dangelmaier, M., Marberger, C., Wenzel, G., & Widlroither, H. (2004). A platform for simulation and evaluation of driver assistance and information systems. In *Proceedings of Driving Simulator Conference – Europe 2004,* Paris.

Dekker, S.W.A. (1996). Cognitive complexity in management by exception: Deriving early human factors requirements for an envisioned air traffic management world. In D. Harris (Ed.), *Engineering Psychology and Cognitive Ergonomics, Volume I: Transportation Systems* (pp. 201-210). Aldershot, UK: Ashgate Publishing.

Dekker, S.W.A., & Woods, D.D. (1999). Extracting data from the future: Assessment and certification of envisioned systems. In S. Dekker & E. Hollnagel (Eds.), *Coping with Computers in the Cockpit* (pp. 131-143). Aldershot, UK: Ashgate Publishing.

Derefeldt, G., Skinnars, Ö., Alfredson, J., Eriksson, L., Andersson, P., Westlund, J., Berggrund, U., Holmberg, J., & Santesson, R. (1999). Improvement of tactical situation awareness with colour-coded horizontal-situation displays in combat aircraft. *Displays 20,* 171-184.

Endsley, M.R. (1995). Toward a theory of situation awareness in dynamic systems. *Human Factors, 37*(1), 32-64.

Frazer, N., & Gilbert, N.S. (1991). Simulating speech systems. *Computer Speech and Language, 5,* 81-99.

Goodrich, M., & Boer, E. (2000). Designing human-centred automation: Trade offs in collision avoidance system design. *IEEE Transactions on Intelligent Transportation Systems, 1*(1), 40-54.

Hart, S.G., & Staveland, L.E. (1988). Development of NASA-TLX (Task Load Index): Results of empirical and theoretical research. In P.A. Hancock & N. Meshkati (Eds.), *Human Mental Workload.* Amsterdam: Elsevier Science Publishers.

Kovordanyi, R., Ohlsson, K., & Alm, T. (2005). Dynamically developed support as a potential solution to negative behavioral adaptation. In *Proceedings of IEEE Intelligent Vehicle Symposium,* Las Vegas, NV (pp. 613-617).

Kovordanyi, R., Alm, T., & Ohlsson, K. (2006). Night vision display unlit during uneventful periods may improve traffic safety. In *Proceedings of IEEE Intelligent Vehicle Symposium* (pp. 282-287). Institute of Industrial Science, Tokyo University.

Mischel, W. (1981). *Introduction to personality* (3rd ed.). New York: Holt, Rinehart, and Winston.

Smiley, A. (2000). Behavioural adaptation in safety and intelligent transportation systems. *Transportation Research Record 724,* 47-51.

Spence, C., & Driver, J. (Eds.). (2004). *Crossmodal space and crossmodal attention.* Oxford University Press.

Strobl, M., & Huesmann, A. (2004). High flexibility: An important issue for user studies in driving simulation. In *Proceedings of Driving Simulator Conference – Europe 2004,* Paris.

Chapter XIII

The Role of Simulation in Business Process Reengineering

Firas M. Alkhaldi, Arab Academy for Banking and Financial Sciences, Jordan

Mohammad Olaimat, Arab Academy for Banking and Financial Sciences, Jordan

Abdullah Abdali Rashed, Saba University, Yemen

Abstract

This chapter discusses the importance of business process simulation, while illustrating the relationship between business process reengineering (BPR) and change management, it focuses the discussion on the role of simulation in supporting BPR and the effect of simulation on business environment related skills, business management related skills, leadership related skills, employees empowering level, process improvement, ethical issues, and stakeholders' management skills. The chapter discusses the value of simulation in implementing reengineering strategies and argues the future challenges of business process simulation and describes the limitations of simulation technology in reengineering business processes. Finally, it concludes with a discussion of the characteristics of successful simulation and simulation applications.

Introduction

Business Process: Definitions and Concepts

The logic of business is to create an advantage and/or utilize an opportunity, given this context; it implies the necessity to identify driving forces in order to fully exploit this idea. In general, one or more of the following issue(s) has the tendency to drive any probable business improvement:

- **Customer:** His/her requirements, culture, expectations, consumerism and even his/her feedback on the final product/service may enforce an organization to change its policies in order to gain their satisfaction, since low satisfaction will negatively impact product promotion.

- **Cost:** Basic notion within business logic for both seller and buyer, and the complex side of this logic appears when this perception is related to quality sensitivity.

- **Competition:** Results from micro and macro business environment, that is, market status, legal issues, consumerism situations, and so forth.

The question here is how organizations can remain competitive and, protecting itself from increasing competition threats at the same time dealing with its revenue from costly operations, attract more customers? Surely the answer to this question is not easy; the question here links the company assets (resources) as inputs, how to treat these assets (processing) and the outcome of the business operations, where the acceptance of the product by customers echo its success. Therefore it can be noted that processing operations are stressed, since it determines the success or failure of any product. Accordingly organizations revise their processes so as to maintain their competitiveness.

Prior to carrying out the hows of redesign and improve organization processes, it is necessary to demonstrate some process definitions and the sagacity behind each one of them. There are many definitions of process; this is due to viewpoint, background and trends of the researcher as well as the market common strategies; that is, push/pull strategies, where the adopted strategies demand considering certain outlooks and neglecting others. Each definition considers one or more of the following perspectives: input (resources), activities, output (product/services), and customer and organization objectives. For example, Pall (1987) expressed process as arranging different organization resources (for example: people, materials, energy, equipment, and procedures) reasonably to accomplish work activities leading to specific work product. Correspondingly, Davenport & Short (1990) described it as a collection of sensibly interrelated tasks executed to attain a certain business product. Moreover, Harrington (1991) stated that business process is making use of the organization inputs by collection of judiciously interrelated tasks that facilitate achieving the organization's goals. Omrani (1992) argue that process is the result of cycle of activities that are collectively taken to attain business goals. Likewise, Talwar (1993) defined the process as a series of identified activities implemented to achieve a specified type of outcome. Hammer and Champy (1993) considered the customer perspective when they defined the process as a collection of activities that, all in all, generate a consequence of importance to a customer. The next definition stressed the process boundaries; this characterization was presented by Davenport (1993) who defines process as "an organizing work activities across the place, with a start, an end and obviously identified inputs and outputs." Earl (1994) characterized process as a lateral or horizontal form that sums up the interdependence of tasks, roles, people, departments and, functions required to supply a customer with a product or service. Ferrie (1995) defined processes as being a definable set of activities that form a known foundation. Finally, Saxena (1996) explains business process as a set of interrelated work activities characterized by specific inputs and value-added tasks that produce specific outputs.

Change Management Approaches: Comparative Study

There are many approaches to change management, in the following discussion the researchers will focus on total quality management (TQM), business process reengineering (BPR), and knowledge management.

Total Quality Management

Total quality management (TQM) can be seen as a non-ending process to achieve necessary quality improvement, which needs effort and time, furthermore leading to the emergence of these efforts properly, which means more degree of commitment and support from workers and management. It is necessary to note that TQM is a novel way to do business, acting as a map to guide organization working processes where TQM engages deeply in cultural change. Juran (1986) developed a quality management framework, in which customers' needs are the focus of TQM, for whom value is delivered through well known, appreciated and administered processes. According to Sink (1991) each process consists of five elements: providers, inputs, value-added processes, outputs and, customers. Processes must be distorted and different forms of information must be considered to support decision-making in the new environment (Olian & Rynes, 1991). In the context of the above process components, management is concerned with improving input, output, customer views, and process quality assertion, in addition to tending providers' management. Accordingly, TQM concepts are activated by controlling performance of the above components (Sink, 1991). To organize TQM efforts, a systematic methodology is necessary to deal with different problems and utilization of chances in a controlled approach, which binds all the efforts to the common target. The quality target ought to be converted into plans, specifications, and measurements, all of which are management's liability (Bonser, 1992). Quality information is a vital element of the TQM infrastructure (Godfrey, 1993). Providing the precise information is critical to support the comprehensive TQM efforts. In short, total quality should be managed based on facts rather than on instinct or feel (Garvin, 1991). Quality management requires data regarding such factors as consumer, product and service performance, operations, market, competitive comparisons, suppliers, internal processes, employees, and cost and finance. To deal with different TQM issues, decision-makers need quality information in which support his/her decision.

Therefore, in a TQM environment, effective processing and dissemination of quality information is a vital role in successful operations, especially if we know that decision quality cannot consistently rise above the quality of information upon which the decisions are based. To support TQM efforts, the organization's information base must be comprehensive and easy to find and use. Poor and inconsistent data will divert efforts by focusing on problems of data quality rather than on the quality of the organization's outcomes.

Business Process Reengineering

The concept of business process reengineering (BPR) was first introduced by Hammer in 1990. BPR has many definitions; Hammer (1990) defined BPR as making use of new information technology to fundamentally redesign business processes to attain remarkable progress in performance. Conversely, Goll (1992) describe it as the full transformation of a business; an unconstrained restructuring of all business processes, technologies, and management systems, as well as organizational structure and values, to improve performance throughout the business. Ahadi (2004) stated that from the practitioner point of view there are five building blocks that are obvious to shape the fundamental topics that characterize BPR: radical or at least considerable change; business process centered analysis; achieve major goals or dramatic performance improvements; information technology is a critical enabler; and organizational changes are a critical enabler. Several organizations have achieved success from their BPR attempts by controlling costs and achieving breakthrough performance in a variety of parameters, such as delivery times, customer service, and quality. However, not all companies that undertake BPR effort achieve their intended results. BPR has great potential for increasing productivity through reduced process time and cost, improved quality, and greater customer satisfaction, but to do so it must be implemented and managed in the best interest of customers, employees, and organizations.

Knowledge Management

Knowledge management (KM) is a process of elicitation, transformation, and diffusion of knowledge throughout enterprise so that it can be shared and thus reused (Turban et al., 2004). At the organizational level, knowledge value is limited if not shared. KM can also be seen as the process of generating value out of organization's intangible resources (Liebowitz, 1999, 2000; Liebowitz & Beckman, 1998). Grant (1996) argues that the ability to integrate and apply dedicated knowledge of organizational affiliates is essential to firm's ability to create and maintain competitive advantage. Therefore, the focus of knowledge management is how to share knowledge and how to create value added benefits to the organization. On the other hand, some organizations have an aged workforce and this may illustrate the necessity for sharing the skills of those experts. Knowledge management is a process of capturing, creation, codification, communication and capitalization of knowledge throughout an enterprise so that it can be utilized effectively and efficiently, where the authors refer to the above process as the KM 5Cs model. During the knowledge capture phase, searching for several sources of knowledge that are necessary and related for performing the work using analogy, metaphor, models, brainstorming, and similar mechanisms are conducted. Knowledge creation phase involve conducting research activities, verification and validation process consecutively to discover the knowledge in the organization, and utilizing experiences and routine procedures to prepare an appropriate culture and systems. Knowledge codification phase includes classification and categorization of existing knowledge in the organization according to its nature into categories, such as administrative, technical, financial, and so forth, knowledge of the organization to reflect what is actually known and done by the organization, and refining and filtering knowledge of the organization in order to access the most significant knowledge. On the other hand, the knowledge communication phase represents the process of considering sources, nature, and attributes of knowledge when transferring and sharing in the organization, encouraging and enhancing the culture of knowledge sharing among organizational members, anytime, anywhere, access when it is needed, and determining who can transfer knowledge. Finally, the knowledge capitalization phase is about investing and utilizing organizational knowledge in an innovative way, improving the existing methods of work practices leading to encouraging individual competitiveness and effectiveness. Enhancing the feeling of individual responsibility bring about changes in organizational culture. Knowledge capitalization is the process of finding new practices for performing organizational work to improve organization performance, balancing cost-benefits, enhancing creativity in decision-making process and problem solving, utilizing the embedded knowledge in order to direct the future behavior of organizational members, evolving organizational knowledge through the feedback, and evaluating the outcomes of organizational knowledge through the services that are offered by the organization.

Liebowitz (2001) discussed challenges that influence knowledge management scheme success in any organization. He argued that organization culture in reference to the required degree of knowledge sharing and how to build and enhance knowledge sharing is a vital role for successful knowledge management; furthermore, seeing knowledge management as part of the strategic vision of the firm. Maintaining knowledge repositories and security are among these challenges. He claimed that user-friendly design, trouble-free knowledge management programs are necessary; and finally, employees have to be motivated, that is through incentives, so that knowledge sharing will likely be accomplished.

Why Simulation?

In the preceding sections, it was argued that the main benefit of BPR is to recognize and put into practice alternatives to the organization that desires to develop its activities and/or processes. This task might be easy when the processes that shape business cycles are static and trouble-free, but what is the solution when:

1.   The processes are of dynamic nature (i.e., customer-based businesses).

2.   The global market is the goal of the organization (i.e., competition conditions).

3.   No time to conduct trial and error.

4.   The company desires to manage customer expectations and outcome predictions that may result from radical change.

5.   The direct execution of all alternatives will be unsafe.

The rate of failure in reengineering projects is over 50%: one of the major tribulations that has a say to the breakdown of BPR projects is the need for tools for evaluating the sound effects of a considered solution before execution (Hlupic & Vreede, 2005).

In the following section the authors will illustrate simulation definitions, the environs under which simulation is advised, the impact of new environments on simulation, the factors that shape the extra value of simulation, and finish with a discussion of the significance of simulation from a business viewpoint.

What is Simulation?

In the context of BPR, simulation can be regarded as a tool to aid the decision-maker to assess the on-hand alternatives, subsequently employing the most appropriate (not necessarily the appropriate) alternative as a tolerable solution for the problem under consideration. As BPR is changing approach, then there will be an as-is model which will brings to light the opportunities on hand, rewards that exist, the level of performance, well-built/frail processes, and so forth, and, on the other hand, a to-be model processing the challenging sides and infrequent potential. Finally, the judgment will exceed through the simulation panel to include, remove, or adjust any process that may guide to client approval. Following the simulation team observations, the client will appraise the model, which will be the portrait of the future according to his/her requirements. So simulation is the attempt to bridge the gap between the client requirements and the anticipated actual status from one side and bridging the gap between the decision-maker and his/her facilities before execution, when all alternatives may be costly. From the above argument, it may be concluded that simulation might result in information that could have different meanings according to the receiver, who may utilize it differently according to his/her requirements.

When to Simulate?

Simulation, as in any activity, may require time to be achieved, financial support is also a matter to bring the right outcome; moreover, specialists who need training and whose passable level of loyalty to the organization to work within the organization margins of benefit is needed. So the question arises as to when simulation is advised? To answer this question we must return to the process philosophy. Pressmen (2005) notes that each business system is composed of one or more business processes, and each business process is defined by a set of sub-processes. In the context of this notification about the hierarchical nature of business system, two related issues may be emphasized:

1.   Complexity increases as we move descending in the hierarchy.
2.   Risk increases as we move upward in the hierarchy.

Considering complexity, the combination between the business processes from one side and between the tasks that form these processes/sub-processes from another side play a major role to determine whether simulation is proper or not. As combination level increases, simulation has to be conducted. On the other hand, when it is difficult to divide the main processes (at the top of the hierarchy) and it is difficult to decide the appropriate alternative based upon the client requirements, simulation has to be conducted. Until now the business process/sub-process as a whole has been considered, but what about the process itself? During the introduction to this section, dynamism, globality, time factor, and expectations management were underlined, all of these factors drive the world toward process oriented adoption, which enable the client to be the controller of the market along with its changing requirements, where the process itself needs to be dynamic. As this dynamism increases, the need for simulation is increased. Simulation in general engaged in situations where the arbitrary content does matter and cannot be modeled by other methodical techniques. Cheng (1992) viewed simulation as a tool of final option to be engaged just once further methods are ruled out. He cited the high computational expenditure and time and effort required in constructing models as drawbacks of the simulation approach. On the other hand, Swain (1993) and Bhaskar (1994) recommended that the simplicity of model structure and price economies in computing originate simulation as a tool of selection for modeling complex systems and validating analytical models before going on to optimization. In the following section we will study the bond between new environment and simulation.

New Environment and Simulation

In the previous subsections it was distinguished that the business chain becomes process-oriented rather than production-oriented due to client enablement in the new economy, where the opportunities and advantages are disseminated among big as well as undersized organizations, as well as the risks affected both sides in the new economy. Enterprise resource planning (ERP) was a solution for many organizations, but this solution has a limited capacity to meet the repeated varying requirements that are the core success in the new economy. ERP may be a good proactive solution, but the new environment needs proactive and reactive or even rapid reactive solutions that sustain an accurate decision-making process as changes take place. It must be concerted that simulation is not substitute to ERP solutions, but hold and function with these solutions side by side to support decision-making.

Simulation Areas and Applications

Dynamism, frequent changes, Internet settings, and globalization are some of several issues that enforce using simulation in novel economy. One of the salient characters of the new economy is the diversity of applications that can be conducted within such huge environment. It can be said that simulation is a tool that can be used in most business applications; this is due to:

- Applications diversity: in the new economy companies found themselves enforced to conduct many profitable activities especially small to medium companies where the rivalry of large companies in a certain activity may bound their ability to survive or even to go on in the market.

- Unsafe environment: in conventional business world ethics, security, payment methods and similar issues can be managed to an accepted limit. In the new economy, these issues are hard to control to the same limit in traditional business world; besides, the frequently changing business surroundings formulate simulation as a preferred solution to control these matters.

- Complexity: complexity goes up rapidly due to the association of new relationships, new partners, new processes, supply chain improvement and so on.

- The human behavior role in business triumph is going up, which means the need to predict its responsive behavior on the way to change. All these aspects are directly related to the cost issues, which step up the business activities.

Hugen et al. (2003) claimed that the upcoming day's simulation technology would get in touch with more and more applications. In the same context, Jain (1999) stated that simulation will become the way to do business in the future: he argued that making use of simulation modeling will extend from traditional applications in manufacturing and the logistics side to business processes and interactive simulation applications in training and sales. The following are some examples of applications and areas where simulation can be valuable:

1. Manufacturing: simulation can support successful manufacturing regarding product design, layouts, flow of materials, manufacturing processes and even the organization structure.

2. Conflict management: simulations enhance decision-making skills for these areas in supporting efforts for conflict evasion and conflict declaration (Oren, 2002).

3. Training: simulation can enhance an organization's efforts to train its staff or new employees about the significance and practice of cooperation or even use simulation as training system, making use of machine learning theories (Oren, 2002).

4. Judgment of manpower requirements.

5. Competitive market analysis.

6. Production planning and inventory control.

7. Corporate planning.

8. Man-machine interaction.

And other diverse applications and areas in which simulation can be used effectively.

Simulation ... Simulation

Up to now, simulation concept, the condition under which simulation is informed, and the relation between novel environment and simulation were established. The question arises: where is simulation in the information and knowledge world? When someone notifies you that the weather is cloudy then he may have knowledge different from that of you (see *Figure 1*).

Ali notifies Ahmad that the climate is cloudy; this means (knowledge) to some extent to Ahmad differs from that of Ali, where this meaning is associated to the necessities of each one, which may or may not be the alike. Figure 1 can be extended to the idea under consideration, that is, simulation. Imagine that the people in the above figure change to be an organization with knowledge resource, there will be area where the knowledge within its resource may mean many things according to each organization, these things result from the organization's diverse requirements, for instance, increasing market share as an information may mean making revenue and rising duty, the options depend on several factors: legal issues, competition status, and so on. In this and similar districts simulation will effectively enhance the decision–maker's efforts to seize the proper decision. An important characteristic of simulation models is their ability to provide quantitative information, providing a foundation for assessment and comparisons of choice decision strategies to settle on whether that choice accomplishes the decision-maker's objectives or not. In the digital settings the business pressures are divided into three categories: market, societal, and technological, each one of these pressures calls for particular reaction types derived from certain requirements of the organization under consideration by making use of diverse responses (alternatives) simulating.

Business Process Simulation

Business logic is based on taking advantage or utilizing opportunity, nevertheless, it can be claimed that advantage or opportunity imply either cost or time in most cases of the business

Figure 1. Knowledge conversion

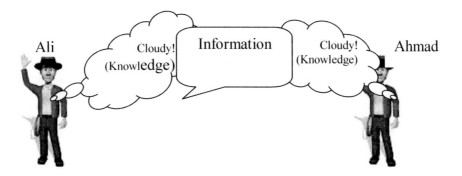

world. The organization may follow many tactics to achieve its objectives, for example, one organization may improve the way to reach its customers (demand pull/supply push), and another one may adjust its working processes, develop the working team or even take on new habits to everyday work. All the above tactics are related strongly to the inquiry mentioned earlier, to be precise, what if?

In the new economy, opportunity or advantage are robustly interrelated to the assets (information, knowledge) of organizations, so it can be argued that simulation aims to bring valuable information within the required time, which will lead at the end of the day to building many scenarios to support the decision-maker, who will agree on the appropriate alternative according to certain requirements. In the following section we will recognize the concept concerning business process simulation.

Business Process Elements

Tumay (1995) claimed that the four basic building blocks of a proper model are:

*   Flow objects: it may be referred to as entities; these are objects that are processed by resources, for example, customer.
*   Resources: these are agents that are used for accumulation value to flow objects; they are to be paid to activities, for example, service representatives.
*   Activities: they are linked by routings to stand for the stream of objects through the simulation model, for example, batching.
*   Routings: label the various types of associates between activities.

Model Performance Measures

Tumay (1996) suggested that the basic model performance measures are cycle time, entity count, resource utilization, and activity cost.

1. **Cycle time:** it is the entire time an entity uses during a process, this time incorporates: value added process time, waiting time, movement time, and so forth. One of the most precious outcomes of business process simulation is the automatic tracking and estimating of cycle time for each entity; furthermore, business process simulation tools present average and maximum values based on the total count of entities that pass through the process.

2. **Entity count**: it comprises the total number of entities that traverse a process or that are still being processes. This signifies that the total entities count is the sum of entities processed plus the entities that are in process. Business process simulation tools routinely track and evaluate entity counts for each type of entity in the model.

3. **Resource utilization**: throughout simulation resources change states from active to idle, from active to inverted. Resource utilization defines the percentage of time that

a resource spends in each condition. The accessibility and transfer of resources shape the share of resources to activities in a model. So the resource uses results to provide helpful information in measuring and analyzing under-utilization or over-utilization of resources.

4. **Activity cost:** a resource is defined in a process model by the number of available units, handling overheads, system expenditure, and predetermined expenses. Activity cost is defined by the resources required to execute it, the time for the activity, and the entities that it processes. Activity cost calculations provide a sensible way for measuring and analyzing the cost of activities.

Simulation Life Cycle

Prior to discussing simulation life cycle, it should be concentrated on the idea that in simulation we are working with a model of the problem and this makes it possible to keep away from possible errors during implementation, such as discovering that the problem under consideration was not entirely covered. Since we simulate requirements that are changing repeatedly, simulation life cycle will be iterative in nature, and this nature entails assessment of exactness (i.e., validation, verification) after completing each phase of simulation study, which makes the reverse switch, is probable. Hlupic & Vreede (2005) suggested that the activities concerning business process simulation include the following steps:

a. Definition of modeling objectives,

b. Definition of model boundaries,

c. Data collection and analysis,

d. Development of business process simulation model,

e. Model testing,

f. Model experimentation,

g. Output analysis, and

h. Recommendations of business process change.

Definition of modeling objectives: any obstacle that binds the capability of making use of opportunity or taking advantage is a problem in the business world; once this problem is acknowledged it will be brought to the table to be solved. So the solution to this problem will be the vision of the upcoming model; for example, evaluating the consequence of entering WTO or evaluating the role of private sector may participate in developing economies. Subsequent to determining the modeling objectives, the question arises of how to achieve these objectives? The answer to this question will be shown through the following discussions.

Definition of model boundaries: within real life there are a lot of problems where solution cost to these problems is high enough to convince the decision-maker to put down them as they are. So it is vital to verify the processes that can be considered before getting the point where we found that the problem wasn't well recognized.

Determining the processes that can be considered depend on the requirements, the objectives of modeling process, and their associations with the processes considered; it also it depends on the availability of the crucial resources: time, experts, or even the information itself.

Data collection and analysis: in the preceding step the availability of necessary resources was revealed. This border line is very important since we study a problem that will be modeled. So experts' ability to mine the information or even data from those experts, and time availability to obtain the associated data is very important factor in the series. There are numerous ways to gather the definite data required: interviews, direct observations, studying documents, and so forth. Following the required data collection, they should be analyzed by means of methodical approach to get to the goal of this step.

Development of business process simulation model: it relates to a simulation model enlargement using a simulation software package. Since simulation life cycle has an iterative nature, the software package should be prepared through an iterative and incremental process, and this entails adopting agile process models where the regularly changing requirements will be easily conducted for example: extreme programming, adaptive software development, starting with a simple model then, with the progress of ideas, this model will be extended and distinguished until an adequate model is obtained.

This is the case if the software is developed in-house, but if it is outsourced then the desirable characteristics are introduced through an RFP document.

Model testing: the point to be emphasized here is that testing must be conducted following each iteration throughout model growth; the purpose of this step is to handle any errors coming in or out of the model, studying the effect of these errors, and recording these errors to be under consideration in the scene. A series of different tests are conducted, each of which has explicit purpose although the final target is to verify that system elements have been properly incorporated and carry out allocated functions.

Model experimentation: formerly, model performance measures, model testing, were discussed; the next step will be for exercising the model by concerning the model performance measures through formal experimentation.

Experiments should be planned in such a way in an attempt to decrease arbitrary errors, and include a broad range of alternatives so that recommendations could be relevant for a variety of organizational units, the experiment should be as simple as possible, and a sound statistical analysis should be applied without making impractical assumptions related to the nature of business processes.

Output analysis: the tip to be stressed here is that, throughout several experiments conducted, major business interrelated key variables are estimated then the consequences explored. Subsequently, statistical tests can be used to verify whether there are noteworthy differences between key model output variables of various experiments.

Recommendations of business process change: simulation of business processes allow the simulation team to revise diverse possible scenarios, which will guide them to the most accurate scenario through which the organization can attain its objectives, followed by formulating the recommended scenario and fire it to the decision-maker who will revise the recommended scenarios resulting in adopting or rejecting the proposal following validating each choice. One of the regular scenarios resulting from simulation is the radical change in the course of actions of conducting certain process(s), that is, business process reengineering. In the following section we will give attention to the role of simulation in supporting BPR.

The Role of Simulation in Supporting BPR

In the previous sections, simulation perception was conversed taking into account the simulation life cycle. The question arises: what to simulate? Pressman (2005) indicated that within business processes people, equipments, material resources, and business procedures are pooled to generate a particular result. Levas (1995) claimed that the long-established sight of business process has been bringing down to a little degree; it is replaced by innovative pattern, where clients, dealer, the stream of work and the communication force supersede the organizational map. Earlier, it was stated that as complexity enhanced, simulation would be preferable; besides, the increase of risk entails that simulation tools be used. It is obligatory to recognize that human issues are fundamental in business cycles, so it is significant to study this factor's effects, trends, and skills. To emphasis that point, Levas (1995) stated: "We will need to move toward process representation and analysis that seeks to capture the unique skills and capabilities of each individual." So it is essential to realize the human capital, which plays a vital role in information and knowledge era. To illustrate the value of employees' role in business process reengineering scheme refer to Paper et al. (2003).

Bridgeland and Becker (1994) asserted that it is critical that a BPR modeling tool provides the way for capturing intellectual assets.

In the preceding sections it was stated that BPR is iterative process, so as long as we can minimize time required for iterations it will lead to successful BPR. Once information technology is involved within business process, the procedure's cycle time will be automatically minimized, but the problem arises where the time losses are related to human understanding, in which it will be a function of their willingness to utilize their skills to accelerate business process principally in new settings, where fluctuations are recurrent. In the following discussions, a variety of domains where simulation can play a vital role in supporting BPR will be considered, these domains include: business environment, business management, leadership, employee empowering, process improvement, ethical issues and stakeholders' management.

Business Environment Related Skills

Within the business environment there are numerous stress places, for example, if the vision of one element of the business cycle is not supported by the operational environment, sometimes even if the members of this environment are aware of the value of this vision, then stress will result. Here and in similar cases simulation as a prognostic tool can work effectively to minimize the probable risks. It must be illustrious that effective strategies can be constructed based on stresses thoughtfulness, which will assist in rising up these stresses that may appear in the upcoming days. Each category of these stresses may be faced with various optional strategies, for example, the way in which the operational environment members believe, act, or are energized, the organizational culture, sociability level, solidarity toward the targets—all of these issues must be well thought out to stimulate critical thinking within frail ends, manage probable reasons behind resistance, and hold the vision upon execution.

By applying simulation practices, the critical bounds, roles organizing, disseminating of responsibilities, and conflict management can be managed effectively. Still, it must be remembered that a dynamic business environment calls for dynamism of the stress places where frequent changes come to pass. Again simulation assists in minimizing the risks resulting from these frequent fluctuations.

Business Management Related Skills

Suppose that a corporation sells abroad a certain amount of its products globally to several markets, suddenly, a crisis take place with one or more of these markets, what is the solution? There are ROI losses; production size implies new storage facilities, which induce extra costs, and so forth. To be proactive, such economic uncertainties must be considered, all probable alternatives must be at the table, simulation seems to be useful tool to care for such economic uncertainties. In the new economy, international agreements are held—that is, WTO, which means new markets—at that moment, the problem is to determine the needs of such markets, and circumstances forced by these contracts. Simulation will support the efforts for balancing market needs, and the forecast of the possible revenues. A vital role of management is to create and facilitate the useful and possible change: as a preliminary stage, you as a manager must ask yourself:

1. Where are the changes required that lead to profitability?
2. What are the barriers that may face that change?

All of these questions represent the aspiration to induce change, and many alternatives are possible: from front line employees, middle management, or the client, the point is what will be the cost of each option? This may depends on the related issues of each option. Simulation is a technique that can be used to minimize the risk that may result from adopting one alternative over others. It is a vital point for management that all business procedures must be kept within the business cycle time line (task time), if any trouble appears, many solutions are available: cancel the unnecessary tasks, energize the weak points, dismissal of some employees, or even inserting new time management habits fitting the cases under consideration. Once these alternatives are on the table they must be simulated to adopt the appropriate one. Before starting any project the management gives enough time to select the working teams. Team's characteristics must concur with the organization vision to the gainful change.

In the new economy individualism disappears while team work character becomes stronger; the number of teams, individuals forming these teams and odd jobs are examples of what may be simulated to support a decision-maker, who will act together with a first round representation of what will occur later on.

Leadership Related Skills

Within the new economy, organizational culture plays a fundamental role in developing an organization's activities. Within operational environment present diverse workforce settings due to the diversity of their tacit assets and how these assets were fabricated, that is, the training through which they gain their assets, or the individual by whom their beliefs toward organizational culture were affected. Successful leadership enables the related leader to deal with diverse backgrounds and contrasting workforce age groups; for example, in an information era one comes across numerous types of workforces: some believe in experience character, others believes in building new techniques involving IT. Simulation is effective tool that supports efforts to identify the appropriate leadership for a certain type of workforce. Some managers are not aware of the hazard that may result when overlooking the fact that various workforce generations' beliefs ought to to be considered; this phenomenon is related directly to the loyalty level of the workforce to their organization. Practical experience indicates that the leader of any group will form a working model for this group, the members go behind his opinions, trends, and the ways he tracks to solve problems. The frequent fluctuations in the new economy's environment needs a further level of leadership dynamism, which necessitate a leader model of particular characteristics, these individualities must be verified depending on the outcomes of simulation study for performance measures before starting work; for example, a key success character is leadership hierarchy, that is where we will start to lead the team successfully? Frontline, managers, and so forth. Furthermore, one more important issue is to know how to lead, for example, in one case for successful leadership the working patterns must be revolutionized, another case requires taking into account the motivation factor. What's more, necessary communication skills for victorious and proper leadership must be confirmed. Bridgeland and Becker (1994) declared that communication skills encompass two purposes in the context of reengineering: to create an undeniable case for change and to generate a vision of what the company is to be. These skills lend a hand in periodic meetings with customers by serving the leader to bridge the gap between customer requirements and the organization products. So, effective leadership affects directly the revenue model of the organization. By simulating the surroundings within which the leader will work, the exceeding tips can be branded and the right personality can be picked. Simulation can effectively support an organization willing to study this type of skills.

Employees Empowering Level

According to Levas (1995) within the information era the worth of human being is not considered in unit\hours but in their involvement to upcoming victory. Bridgeland and Becker (1994) argued that the high crash speed of BPR schemes is due to inborn and ordinary variance of attention among workers and the organization as whole. The employee responsibility in potential success is linked strongly to the level this employee is motivated and recognized. In the preceding sections, workforce backgrounds and generations' character were discussed; this discussion leads to the thought that there are diverse groups of employees to be considered in the reengineering process. The needed efforts to systematize each class to change must be considered, as well, the communication skills to rejuvenate employees and communication skills for front line employees who deal directly with customers must be judged. Finally, each group of employees will require a certain type of management, which will forward the group efforts to profitable change. By simulating everyday jobs, roles, customer's culture, worker's qualifications and skills and then correlating them to the performance measures, then the required empowering level is identified and the incentives range are recognized.

Process Improvement

BPR is an approach that is generally used to drastically modify the process of coming with and creating novel and improved traditions to run a business (Levas, 1995). But the question that arises here is: What are the processes that will be modified? To answer this question various dimensions have to be taken into account, that is, organization vision and priorities, customer requirements and satisfaction, and so forth. Within the activities of business process progress, information resources and ways for information gathering must be identified. Considering "as is" a model is not enough for BPR use and it will go in front to risky "to-be" model (Levas, 1995). It is noted that customer requirements are ambiguous in most cases and they are changing repeatedly. Since simulation is a predictive tool, it can be used to minimize the risks that may result. To demonstrate the value of the simulation modeling efforts for the assessment of current performance levels and quantifying possible improvement, please refer to Hulpic and Vreede (2005).

Ethical Issues

Bhaskar (1994) asserted that employees have unstable level of influence and control, and use these personalities to carry or defend against the accomplishment of an economically wise reengineered process. For successful BPR it is essential to place the potential options of ethical matter sound effects at the table. As discussed earlier, dissimilar workforce age groups of employees, which have different settings and different viewpoints of how the business work runs, must be well thought-out. Ethical issues are not only distress employees, but also the client must be taken into account. The culture of the client, his/her consumerism, and his/her behaviors toward business products must be regarded. Simulating the interior and exterior settings will facilitate realizing the trends of individuals as well as organizations (competitors or friends) toward ethics. It appears that simulation can support efforts that are dedicated to revise ethical issues consequences on BPR.

Stakeholder's Management Skills

It was asserted earlier that the construction of a vision of what the company is to be is one of communication skills functions. Stakeholders are all the parties (individuals or groups) who deal with the change; these include the employee and managers (discussed early), clients, community, government, and so forth. In most cases stakeholders are concerned with what the company is to be. Unless you as manager can win over the stakeholders, especially effective ones, with the upcoming portrait of the organization you will lose their support.

Each one of the stakeholders has a certain intuition regarding the organization vision and goals, this intuition relays robustly on their stakes of the change success. If the manager has the ability to recognize these visions he/she can prepare to the proper means to get their support. By determining the improvement level, procedures that need to be modified, performance measures, and the roles and responsibilities, one can readdress a certain group of stakeholder's influence to support the change not to resist it, and arrange to overcome this resistance if it take place. Simulation can support this manager in the effort to win in his/her battle. At the end of this section, it must be indicated that simulation supports diverse business process reengineering domains and facilitates:

- Minimizing risks.
- Desired change management.
- Expectation management.

Besides that, it is a tool that enables a decision-maker to model human responses either within the business cycle (employee) or those who have an effect on this cycle (external stakeholders). These entire capabilities enable a decision-maker to construct the right "to-be" model, which leads to effective execution of changes desired. The above discussions bring to light the role of simulation in supporting BPR, which is the most complex and painful period of transition to competitiveness.

In the following case study we will consider a comprehensive example for the role of simulation in business process reengineering.

Case Study: Team-Based Human Resource Planning Model

Cheng et al. (2005) proposed a team-based human resource planning model (THRP) with a center of attention on construction management process reengineering, where a team approach was applied in reorganizing the structure of a company to smooth the progress of newly designed process, and simulation was used to forecast labor power aptitude for the new organization after business process reengineering. The THRP method aims to determine the maximum loading of projects that can be held by the original labor power, in addition to identifying the range of labor power needed for the upcoming project loadings. The THRP method includes four phases, namely, process reengineering, data preparing, human resource allocation, and simulation. Using the THRP method, a company cannot only design a team-based organizational structure, but also assign human resources based on cross-functional processes. Furthermore, human resource utilizations before and after process reengineering can be assessed to evaluate the process reengineering practicability. Hence, optimal labor power can be assessed, and human resources allocated to fit the changing processes and situations of the growing business. *Figure 2* shows THRP model.

Figure 2. The THRP model

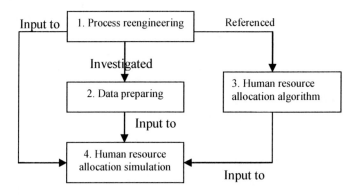

To validate the soundness and feasibility of the THRP model Cheng et al. (2005) provided a case study. The THRP model was employed to assist "Company A" with reengineering of the construction planning process and evaluating the rationality of human resource use in the reengineered processes.

Process Reengineering Phase

Figure 3 shows the functional organization structure of "Company A." In the figure, the construction planning process was executed by the construction planning department. To execute different specific functional goals, the department was divided into three functional task groups. The planning process was enabled and split into (1) bidding/contracting, (2) estimation/budget and (3) purchasing sub-processes, which were carried out by the related task groups sequentially. However, due to the fact that task groups were accountable only for partial goals of the construction planning process, the three task groups hardly supported each other and some of them experienced overload leading to inefficient human resource deployment in functional organizations. Taking into account the ineffectiveness of previous construction planning, the manager had chosen the construction planning process to be reengineered.

Process Representation

The major rationale of processes representation is to develop a systematic definition of these processes. Two tasks have to be carried out:

Figure 3. The organizational structure of Company "A" before BPR

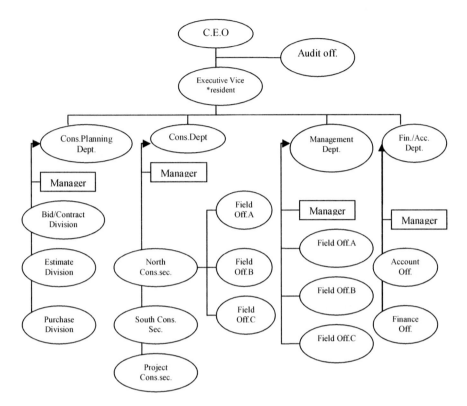

Task 1. Activity Data Collection

To map a business process model from function-based to process-based organization, detailed information on the activities of functional departments is required. E-EPC diagram of the Architecture of Integrated Information System (ARIS) process modeling tool was used to outline both existing and the newly designed management processes of a construction company.

Task 2. Process Modeling

This is accomplished based on the collected data using a modeling tool, for example, the Architecture of Integrated Information System (ARIS) modeling tool.

Process Evaluation

The intention here is to analyze the rationality and effectiveness of process-based on the e-EPC process diagrams created in the process representation step. For example, hidden problems within the original bidding/contract processes were located as follows:

1.   Only quantity surveys were implemented with computer-aided software while all other activities were completed manually; as a result, the activities efficiency and the labor power utilization were limited.

2.   Job loading was not proportional to task divisions due to the magnitude of various construction projects. Actually, active assign of proper labor power actively to accurate groups in a functional organization structure is not easy.

3.   No valid history reference data on material and labor was on hand to assist the estimator so that cost estimations are not up to snuff to correspond with current prices.

4.   The time consumed by document circulation increased because all documents were required to be submitted to the managers.

Process Redesign

To fix the process defects, the process model was redesigned on process design principles and a new process model created. A cross-functional information system of cost estimation system was applied. Some activities could be integrated and executed by cost estimation system.

Figure 4. The organizational structure of company "A" after BPR

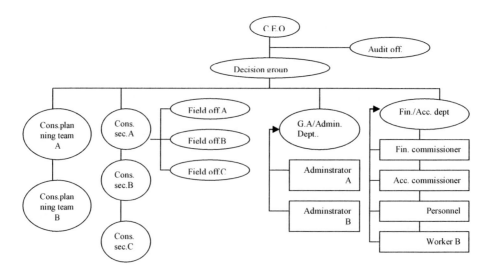

Team-Based Organization Creation

A team approach was applied to join multiple division functions into one unit which could facilitate functional interfaces and parallel process activities. *Figure 4* shows the team-based organization structure of Company A corresponding to the new construction planning processes. The primary difference between the previous structure and the team-based structure is the appearance of the decision group on one hand and the two construction planning teams who are accountable for the entire construction planning process.

In the adjusted organization structure, the construction planning team is flexible in number depending on work loadings. As a reference for human resource allocation decision-making, an evaluation method with simulation tools was used to reveal the relation between work loading and the labor power of work teams so that the managers could assess the capability of labor power allocated in a team was used.

Preprocessed Data Phase

In the data preparation phase, the values of project parameters and process parameters were estimated according to the historical data of projects and activities of Company "A." *Table 1* shows the project parameters of the Company "A."

The process parameters for both the original and the newly designed processes were estimated and then evaluated for labor power in activities, for probability distribution functions, and for durations with 95% confidence level.

Human Resource Simulation Phase

In this section, to validate the feasibility of the simulation system, the "construction planning process" of Company "A" was taken as an example. The current process model within the original organization structure was developed and simulated. Comparing the simulation results with the performance of the current process model, the simulation system was validated. The results of the model validation were very close to the real situations. For sixteen

Table 1. Project parameters of Company "A"

Project parameter	Function/Value
Start time of project	Uniform distribution
Success ratio of bidding	20%
Number of subcontracts	25
Required lab power	7 employees

projects, the maximal project loading of seven workers was estimated by the simulation system with the test model. This result matched the real capability in the range of 15 to 18 projects estimated by the mangers of Company "A." Moreover, the average of idled labor power also was close to the real situation as evaluated by the managers with experience.

After the model test, the newly designed construction planning process was modeled in the eM-Plant system and evaluated with two situations: (1) the number of members of the construction planning team remained constant at seven, which was the total labor power of the construction planning department; (2) the project loading was taken as a function of the number of team members, and the combinations of labor power and project loading were estimated.

The main purpose of the first simulation is to estimate the potential project loading that current labor power can take. Therefore, the value of the current number of personnel available was set at seven workers. Moreover, each team consists of one senior engineer and at least one junior engineer, so there were eight possible labor power allocation alternatives for seven workers. *Figure 5* shows the allocation alternatives and the amount of projects which were simulated.

As applying the forward scheduling method, the success ratio was always higher than 90% when the project loading was less than 25 projects. It decreased to 83% (less than the threshold of 90%) as the loading increased to 26 projects. Therefore, the maximum project loading for seven members was evaluated at 25 projects. Likewise, as the backward scheduling method was applied, the success ratio was always higher than 90% when the project

Figure 5. The possible labor power deployment alternatives tree for C.P.T with NAP==7, NAP: number of personnel available, AOP: amount of projects

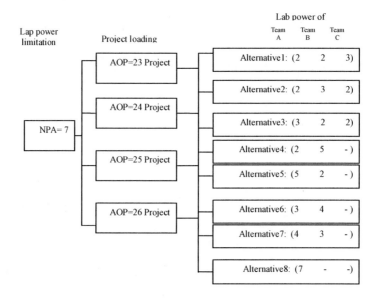

loading was less 40 projects, but decreased to 80% as the loading increased to 41 projects. Therefore, the maximum project loading for seven members proved to be 40 projects when the backward scheduling method was applied.

Because forward scheduling deploys the maximum resource at the beginning of a project to accomplish the process as soon as possible, labor power capability might be limited by more resource conflicts than in the backward scheduling method. Therefore, it was guessed that optimum project loading falls between the two above referenced extreme cases. Based on this result, Company A's labor power margin was adapted to a minimum of 25 projects and a maximum of 40 projects.

In addition, the outputs of the simulation exhibit positive evidences of the advantages of process reengineering. That is, by comparing simulation results between the original process and the new, integrated construction planning process, we can see the labor power capability increasing from 16 projects to 40 projects, and idled labor power decreasing obviously due to the integration of functions.

For the second situation, this study extended the number of personnel available from 2 to 15 workers to realize the relation between labor power and project loading. In *Figure 6*, one NPA value corresponds to two project loadings; one is backward scheduling result and the other is forward scheduling result. As each NPA has been simulated, two curves could be finally sketched which present the relation of labor power and project loadings. Based on *Figure 6*, not only the capability range of specific labor power, but also the range of labor

Figure 6. Simulated output curves of the new construction planning process

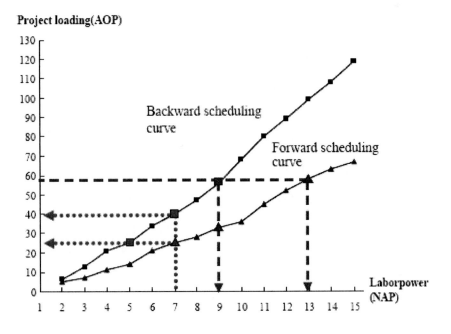

power requirement for specific project loading can be identified. Therefore, managers can estimate the efficiency of labor power or the amount of it needed based on the number of projects.

Successful Simulation

In the prior section the role of simulation in supporting BPR was discussed. Now, how to achieve successful simulation? According to Sadowski et al. (1999) successful simulation is one that brings functional information at the proper time to bear an evocative decision. Aguilar et al. (1999) acknowledged that the victory of a simulation project fundamentally depends on four factors: the process characteristics, the objectives of simulation, the quality of input data, and the management of expectations. The key aspect of presenting the functional information is to look at it from the viewpoint of the audience for which it is intended.

- What they need to know
- Why they need to know
- Anticipate the questions they may ask
- Their view of the system
- Take time to introduce them to how you look at things and investigate how this varies from their point of view.

In the next discussion four aspects that affect the success of simulation will be underlined.

Communication

Simulation models can be viewed as a problem recognition rather than a problem solving tool (Hlupic & Vreedi, 2005). In order to superiorly identify the problem it is essential to communicate the situation under consideration; if the situation communicated properly the problem to be solved will be recognized, defined and then appropriate formula for the problem will result. Through communication the relations between the situation problem and simulation intention will be defined leading to the implementation of the acceptable action. Robinson et al. (1998) claimed that the classic approach to signify human decision-making in a simulation model is to elicit the decision regulations from the decision-maker. This means that the communication channels with decision-makers are required to be sustained to identify the superlative match between different scenarios resulting from simulation, along with organization goals that will provide a solution approved by the decision-maker. Internal and external organizational environments have to be communicated; for example, employees, organizational chain of command, encouragement arrangement, market, customers, and competitors, are a number of issues have to to be regarded. Throughout communication the influence of diverse variables can be appraised, and the significance of each one can be labeled. One more party that must be communicated is the working team, that is, the team who is accountable for simulation process; this team must be supervised and led, as well as guided properly. Communication can be considered as the foundation from which either you tag along a mistaken path—meaning the risky one—resulting in organization failure or you tag on the precise path, which guides to success. The more properly and earlier that you handle the problem, the sooner that you can initiate your project (simulation). There are numerous ways to communicate with the above parties: interview is appropriate technique especially with top management, as well as observation of business process, besides several other techniques that can be used for this purpose. JAD sessions can be conducted to communicate all these parties, these meetings can be wise for eliciting requirements, brainstorming to analyze initiatives, collaboration and thorough analysis.

Complexity

In the former arguments it was acknowledged that simulation is chosen as complexity and risk of as-is as well as to-be models grows to be high. It is of the essence for simulation panel to go behind planned approach in their work, one key inspiration behind the use of such an approach is to administer the model complexity in order that the model can be effectively confirmed and certified (Balci, 1990). In the preceding aspect, that is, communication, the value of organization environments was discussed; within these settings we deal with employees, where each employee has personality venture, incentive style and grand (Bridgeland & Becker, 1994). What's more, by way of market instability with different input variables and the significance of each one to the business objectives and working business processes, simulation team selection plays a fundamental role. The bottleneck in the on-hand resources has to be considered; for example, time, people and tools. As this estimation encompass high accuracy, the decision-making value of business process simulation will be further be significant. Abstraction level is imperative for reducing complexity level in a simulation model, which then reduces essential time and resources dedicated to this effort. At the initial stages of simulation, it is noteworthy to gather as much information as possible: Benjamin et al. (1998) believed that it is crucial to start with low level of abstraction which imply huge amount of information. In addition, a system must be decomposed into subsystems to overcome its complexity, which means planning simulation project at proper level of details (divide and concur principle). Complexity reflection has to be linked to task worth analysis, that is to say, the tasks that represent keystones must be taken into account throughout the simulation work. On the other hand, the cost of each level of abstraction, that is, time, efforts, and resources analysis looked-for to accomplish the available and needed amount of information, in addition, the cost result from quality analysis must be thought about.

Dynamism

Hlupic & Vreedi (2005) emphasized that a simulation model can be easily customized to go behind modification besides utilizing it as a decision support tool for unbroken progress. The majority of these days' businesses are dynamic in nature and yet their objectives are habitually unpredictable due to the novel environment economy necessities. Significant assistance is the reality that experiments with business models can be reiterated, which facilitates the assessment of the sound effects of changes (Hlupic & Vreedi, 2005). In the framework of the above interpretation, it is essential to review the resulting model meticulously by the simulation team, focusing on the business aspirations that guide the simulation project. The effects of any adjustment call round the model have to be agreed upon between simulation team and business owner (or its agent); this will go ahead to enhanced communication and acquiring the necessary support of decision-maker. During the employment, new resources (time, personnel, etc.) may be required to keep on simulation project, in these cases action and reaction will not work, it is important to be proactive for these unexpected events. To overcome these and similar cases, it is necessary to have an expectations agenda and cost margin to go on with the project safely. Sadowski et al. (1999) argued that all the way throughout the project, suppleness is significant as circumstances come to pass, such as scope adjustment, difficulty with data gathering or professionals' accessibility. The simulation team has to look for innovative traditions to decipher evils.

Scope

As Sadowski et al. (1999) stated, "The most common types of misguided simulation studies are these where the scope is too ambitious or ill defined." In the two previous sections, task value analysis and cost analysis were pointed out as components of complexity factor examination, if the scope of simulation is amorphous everything appears to be significant according to the business under consideration. The scope stands for the boundaries of the work, Sadowski et al. (1999) asserted that the project originator must know what to wait for from simulation and the simulation team must know what is anticipated from them. So expectations management is one of the restrictions that affect simulation sensation. On the other hand, it is important that simulation concepts (for example, cycle time) have obvious denotation according to the related work. In the earlier discussions it was stated that successful simulation conveys precise information at the exact time to support an accurate decision; delivering information mostly depends on the individual, this information incorporates business rules and working processes, so the right individual is critical periphery in simulation project. One more boundary is the business objectives by which simulation work is bounded. Besides, business logic is a crucial character that is related to overall business settings that shape the margins of simulation plan. In order to obtain a proper scope a structured walkthrough must be accomplished principally with the dynamism attribute of new business environment.

Challenges and Limitations of Business Process Simulation

In the previous sections it was acknowledged that simulation is a predictive tool used to minimize the effects of probable risks. While simulation has numerous capabilities and characteristics that make it the best way to forecast the project future under certain circumstances, it has a lot of pitfalls:

1.   Simulation is experience sensitive: the simulation consequences are strongly affected by the experience of simulation team, for example, studying diverse workforce generations necessitates adequate background in this field.

2.   Simulation is the superlative option when the problem complexity is high: in straightforward cases decision-makers can use many other tools, for example, linear programming. If the problem complexity is not resolved accurately, the cost and efforts will not be easily justified when the stakeholders inquire about them.

3.   Simulation is time sensitive: Hulpic & Vreede (2005) argued that as business surroundings and evils are turn out to be more complex, the time available to deliver solutions is more limited. Dynamic nature, new environment, and opportunities, become main concepts in business logic which is restricted in most cases by cost/benefit issues, so the rapid decisions necessitate that simulation timeline to be as minimum as possible.

4.   Simulation is sensitive to "on-hand information:" unless the simulation team can effectively elicit the right information, the simulation consequences will be completely or at least partially distant from the acceptable solution, so it is necessary to engage the right people who have the right information; this obliges selecting the effective roles in the organization.

5.   Simulation scope (as discussed in pervious section): this issue is in command of results by project limitations, business objectives, and so forth.

Hulpic & Vreede (2005) confirmed that the number of stakeholders, model building time and management awareness are the key challenges of simulation:

1.   Number of stakeholders: the simulation team need to communicate properly and efficiently with all roles in the organization. They communicate to top management to understand the objectives of the business, with employees to recognize the proper methods to empower each one of them and to know how to plan the daily habits of each one of them; to revise staff age group styles diversity and even working processes must be communicated to study the running process stream. All of these roles are carried out by people, therefore those people have to be involved before the simulation life cycle starts, which means more time, efforts and costs.

2.   Model building time: as discussed previously, the time is inadequate in the new economy to think in all business situations. Unfortunately, simulation needs time, efforts, and

human resources, and if simulation is not justified by the problem complexity along with its consequences it is favored to use a different tool.

3.   Management awareness: management support is crucial in any activity success, in simulation case, it is necessary that management be aware of the advantages of simulation, the value of its outcomes, its responsibility in recognizing leadership skills, management skills, and so forth.

Besides these challenges, the culture is an additional challenge that must be concerned: at individual and national levels, this signifies the change acceptance level within the cognitive limit of the target society. Again, religion, habits, and traditions may make the predictive tool unable (even within a certain limit) to simulate the future. In order to overcome these challenges it is necessary to:

1.   Select the appropriate simulation team.

2.   Keep it simple: start with simple and known things on the road to further complex and unknown ones.

3.   Divide and concur: exhaustive working may lead to disseminating simulation efforts; hence it is better to divide the work among simulation team.

4.   To overcome time-consuming communication it is recommended to hold meetings for all the stakeholders within one session as possible; that is, JAD session.

5.   It is recommended that agile methodologies are used to develop simulation software package, since agile approaches concern frequent changing, customer involvement, and so forth.

6.   It is necessary to improve simulation team communication skills taking into account the cultural matters.

At the implementation level, the following are some issues that must be considered (Berio & Vernadat, 1999):

1.   Provide a general model representation for the user (user interface).

2.   Incremental modeling (concept of domains), because modeling is an iterative process.

3.   Implementation of the view mechanism (for the management of, model complexity).

4.   Distinction between material, information and control flows (for instance, using different graphical representations or notations).

5.   Functional hierarchy for the decomposition of domain processes (top level), business process (intermediate level), and enterprise activities (lowest level).

6.   Distributed simulation capabilities (to emulate concurrent processes and stand alone resource behaviors).

Summary

The logic of the business world is to create an advantage and/or utilize an opportunity.

Customer, cost, and competition are the issue(s) that have an aptitude to drive any plausible business improvement. There are many approaches to change management; in the preceding discussion in this chapter the authors focused on total quality management (TQM), business process reengineering (BPR) and, knowledge management. Considering the rate of failure in reengineering projects was over 50%, one of the major tribulations that have a say to the breakdown of BPR projects is the need of tools for evaluating the sound effects of a considered solution before execution.

Complexity and risk are the main issues have to been emphasized when simulation activity is to be conducted. Flow objects, resources, activities, and routings are the basic building blocks of any proper model. Cycle time, entity count, resource utilization and activity cost were illustrated as the measures for any model performance. The chapter illustrated the model experimentation, output analysis, and recommendations of business process change as the major business process simulation steps. The number of stakeholders, model building time and management awareness were positioned as the key challenges of simulation. Throughout the preceding in this chapter the discussion on the effect of simulation on business environment related skills, business management related skills, leadership related skills, employees' empowerment level, process improvement, ethical issues, and stakeholders' management skills were discussed. Finally, for successful simulation communication, complexity, dynamism and scope must be considered and then simulation can be used in diverse applications and areas as outlined in this chapter.

References

Ahadi, H.R. (2004). An examination of the role of organizational enablers in business process reengineering and the impact of information technology. *Information Resources Management Journal, 17*(4), 1-19.

Aguilar, M., Rautert, T.,& Pater, A.J.G. (1999). Business process simulation: Fundamental step supporting process centered management. In *Proceeding of the 31ˢᵗ Conference on Winter Simulation: Simulation - Bridge to the Future,* Phoenix, AZ, U.S. (pp. 1383-1392). New York: ACM Press.

Balci, O., (1990). Guidelines for successful simulation studies. In *Proceedings of the 22nd Conference on Winter Simulation,* New Orleans, LA, U.S. (pp. 25-32). Piscataway, NJ: IEEE Press.

Benjamin, P., Erraquntla, M., Delen, D., & Mayer, R., (1998). Simulation modeling at multiple levels of abstraction. In *Proceedings of the 30ᵗʰ Conference on Winter Simulation,* Washington, DC (pp. 391-398). Los Alamitos, CA: IEEE Computer Society Press.

Berio, G., & Vernadat, F.B. (1999). New developments in enterprise modeling using CIMOSA. *Computer In Industry, 40*, 99-114.

Bhaskar, R.,(1994). Analyzing and re-engineering business processes using simulation. In *Proceedings of the 26th Conference on Winter Simulation,* Orlando, FL, U.S. (pp. 1206-1213). San Diego, CA: Society for Computer Simulation International.

Bonser, C.F. (1992). Total quality education? *Public Administration Review, 52*(5), 504-512.

Bridgeland, D., & Becker, S. (1994) .Simulation satyagraha, a successful strategy for business process reengineering. In *Proceedings of the 26th Conference on Winter Simulation,* FL, U.S. (pp. 1214-1220). San Diego, CA: Society for Computer Simulation International.

Cheng, T.C.E. (1992). Computer simulation and its management applications. *Computer In Industry, 20*, 229-238.

Cheng, M., Tsai, M., & Xiao, Z. (2005). *Construction management process reengineering: Organizational human resource planning for multiple projects.* Article in press.

Davenport, T.H., & Short, J.E. (1990). The new industrial engineering: Information technology and business process redesign. *Sloan Management Review, 31*(4), 11-27.

Davenport, T.H. (1993). *Process innovation: Reengineering work through information technology.* Cambridge, MA: Harvard Business School Press.

Earl, M.J. (1994). The new and the old of business process redesign. *Journal of Strategic Information Systems, 3*(1), 5-22.

Ferrie, J. (1995). *Business processes – a natural approach.* ESRC Business Processes Resource Centre, University of Warwick. Retrieved from http:\\bprc.warwick.ac.uk\forum1.html

Garvin, D.A. (1991). How the Baldnge Award really works. *Harvard Business Review*, pp. 80-93.

Godfrey, A.B. (1993). Ten areas for future research in total quality management. *Quality Management Journal, 1*(1), 47-70.

Goll, E.O. (1992). Let's debunk the myths & misconceptions about reengineering. *APICS Magazine*, pp. 29-32.

Grant, R.M. (1996). Prospering in dynamically: Competitive environment: Organizational capability as knowledge integration. *Organization Science*, pp. 375-387.

Hammer, M. (1990). Reengineering work: Don't automate, obliterate. *Harvard Business Review, 68*(4), 104-112.

Hammer, M., & Champy, J. (1993). *Reengineering the corporation.* New York: Harper Collins Books.

Harrington, H.J. (1991). *Business process improvement.* New York: McGraw-Hill.

Hlupic, V., & Vreede, G. (2005). Business process modeling using discrete-event simulation: Current opportunities and future challenges. *International Journal of Simulation and Process Modeling, 1*(1/2), 72-81.

Hugan, J.C., Banks, J., Lendermann, P., Mclean, C., Page, E.H., Pegden, C.D., Ulgen,O., & Wilson, J.R. (2003). The future of simulation industry. In *Proceedings of the 35th Conference on Winter Simulation,* New Orleans, LA, U.S. (pp. 2033-2043).

Jain, S. (1999). Simulation in the next millennium. In *Proceeding of the 31st Conference on Winter Simulation: Simulation - Bridge to the Future,* Phoenix, AZ, U.S. (pp. 1478-1484). New York: ACM Press.

Juran, J.M. (1986). The quality trilogy. *Quality Process*, pp. 19-24.

Levas, A., Boyd, S., Jain, P., & Tulskie, W. (1995). Panel discussion on the role of modeling and simulation in business process reengineering. In *Proceedings of the 27th Conference on Winter Simulation,* Arlington, VA, U.S. (pp. 1341-1346). New York: ACM Press.

Liebowitz, J., & Beckman, T. (1998). *Knowledge organizations: What every manager should know*. Boca Raton, FL: CRC Press.

Liebowitz, J. (1999). *The knowledge management handbook*. Boca Raton, FL: CRC Press.

Liebowitz, J. (2000). *Building organizational intelligence: A knowledge management primer*. Boca Raton, FL: CRC Press.

Liebowitz, J. (2001). Knowledge management and its link to artificial intelligence. *Expert Systems with Applications*, *20*,1-6.

Olian, J.D., & Rynes, S.L. (1991). Total quality work: Aligning organizational processes, performance measures, and stakeholders. *Human Resource Management*, *30*(3), 303-333.

Omrani, D. (1992). Business process reengineering: A business revolution? *Management Services*, *36*(10), 12-15.

Ören, T.I. (2002). Future of modeling and simulation: Some development areas. In *Proceedings of the 2002 Summer Computer Simulation Conference,* San Diego, CA, U.S.

Pall, G.A. (1987). *Quality press management*. Englewood Cliffs, NJ: Prentice-Hall.

Paper, D., Tingey, K.B., & Mok, W.(2003). *The relation between BPR and ERP Systems: A failed project*. Hershey, PA: Idea Group Publishing.

Pressman, R.S. (2005). *Software engineering: A practitioner's approach*. New York: Mc-Graw Hill.

Robinson, S., Edwards, J.S., & Yonfa, W. (1998). An expert systems approach to simulating the human decision maker. In *Proceedings of the 30th Conference on Winter Simulation,* Washington, DC (pp. 1541-1546). Los Alamitos, CA: IEEE Computer Society Press.

Sadowski, D.A., & Grabau, M.R. (1999). Tips for successful practice simulation. In *Proceedings of the 31st Conference on Winter Simulation: Simulation – A Bridge to the Future (Vol. 1),* Phoenix, AZ ,U.S. (pp. 60-66). New York: ACM Press.

Saxena, K.B.C. (1996). Reengineering public administration in developing countries. *Long Range Planning*, *29*(5), 703-711.

Sink, D.S. (1991). The role of measurement in achieving world class quality and productivity management. *Industrial Engineering, 70,* 23-28.

Swain, J.J. (1993). Flexible tools for modeling. *OR/MS Today,* 72-81.

Talwar, R. (1993). Business reengineering: A strategy-driven approach. *Long Range Planning, 26*(6), 22-40.

Turban, T., King, D., Lee, J.K., & Viehland, D. (2004). *Electronic commerce: A managerial perspective.* 3/E Prentice Hall.

Tumay, K. (1995). Business process simulation. In *Proceedings of the 27th Conference on Winter Simulation,* Arlington, VA, U.S. (pp. 55-60). New York: ACM Press.

Tumay, K. (1996). Business process simulation. In *Proceedings of the 28th Conference on Winter Simulation,* Coronado, CA, U.S. (pp. 93-99). New York: ACM Press.

Chapter XIV

Virtual Reality and Augmented Reality Applied to Simulation Visualization

Claudio Kirner, Methodist University of Piracicaba, Brazil

Tereza G. Kirner, Methodist University of Piracicaba, Brazil

Abstract

This chapter introduces virtual reality and augmented reality as a basis for simulation visualization. It shows how these technologies can support simulation visualization and gives important considerations about the use of simulation in virtual and augmented reality environments. Hardware and software features, as well as user interface and examples related to simulation, using and supporting virtual reality and augmented reality, are discussed, stressing their benefits and disadvantages. The chapter intends to discuss virtual and augmented reality in the context of simulation, emphasizing the visualization of data and behavior of systems. The importance of simulation to give dynamic and realistic behaviors to virtual and augmented reality is also pointed out. The work indicates that understanding the integrated use of virtual reality and simulation should create better conditions to the development of innovative simulation environments as well as to the improvement of virtual and augmented reality environments.

Introduction

Simulation visualization aims to convert data and behavior of a system being simulated to user-friendly, understandable information. Using this information, the user can analyze the system and make decisions about the real system, related to dimensioning, parameters, behaviors and other system features.

In a first stage, the great amount of data generated by simulation was hard to analyze. With the use of multimedia, results from simulation were converted to graphics, like 3-D charts and other representations, showing colors, sizes and 2-D animations, simplifying the data analysis. However, the screen size imposed restrictions to the data amount that was possible to be visualized. Virtual reality applied to data visualization broke the limit of screen size and let the user view data in 3-D environments generated by computer. Besides the advantage of spatial view, people now can represent objects as they really are. In multimedia environments, the objects were represented by symbols, such as buttons, menus, icons, and so forth.

Virtual reality allows the user to analyze data and see the behavior of an installation being simulated, represented as a 3-D model.

The main difficulty of virtual reality is placing the user into the virtual environment, so that he/she can navigate, inspect and interact with the simulated environment. There are two ways to solve this problem:

- Using an avatar, like an intelligent agent, who knows the virtual environment and is able to receive commands from the user and execute actions in the virtual environment. In this case, it is necessary to use multimodal interactions and artificial intelligence in the virtual reality environment.

- Using augmented reality, which allows bringing the virtual environment to the user space, where he/she dominates and does not need special devices to interact with virtual objects. However, the use of multimodal commands is recommended, since the user uses the hands to manipulate objects and requires another way, like voice, to issue commands to the system, such as grab, release, and so forth.

In this way, this chapter discusses how virtual reality and augmented reality can be used to support simulation visualization, analyzing hardware and software aspects in each case. It emphasizes interaction techniques, since this subject is essential in the decision-making process.

Although the simulation visualization is the main focus of this chapter, it also makes considerations on the use of simulation in virtual and augmented reality environments, giving more realistic appearance and behavior to them.

This chapter also presents case studies on the application of virtual and augmented reality in simulation environments.

Modeling, Simulation, and Visualization

Modeling, simulation and visualization of systems depend on system concept defined as an entity composed by parts that interact among themselves and with the external world. A system can be: a factory, a computer, the computational application, a body, an atomic structure, a group of planets, a project, and so forth. To understand the functioning of a system, it should be evaluated directly, when possible, or through the use of simulation. Many times, simulation is the only way to express the behavior of a system when considering extreme situations involving characteristics, such as very small, very big, very fast, very slow, very expensive, dangerous, and so forth. Before performing the simulation, the user needs to create a model, which acts as the real or theoretical system, as shown in *Figure 1*.

Modeling

The simulation models are structures that represent real systems or projects and behave as they would. They can be classified as behavioral, functional and structural (Hein, 2000).

Other classifications focus on the models in distinguished or detailed form. Fishwick (1995) emphasizes the symbolic modeling and state-based modeling methods, using events and states. These models include the following types: conceptual, declarative, functional, constraint, spatial and multimodels. Maria (1997) states that a model for simulation is a mathematical model that can be: deterministic or stochastic, depending on the data; dynamic or static, depending on the consideration of time; continuous or discrete, if the time is continuous or considered in discrete points. The choice of the model depends on a tradeoff between realism and simplicity. If it is excessively realistic, the model will be complex. If it is excessively simple, the model will not reflect the functioning of the real system or the expected functioning of the project.

Figure 1. Real system and simulation

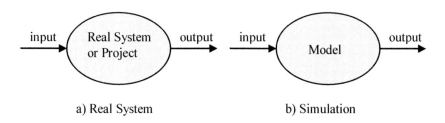

a) Real System   b) Simulation

Simulation

Simulation can be defined in several ways. Fishwick (1995) defines computer simulation as a process of designing a model of real or theoretical system, executing it in a digital computer and analyzing the output data. According to Bellinger (2004), simulation is the manipulation of a model in order to operate with compression or expansion of time and/or space, allowing the perception of its interactions. Maria (1997) defines simulation as the operation of a system model, which can be reconfigured and experimented, as if it was the real system, allowing the performance evaluation of the existing or proposed system.

Therefore, simulation intends to implement models of real or proposed systems, aiming to understand them, analyze their behavior and improve them. The diagram of *Figure 2* shows the steps of the simulation process.

Starting at the real or proposed system, defined here as the object of study, the specialist develops a model whose behavior is similar to the object of study and, at the same time, not very complex. The simulation starts from a set of input data, generating output data or results. The simulation data set, including input and output data, is then analyzed by a specialist, who can adjust the model. When the specialist obtains the intended behavior, the real or proposed system can be modified and adjusted for the improved or validated situation.

There are several types of simulation, which can be classified according to a taxonomy based on three properties: behavior, presence of time and type of values (Sulistio, 2004). Behavior indicates the way of occurrence of events during the simulation. Presence of time indicates if time is a significant variable or not. Type of values is related to the variation zones of the values of the simulated entities.

Figure 2. Simulation process feedback

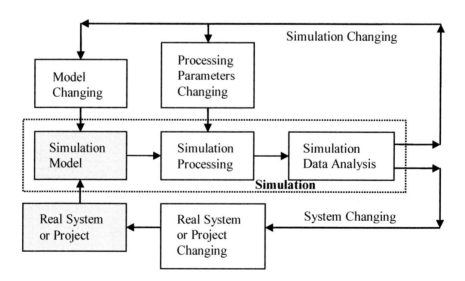

Figure 3. Simulation taxonomy emphasizing discrete event simulation

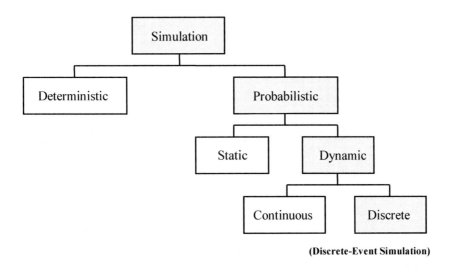

Thus, the simulation can be: deterministic or probabilistic, depending on its behavior; static or dynamic, depending on the presence of time; and discrete or continuous, depending on the type of values. *Figure 3* illustrates this taxonomy, emphasizing discrete event simulation, which is one of the most popular types of simulation.

Discrete event model is based on the concept of state, events, activities, and processes. Time is an important element, since the state changes in discrete times called event times. An event is an occurrence, which changes the state of the model and can trigger new events, activities or processes. An activity represents a period of time during which actions are executed. A process is a set of events and contains a group of activities. The modeling of an application uses one of the following approaches: event schedule, activity scanning or process interaction.

A discrete event simulation (probabilistic, dynamic and discrete) tends to be computing intensive. Its execution uses fixed-increment time advance or next-event time advance. Briefly, it can be said that a discrete event simulation functions as a repeated sequence of the following actions:

- Determining which type of events will occur next;
- Placing the simulation clock to the same time of the next event;
- Updating the statistical variables, if necessary;
- Executing actions associated with the current event;
- Scheduling a time for the next occurrence of that type of event.

A very important aspect of simulation is the user interface, since it determines how the user interacts with the simulation tool. Visual user interfaces are extremely helpful, because it offers better conditions of interaction and easiness of use, which makes it friendlier. In this case, the visualization of the system, its behavior and its input and output data have fundamental importance.

Visualization

During the simulation process, the input and output data must be presented in different ways to different people, depending on their functions in the use of the tool. These people can be simulation specialists, planners, managers, and so forth, so that their interests in the results differ in each case. Several visualization techniques have been used to show the simulation data, including charts, layout plans, 2-D and 3-D animations, and virtual and augmented reality. With the technological advances, the systems have become more complex, generating bigger volume of data, but, at the same time, the computational platforms have become more powerful, making possible the execution of the simulation and the visualization of those systems. On the other hand, users have also become more demanding, requiring more interactive interfaces with realistic visualization. As a result, the visualization techniques with 3-D animation, virtual reality and augmented reality are acquiring space in the simulation process.

The visualization in the simulation process, shown in *Figure 4*, involves an interactive feedback related to its adjustment to the needs of different users.

Figure 4. Visualization in simulation process

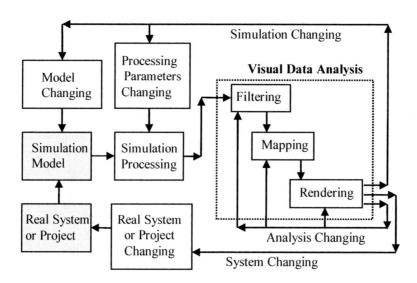

In *Figure 4*, the visual data analysis module is composed of three processes: filtering, mapping and rendering, to be applied to the simulation data, involving inputs and outputs (Lang, 2003). The filtering process is responsible for selecting or sampling data and producing deducted information. The mapping process transforms data into abstract visual representations, leading to a collection of geometric primitives (triangles, lines, points, clouds, etc.) as well as surface properties such as textures and materials. In the rendering process, the scene description, lighting information and camera positions are used to generate and display the images of the scene. Virtual reality and augmented reality need specific mapping and rendering processes, supported by the authoring systems of the 3-D environments.

Virtual Reality Applied to Simulation Visualization

The visual analysis of simulation data has intensively used multimedia to show graphics and diagrams, related to simulation results and behavior. Multimedia is defined as the integration, controlled by computer, of graphical texts, images, video, animations, audio and other media that can represent, store, transmit and process digital information (Marshall, 2001). Although multimedia is powerful and simple to use, it limits the visualization to the computer screen (which is 2-D). This space can be improved by the use of multiple overlapped or spread windows. The computational multimedia platforms require specific processing executed in specialized sound and video boards, besides internal and external channels with high capacity of transmission.

With the advance of computer technology, particularly in computer graphics, graphical interactive 3-D computation became feasible, allowing the breaking of the barrier of the screen and showing more realistic three-dimensional representations and behaviors of systems. Virtual reality is one of the technologies that allow the implementation of interactive 3-D visualization without restrictions and in real-time.

Definitions of Virtual Reality

According to Capps (2003), virtual reality is "a computer-generated, 3-D spatial environment in which users can participate in real-time." Complementarily, Burdea (1994) states that: "Virtual reality is a high-end user-computer interface that involves real-time simulation and interactions through multiple sensorial channels."

Despite the fact that virtual reality also uses multiple media, its primary focus is on the user interaction and the generation of 3-D images of the environment, in real-time. In order to do that, the main characteristic of the computational platform, appropriated for virtual reality applications, is the capacity of graphical processing used to generate images in real-time, complemented with non-conventional devices for the interaction in the 3-D environment.

A comparison between multimedia and virtual reality can be synthesized in the following way:

- Multimedia works with captured or pre-processed images; it prioritizes the quality of the images; it demands high capacity of transmission; it uses techniques of data compression; it acts in a 2-D space; and it works with conventional devices.

- Virtual reality works with images generated in real-time; it prioritizes the user interaction; it demands high processing capacity; it employs graphical computation techniques and resources; it acts in 3-D space; and it supports special devices.

As well as in multimedia, virtual reality needs the user to be transported into the application domain (virtual environment), which usually requires adaptation and training.

Types and Components of a Virtual Reality System

Based on the user's sense of presence, virtual reality can be classified into two types. When the user is partially transported into the virtual world through a window (screen or projection, for example), virtual reality is considered non-immersive. When the user is completely transported into the application domain through multisensorial devices (such as helmet, cave and their related devices, for example), which capture his/her movements and behavior and react to them, stimulating the sensation of presence inside of the virtual world, virtual reality is immersive. *Figure 5* shows examples of these two basic types of virtual reality. With the technological evolution, new devices may appear, but this basic classification remains.

A virtual reality system has two basic components: hardware and software. Hardware encompasses the input devices, multisensorial displays, processors and networks. Software includes simulation and animation controllers, authoring tools, database of virtual objects, interaction functions and input and output interfaces. In the next three items, hardware, software and computer network supporting virtual reality are discussed.

Figure 5. Virtual reality: non-immersive/immersive

    a) Non-imersive VR with monitor   b) Imersive VR with HMD

Hardware

Virtual reality hardware involves several input devices that help the user to communicate with the virtual reality system. Some those devices are: tracker, glove, 3-D mouse, keyboard, joystick, voice recognition device, and so forth.

Displays are sensorial output elements that involve more than the vision, such as visual displays, audio displays and haptic displays.

Processors are important elements of the virtual reality system that have benefited from the technological advances and trends of the videogame market, leading to complex three-dimensional applications. They involve both the main processors and the support processors in graphical, sound and other boards of specialized processing. Moreover, hardware can involve parallel processing and supercomputer environments.

Software

Virtual reality systems are complex and include real-time interactions among many hardware and software components. Virtual reality software can work in the preparation phase of the system, as authoring software for 3-D environments, or in the execution phase, as run-time support.

The authoring software involves: languages, such as VRML (Web3D Consortium, 2006a) and X3D (Web3D Consortium, 2006b); graphical libraries, such as OpenGL (ARTLab, 2006); toolkits based on Python and C/C++ libraries, such as Vizard (WorldViz LLC, 2006); and graphical toolkits, such as VizX3D (Virtock, 2006) and EonStudio (Eon Reality, 2006). The virtual environment development comprises 3-D modeling, preparation and manipulation of textures, sound manipulation, construction of animations, and so forth.

As run-time support, the virtual reality software must include the following actions: interact with the special devices; look after the user interfaces; support visualization and interaction; control the virtual environment simulation and animation; and perform the network communication for remote collaborative applications.

In some cases, the virtual reality software needs to be complemented with other resources, as it occurs with the VRML language, which must be integrated with Java language through EAI interface, to allow the development of systems with more powerful interactions and network communication. In other cases, the virtual reality software already has these optional resources or modules that are enough for complete use, such as Vizard and the EonStudio systems.

Computer Network

Computer networks are being incorporated in virtual reality applications, mainly because of the growth of Internet resources and the tendency of expansion in the use of collaborative work in several areas. However, since virtual reality does not demand traffic of images in the network, although it uses sporadic texture downloads, the necessary throughput is very low. This gives a computer network conditions to accommodate hundreds or thousands of users in collaborative applications. The network performs application download in the beginning of the execution, and makes the communication of short streams of data and positioning related to virtual objects of the scene along the execution. Moreover, to still decrease the traffic of data in the network during the system execution, there are techniques used to save traffic, such as dead reckoning and level of detail. The dead reckoning technique allows the network application to generate positions by calculation, sending data only when the calculated one differs from an error value related to real position tracked in a node. The level of detail is very useful in cases of dynamic download of parts of the virtual world, so that, depending on the distance of the user, the virtual objects to be downloaded can be simplified.

Interaction in Virtual Environments

Interactions in the virtual environment are part of the system interface, involving the interface with the devices and the interface with the user. The interface with the devices encompasses the hardware resources, such as the devices and its connections, besides the control software, called driver. The interactions occur through the use of the devices. The user interface includes the actions performed in the 3-D environment. The user can simply observe the behavior of the animated simulated virtual environment, having a passive experience, or be an agent of the system, intervening with its functioning.

The user's interactions comprise navigation, selection, manipulation, and system control (Bowman, 2005). The navigation is related to the movement of the user inside the virtual environment. It involves travel, which consists of the mechanical movement in the environment, and wayfinding, which is the cognitive component of the navigation. Travel is used for exploration, search and maneuvering and involves selection of direction, target, velocity, acceleration and actions such as: start to move, indication of position and orientation, and stop moving. Wayfinding is a decision-making process that allows the establishment of the way to be followed. It depends on the user spatial knowledge and behavior and on artificial cues, such as maps, compasses, signs, reference objects, artificial landmarks, trails, audio and olfactory cues, and so forth.

The selection consists of the choice of a virtual object to be manipulated. It involves three steps: indication of object, confirmation and feedback. Indication is usually made with the fingers or the hands driving some input device. It can occur by occlusion, object touching, pointing or indirectly. The system must show the selection through visual, auditory or haptic elements, such as changing color, blinking, emitting sound, emitting reaction, and so forth. The selection must be confirmed to be effective, which can be done through an event, such as mouse click, pressed key, gesture, voice command or some other action. It is important to have a feedback indicating that the action has occurred.

The manipulation of a selected object consists on the alteration of its position, through translation or rotation, or of its characteristics, involving scale, color, transparency and texture. The selected object can also be erased, duplicated, deformed or modified by other actions.

The system control consists of issuing of user commands to be executed by the system. The commands can be issued through graphical menus, voice commands, gesture commands or specific command devices.

Processing of Virtual Reality

A virtual reality system has several processing modules intended to treat inputs, simulation, animation, and rendering, using database information of the virtual environment and the user interface (Isdale, 2000). *Figure 6* shows a simplified diagram of the processing of a virtual reality system.

A processing cycle can be summarized by the reading of data from input devices, the execution of the simulation and animation, and the sensorial rendering. Sensorial rendering includes visual, auditory and haptic rendering. As the system executes in real-time, the time between the reading of the input data and the respective rendering is called latency time or reaction time of the system. To avoid discomfort to the user, the usual latency time must be, at most, 100 milliseconds, which implies in a frame rate of, at least, 10 pictures per second for the visual rendering.

Virtual Reality in Simulation

Virtual reality can be used to visualize data and behavior of a simulation, as well as the simulation can implement specific behaviors in a virtual reality system. These two characteristics are summarized as following.

Using Simulation in Virtual Reality Systems

Normally, a virtual reality system tries to reproduce the realism of the real world or the theoretical behaviors defined by the designers. Many properties or behaviors of the real world, such as fog, collision and gravity, can be found in virtual reality authoring software. These

Figure 6. Processing of virtual reality system

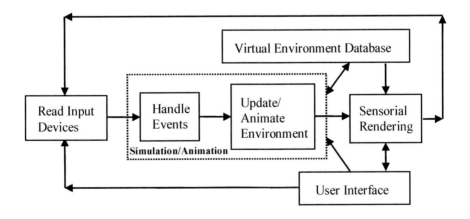

properties and behaviors are obtained through specific simulations that can be performed in the virtual reality software by actions on certain objects or on the entire environment.

Some examples of simulated behaviors usually available in virtual environments are: object movement; collision detection and reaction; physical simulation; and so forth. Although they make the virtual environments more realistic and stimulant, these behaviors require intensive processing, demanding more powerful computers to avoid degradation during the system execution.

Using Virtual Reality for Visualization of Data and Behaviors in Simulation

In this case, virtual reality is used in a module to support the simulation, which is useful for the visual analysis of data and behaviors. The module can be complex and powerful, presenting functions for a realistic, interactive visualization of the simulated environment. The virtual environment, representing the simulated environment has to be previously modeled by a virtual reality authoring tool and, after, to be integrated to the simulator through a special interface. The virtual reality module must also have a user interface which makes possible the interactive visualization and the alteration of simulation parameters.

The virtual reality module can be separate or integrated to the simulation process. If it is separate, it will communicate with the simulator through commands and parameters passing. The time and events management will be in the simulator. If it is integrated, the virtual reality module will have a part of its events used in the simulator, as it occurs when details of the animation, carried through the virtual reality, are controlled by the simulation. In this case, some events and timing parameters of the animation will be changed with the simulator.

A Case Study on the Use of Virtual Reality in Simulation

This case study consists of the simulation of an automatic machine to bottle milk that carries through the functions of filling up, covering and labeling. The implementation brings some challenges for simulation, animation and visualization, such as synchronization between the execution stages, hierarchical animation, techniques to obtain cyclical animation, and complex animation to simulate liquids.

When a simulation supported by virtual reality is created, it is necessary to define, at first, if the focus of the work will be the graphical quality or the application performance. In this case, the simulation performance was prioritized due to the great amount of connected animation that would have to be synchronized. Moreover, it was chosen a simulation integrated with virtual reality, to take advantage of the times and events of the animation.

First, the simulation environment was created. It was composed of a room with three walls, floor and ceiling. Some lightening effects had been included in the textures, following a technique very used in computer games. The plant has a mat to load the bottles and a superior part to fit the components of the machine.

The system executes as an infinite production line. The cycle starts with the input of an empty bottle in the left side of the machine. After accomplishing the three production stages (filling up, covering, and labeling), the bottle is dispensed in the right side of the machine, thus finishing the process.

So that the described animation had a perfect functioning, several animation techniques had been studied and implemented. The main techniques used in the implementation are cyclical animation, opacity animation, and animations in the structural grid of objects, presented as follows.

Cyclical Animation

These animations are the basis for the execution of the simulation in this case study. In this case, the animation cycle is restarted each time that the animation is accomplished. The

Figure 7. Cyclical animation of a milk bottle

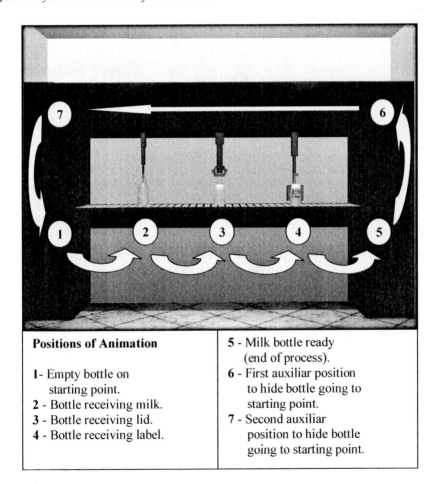

Positions of Animation	5 - Milk bottle ready (end of process).
1- Empty bottle on starting point.	6 - First auxiliar position to hide bottle going to starting point.
2 - Bottle receiving milk.	
3 - Bottle receiving lid.	7 - Second auxiliar position to hide bottle going to starting point.
4 - Bottle receiving label.	

most interesting example is about the bottles that pass through all the production line and, after that, they come back to the starting point, beginning the cycle again. It was necessary to implement an imperceptible return to the starting point. Then, the bottle, when coming back, appeared in a fast way, however perceivable, crossing the mat. Keys had been created to locate the bottle in return to the starting point, imperceptibly. *Figure 7* illustrates this process.

To create the illusion of bottles continuously entering in the left side of the machine, there were used four units of bottle, one for each visible stage of process and another one to complement the illusion. As the animation cycle is the same for all bottles, the controller of the first bottle was instantiated in the other bottles. Besides, as each animation had to start in a pre-defined instant time of the execution cycle, an option of the time generator was used that allowed one to define the starting point of the animation cycle. In this way, it was obtained the cyclical animation of the bottles, in a relatively simple and efficient form.

Opacity Animation

In some situations, it can be very useful to hide objects up to one definitive instant in time and, after that, make them appear in another instant in time. The simulation of milk bottling uses this technique in most objects, which made the creation process easier.

Animation in the Structural Grid of Objects

The animation of the structural grid of objects was implemented with VRML. It allows the modification of the structure of an object, point to point, which generates interesting effects. For this, the controller uses an object in different poses, interpolating them over time to obtain the animation. The simulation used this technique for the animation of the milk bottle. Five poses of the milk bottle during the filling up phase were created. These poses consist of the same object modified in its form, according to the necessity. The final model was designed first with all the details, and, later, it was adjusted for the less detailed forms. *Figure 8* exemplifies the five poses generated.

With the poses implemented, the animation controller was added and the time intervals between a position and another one were adjusted, keeping the animation light. The use of this technique enabled to create the milk bottling, which could be done using any shape of container, as long as it were respected the previously discussed parameters.

The interaction with the system is obtained, initially, by navigation. To do that, some cameras were created in the scene, with which the users can leave and disconnect themselves, navigating freely and carefully examining the points of interest. *Figure 9* shows the several initial positions of the cameras installed in the simulation.

Therefore, virtual reality allows the users to familiarize themselves with the machine, besides carefully observing the time of each cycle, visualizing the process in angles impossible to be seen in the real machine.

Figure 8. Milk interpolation used to fill the bottle

Figure 9. Viewpoints of the simulated process

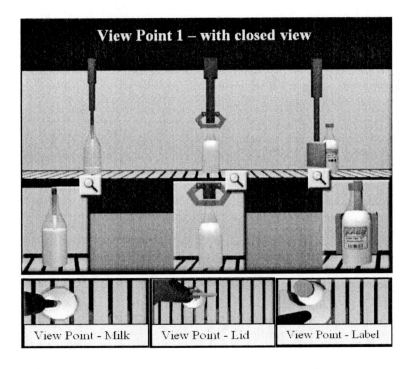

Augmented Reality Applied to Simulation Visualization

The advance of multimedia and virtual reality, complemented by the increasing power of computers, allowed the real-time integration of video and interactive virtual environments. In addition, the increase of bandwidth of computer networks also have been influencing positively the evolution of multimedia, making possible the efficient transference of images and other information.

Augmented reality, enriching the physical environment with virtual objects, benefited from this progress, making viable a series of applications of this technology, both in sophisticated and popular platforms.

Differently of virtual reality, which carries the user into the virtual environment, augmented reality keeps the user in his/her physical environment and carries the virtual environment to the user space, allowing more natural interaction with the virtual world, without the need of specific training or adaptation. New multimodal interfaces are being developed to facilitate the manipulation of virtual objects in the user space, using the hands or simple interaction devices.

Definitions of Augmented Reality

Azuma (1997) defined augmented reality as systems that present three main characteristics: it combines real and virtual scenes; it works interactively in real-time; and it is registered in 3-D.

In summary, augmented reality can be defined as the improvement of the real world with computer-generated data in real-time, using technological devices. However, there is a broader context defined as mixed reality that combines the real world with the virtual world using computational techniques, as shown in *Figure 10*, adapted from the reality-virtuality continuum proposed by Milgram (1994).

In the mixed reality environment, augmented reality occurs when virtual objects are placed in the real world. The user interface is in the real environment, adapted to visualize and manipulate the virtual objects placed in the user space. Augmented virtuality occurs when real elements are inserted in the virtual world. The user interface transports the user into the virtual environment, allowing him/her to see, read and manipulate not only virtual elements but also real ones.

Augmented reality and augmented virtuality are particular cases of mixed reality. Augmented reality uses computational techniques that generate, place and show virtual objects integrated to the real scene, while augmented virtuality uses computational techniques to capture real elements and reconstruct them as realistic virtual objects, placing them inside of virtual worlds and allowing their interaction with the environment. In any of the two cases, the functioning of the system in real-time is an essential condition.

Figure 10. Mixed reality environment (Adapted from Milgram, 1994)

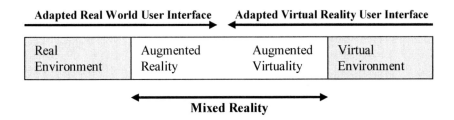

Augmented reality involves four important aspects: high quality rendering of combined worlds; precise registration (alignment in position and orientation) of virtual objects into the real world; interaction between real and virtual objects; and real-time processing.

The augmented reality environment uses multimedia resources that include high quality image and sound, besides virtual reality resources, including the image generation of virtual objects and real-time interaction. Thus, the computational platform for such environment must have the appropriate characteristics for multimedia and virtual reality, such as: processing capacity and media transference (image, sound, etc.); 3-D graphical processing capacity; real-time interaction; and support for non-conventional devices. Nowadays, the computers prepared to process games usually have all of these characteristics.

A comparison between virtual reality and augmented reality is given as follows.

- Virtual reality works only with the virtual world; the user is transferred into the virtual environment; and it prioritizes the user's interaction characteristics.

- Augmented reality has a mechanism to combine the real world with the virtual world; it keeps the sense of presence of the user in the real world; and it emphasizes the quality of the images and the user interaction.

Since augmented reality keeps the sense of presence of the user in the real world, it has a strong tendency to use technological resources invisible to the user in order to keep him/ her comfortable in that environment. Such resources, such as optical tracking, projections and multimodal interactions are being more and more used in augmented reality applications, while virtual reality uses a greater number of special devices to offer multisensorial resources to the users.

Types and Components of an Augmented Reality System

Augmented reality can be classified in two types, depending on how the user sees the mixed world. When the user sees the mixed world with the eyes pointing directly to the real positions with optic scene or video, it means that the augmented reality is of direct vision (immersive). When the user sees the mixed world through a device, as screen or projector, not lined up with the real positions, the augmented reality is of indirect vision (non immersive). Both types are illustrated in *Figure 11*.

In the direct vision, the real world images can be seen directly by eyes or through video and the computer-generated virtual objects can be projected into the eyes, mixed to the real world video or projected in the real scene. In the indirect vision, the real world images and the virtual world are mixed in video and shown to the user.

Figure 11. Augmented reality types based on vision

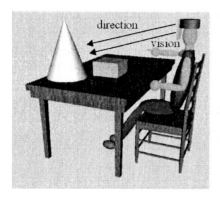

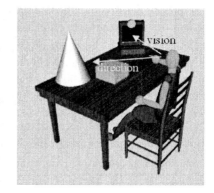

a) Direct Vision – Optical HMD   b) Indirect Vision - Monitor

Figure 12. Augmented reality implemented with direct and indirect vision

a) Direct Vision
(with HMD)

b) Indirect Vision
(Mirror on Monitor)

Augmented reality with direct vision can be implemented with the use of: optical helmets (optical see-through); helmets with micro camera connected to them (video see-through – see *Figure 12a*); handheld displays pointing to the real world; or virtual object projections in the real environment. Augmented reality with indirect vision is implemented with the use of cameras and monitors or projectors. The camera can be placed in several positions, such as: on the user's head, generating a first person view; behind the user, generating a third person view; or in front of the user, directed to him/her, generating a mirror view (*Figure 12b*).

Another way to classify augmented reality (AR) systems is based on the technology of displays, consisting of optical see-through AR; video see-through AR; monitor-based AR; and projector-based AR (Milgram, 1994; Isdale, 2000).

An augmented reality system has two basic components: hardware and software. Hardware involves the input devices, displays, processors and networks. Software includes the virtual reality software together with image markers and mixers, interaction functions and multimodal interfaces. In the next three items, hardware, software and computer network supporting augmented reality are discussed.

Hardware

Augmented reality hardware can use virtual reality devices, although it tends not to obstruct the user's hands, which must act naturally in the mixed environment. Techniques of visual tracking, using computational vision and picture processing are important in this case. For applications in open spaces, GPS is an interesting alternative for tracking. In this last case, the miniaturization of resources and the duration of the load of the battery are relevant elements to guarantee the user's comfort.

The power of main processing and support boards to deal with multimedia and virtual reality must be high enough to guarantee the real-time execution of the following tasks: video treatment; 3-D graphical processing; generation of mixed images; incorporation of sound; haptic execution; multimodal control; scanning of input devices with emphasis in tracking; and so forth.

Software

Augmented reality demands hardware resources and, additionally, it imposes software challenges as more complex and powerful applications are developed. The augmented reality software is used in the stage of system preparation, through tools of mixed environment authoring, and in the stage of execution, as a run-time support.

As an authoring tool, the augmented reality software is used to implement virtual objects and integrate them to the real environment, including some behaviors. It can use auxiliary elements for the capture of position or the same elements of the real scene.

The adjustment of virtual objects in the real space, made during calibration, can be interactive and visual or based on position parameters. Some authoring software comprises frameworks, which allow the preparation and the interaction with virtual objects. Many of them import previously modeled objects generated in languages or libraries as VRML,

X3D and OpenGL, for example. Other authoring software includes simpler tools, generating more complex systems. Examples of augmented reality authoring software are: ARToolKit (Billinghurst, 2006), MRToolKit (Freeman, 2005), Studierstube (Schmalstieg, 2002), Tiles (Poupyrev, 2001), APRIL (Ledermann, 2005), DART (MacIntyre, 2003), MARS (Guvem, 2003), AMIRE (Zauner, 2003), MXRToolKit (Mixed Reality Lab Singapore, 2006), and LibTap (Technotecture, 2006).

As run-time support, the augmented reality software must promote the tracking of the real objects, static and mobile, and adjust the virtual objects in the scene, both for fixed points of view and for in-movement points of view. Moreover, the augmented reality software must allow the interaction of the user with the virtual objects and the interaction between real and virtual objects in real-time. The storage of the scene with the positions and characteristics of the associated virtual objects can be used for the authoring or continuation of the interaction step, starting from a previous situation. The run-time support must also act in the simulation and animation control of the virtual objects placed in the scene; take care of the visualization of the mixed scene; and implement the network communication for collaborative applications. In the same way that the VRML language is considered one of the most popular virtual reality resources, ARToolKit is one of the most popular augmented reality resources.

ARToolKit is a software library, based on C and C++ languages, indicated to develop augmented reality applications. It is based on markers (cards with a rectangular frame and a symbol inside it) so that computer vision techniques can be used to calculate the real camera position and orientation relative to markers, allowing the system to overlay virtual objects onto markers (see *Figure 13*). ARToolKit includes the tracking libraries and complete source code, enabling customization (Billinghurst, 2006).

Figure 13. Augmented reality using ARToolKit

a) Marker Card   b) Virtual Object over the Marker Card

Computer Network

The advantages of augmented reality, especially concerning to the easiness of interaction in the mixed world, make this technology appropriated for local and remote collaborative works. In local face-to-face applications, the users see each other and interact among themselves and with real and virtual objects. However, in remote applications, the scene must be reconstructed in each point of the network. This generates a regular traffic of virtual reality information, complemented with a more intense traffic of multimedia data, including video, textures, sounds, and so forth. In this in case, the network can suffer saturation due to the traffic and the requirements of real-time, restricting the maximum number of simultaneous users in the application to a smaller value than of those ones used in similar collaborative virtual reality applications. Techniques of compression, degradation in image resolution, dead-reckoning, and level of details can be used to implement the collaborative application in low speed networks or to increase the number of users in a network. The same techniques can be used to fit the users in different heterogeneous bandwidth networks, establishing a compatible quality of the application with each user's resources.

Interaction in Augmented Reality Environments

Initially, the augmented reality systems focused on displaying information, without considering how the users would interact with them. Some systems were limited to reproducing, in the augmented reality environment, the graphical interfaces already known in 2-D and 3-D systems, as on-screen menus, recognition of gestures, and so forth.

The interface with devices includes the hardware resources or devices and specialized software such as device drivers that give support for the interactions. Augmented reality tends to use devices that the users do not perceive, giving more naturalness to their actions.

Besides using graphical interfaces, augmented reality systems present two tendencies: exploring different displays and devices; and integrating with the physical world using tangible interfaces (Azuma, 2001).

Alternative displays as handheld and special control devices exploring multimodal interactions are being tested as elements of interaction in augmented reality systems, allowing the inclusion of interaction techniques, such as World In Miniature – WIN (Bell, 2002).

Tangible interfaces allow direct interaction with the physical world through the use of hands or real objects and tools as a real paddle. One of the simplest and most popular ways of implementation of tangible interfaces is obtained by the ARToolKit, having used see-through video. The presence of a marking card in front of the camera places the virtual object associated to it on the marker. The manipulation of the card with the hands also moves the virtual object. Besides the virtual object, sounds can be played when the card enters the field of vision of the camera. Control cards can be implemented to intervene with selected objects of other cards, making geometric alterations, object exchange, capture or duplication, deletion, and so forth. The selection can be made with a control card through physical approach, inclination, occlusion, and other actions. With this, the objects can be modified or repositioned, generating innumerable applications. The Tiles system explores this type of interaction (Poupyrev, 2001).

Another interesting application is the Magic Book (Billinghurst, 2001) that implements interfaces between augmented reality and virtual reality. The project, which uses a physical book showing objects and virtual scenes on its pages, allows the user to use the book outside of the computational environment, enter in the augmented reality environment placing the book in the computational environment, and, finally, enter in the virtual world, hiding the real world video.

Another approach for the development of augmented reality interfaces consists of the use of virtual agents, whose actions are commanded by the user through gestures and voice commands. An agent can move virtual objects so that the user can inspect it, for example. It is possible, therefore, in augmented reality environments to carry through navigation, selection, manipulation and control of the system.

Processing of Augmented Reality

An augmented reality system has some processing modules to deal with the aspects related to virtual reality and real world that carry through the combination of the worlds and assure the user interaction and the interaction between real and virtual objects. The diagram of *Figure 14* shows an overview of the processing modules of a generic augmented reality system.

A processing cycle can be summarized by the following tasks: capture of video and execution of the object tracking; processing of the virtual reality system, including reading of devices and simulation and animation; registration, mixing real environment with virtual reality objects; and sensorial rendering, involving visual, auditory and haptic rendering. As the system works in real-time and must present a latency time equal or less than 100 milliseconds, the involved processing becomes bigger than one considered during the discus-

Figure 14. Processing of augmented reality system

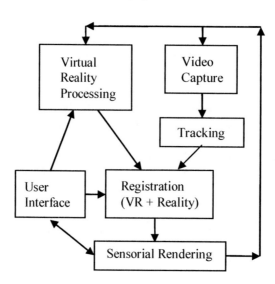

sion of virtual reality systems. Now, the processing of the virtual reality system is part of a broader and more complex set, also involving multimedia techniques.

Augmented Reality and Simulation

Augmented reality is indicated to visualize data and behaviors of a simulation, placed in the physical environment of the user. On the other hand, the simulation can be part of the augmented reality environment, implementing specific behaviors in the virtual objects mixed to the augmented scene.

Using Simulation in Augmented Reality Systems

An important objective of augmented reality is to insert virtual objects in the real world, creating the illusion that the entire scene is real. For this, the simulation is used to make the virtual objects have appropriate behaviors, such as movement, collision, reaction, physical simulation, and so forth. The behaviors do not need to imitate the reality, but they must give, to the synthetic elements, specific properties. As the simulation uses intensive processing, the computational platform must have power enough to execute all the modules in real-time.

Using Augmented Reality for Visualization of Data and Behaviors of the Simulation

Augmented reality can take the simulated environment, its data and its behavior to the user space, allowing their manipulation with the hands using simple tools or multimodal commands. The simulation can also be integrated to the real environment, whose results will be incorporated to the environment, as the simulation of the air traffic, which is shown in a real airport. In these cases, the user interactions with the environment can occur in two ways: the course of the real world can be modified by virtual objects and/or the virtual objects are restricted to operate according to the real world rules (Gelembe, 2005).

The simulated virtual objects can be autonomous (intelligent) or controlled by the user. In both cases, it is necessary for a user interface to establish behavior parameters or to carry through interaction with the augmented reality environment.

Due to the complexity, the trend is that the virtual reality module be separate of the simulation, since the virtual objects will have to be updated and registered in the physical world, in real-time.

A Case Study on the Use of Augmented Reality in Simulation

To illustrate the use of augmented reality in visualization of simulation data, the DataVis-AR tool will be presented as following.

The main objective of the tool is to show, in the physical environment of the user, simulated virtual 3-D scenes, including graphics representing the behavior of parts of the system. Moreover, the tool allows the manipulation of the virtual elements, aiming to give to the user the best viewpoint or the selection of system parts and data to be visualized. The tool can also be connected to a data acquisition system to show graphics representing current data gotten in real-time or historical data stored in a database, running in different speed to be better analyzed by the user.

The tool used marking cards to locate virtual objects on a table, such as maps, highways, toll plazas, graphics of traffic, flow of cars, and so forth.

Initially, the model of the map of the State of São Paulo - Brazil was placed on the table showing some highways and toll plazas. When choosing a toll plaza, the corresponding marker is placed in the environment so that the toll plaza appears on the map. Another marker allows the positioning of graphics receiving data in real-time or reflecting a simulated or sequence historical data on the tax of cars/time (hour or fraction of minutes) in the lanes of the toll plaza. The virtual elements can be located and be visualized in accordance with the user interest. The only care to be taken refers to the positioning of virtual objects in relation to the respective markers, avoiding the risk of occlusion during the manipulation of other markers. *Figure 15* shows the visualization of the augmented reality environment.

The DataVis-AR system was implemented with a modified version of the ARToolKit software. The modification made it possible to create special marking cards, called control cards or paddles. These cards are used to manipulate virtual objects with actions, such as: exchange, capture, duplication, movement, sound activation, alteration in execution time, and so forth. *Figure 16* shows the diagram of processing of the DataVis-AR system.

The simulation module generates simulation and animation data to the VR module to be shown to the user, including values for the graphics and positioning of the cars on the road. The VR Module processes the data and generates the upgraded view of the virtual environment and the graphics. Meanwhile, the system captures the video and tracks the marking

Figure 15. Visualization using DataVis-AR System

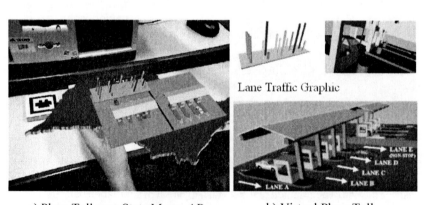

a) Plaza Toll over State Map – AR   b) Virtual Plaza Toll

Figure 16. Processing of DataVis-AR System

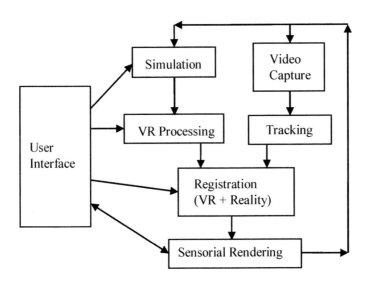

cards spread in the real scene. The registration module places the virtual objects (scenes and graphics) over the markers, creating the augmented environment. The sensorial rendering module shows the augmented environment in the helmet, monitor or projection and activates the corresponding sounds. In the case of the toll plaza, cars sounds are being used to make the scene most real and warning sounds indicate limits in the graphics or point out some scene regions that need attention. Graphical characteristics, such as change of color and pulsation, are also used to indicate warning.

The user can have an auxiliary panel, at the side of the simulated scene, to store virtual objects selected for posterior analysis. In this case, a duplication paddle copies the graphic and transfers it to the support panel. The user can also adjust the virtual graphics and the scene, aiming a better visualization and interaction.

To accomplish changes in virtual objects, reflecting adjustment on the graphics and on the positioning of the cars in accordance with the received data during execution time, some alterations in the ARToolKit were necessary. These alterations permitted the reload of virtual objects during the execution. A VRML code editor was also implemented to perform the alteration of the code, in accordance with the data received from the simulator or the data acquisition system. Thus, it was possible to show dynamic animation and visualization.

The excerpt of code shown in *Figure 17* illustrates the use of comments to indicate, to the editor, the lines that will be modified with the data received from the simulator or data acquisition system.

The DataVis-AR system applied to toll plaza simulation above described is indicated for engineers, planners, and supervisors involved. Several benefits can result from this application, such as:

Figure 17. Excerpt of VRML code showing lines to be changed

```
#VRML V2.0 utf8
# This is an excerpt of the graphic VRML code
  DEF dad_Group5 Transform {
##1 This marker point indicates that the position below will be changed by external data
  translation -10 0 0
   ...
##2 This marker point indicates that the diffuse color below will be changed by external data
         diffuseColor .7 1 .7
...
        geometry Cylinder {
##3 This marker point indicates that the height below will be changed by external data
        height 5.000
##4 This marker point indicates that the radius below  will be changed by external data
        radius 0.300
...
# Lines to be changed with new values (v) replacing the originals
##r1
#    translation v1 v2 v3
##r2
#         diffuseColor v4 v5 v6
##r3
#       height v7
##r4
#       radius v8
##end
```

- improvement of the process of data visualization by the use of a very interactive process, involving the users;

- analysis of delays, throughput, and interference of traffic conditions;

- identification of impacts from changes in the configuration of a toll plaza;

- evaluation of plans for repaving or other construction activities, identifying which minimize the disruption to traffic during construction;

- assistance in designing better toll plazas, including toll collection schemes, and so forth.

Besides, the resulting reduction in traffic overload will provide energy and environmental benefits, as well as economic benefits achieved from probable lower costs for transportation of goods and services.

Conclusion

This chapter presented an overview of simulation visualization using virtual and augmented reality.

Although virtual reality has demonstrated to be an option to overcome problems resulting from multimedia visualization, it brings inside itself other kinds of problems. Virtual reality demands special devices and user training, making its use restricted to users who can count on these components.

A way to minimize those problems is through the use of augmented reality. This technology brings virtual objects to the user space, where the user can manipulate real and virtual objects with the hands, using gestures and/or voice commands. Therefore, augmented reality is a new frontier technology, which goes beyond virtual reality, expanding the advantages of virtual reality and minimizing its difficulties. Augmented reality takes advantage from hardware and software evolution, resulting in powerful processing, fast optical tracking, multimodal interactions, and innovative 3-D graphical interfaces.

With these technological advantages, innovative simulation visualization can be developed, showing simulation results over real places, such as airports, industries, and so forth. It is possible, for example, to test a simulated airplane taking off or landing on a real airport.

Virtual reality and augmented reality, associated with artificial intelligence, promise to be a breakdown technology increasing several application areas particularly simulation visualization.

Acknowledgment

The authors would like to thank Rafael Santin and Arthur Augusto Bastos Buccioli for their valuable contribution in the implementation of the case studies.

References

ARTLab. (2006). *OpenGL: The industry's foundation for high performance graphics*. Retrieved February 09, 2006, from http://www.opengl.org/

Azuma, R.T. (1997). A survey of augmented reality. *Presence: Teleoperators and Virtual Environments*, 6(4), 355-385.

Azuma, R.T. et al. (2001). Recent advances in augmented reality. *IEEE Computer Graphics and Applications*, 21(6), 34-37.

Bell, B., Hollerer, T., & Feiner, S. (2002). An annotated situation-awareness aid for augmented reality. In M. Beaudouin-Lafon (Ed.), *15th Annual ACM Symposium on User Interface Software and Technology* (pp. 213-216). Paris: ACM Press.

Bellinger, G. (2004). *Modeling & simulation: An introduction*. Retrieved February 9, 2006, from http://www.systems-thinking.org/modsim/modsim.htm

Billinghurst, M. (2006). *ARToolKit*. Retrieved February 9, 2006, from http://www.hitl. washington.edu/artoolkit/

Billinghurst, M., Kato, H., & I. Poupyrev, I. (2001). The MagicBook: Moving seamlessly between reality and virtuality. *IEEE Computer Graphics & Applications, 21*(3), 6-8.

Bowman, D. et al. (2005). *3D user interfaces: Theory and practice.* Boston: Addison-Wesley.

Burdea, G., & Coiffet, P. (1994) *Virtual reality technology.* New York: John Wiley & Sons, Inc.

Capps, M. (2003). A definition of VR. In J. Isdale, J., Introduction to Virtual Environment Technology, Paper presented at *IEEE VR 2003.* Retrieved February 09, 2006, from http://vr.isdale.com/IntroVETutorial.2003.color.pdf

Eon Reality, Inc. (2006). *EonStudio.* Retrieved February 09, 2006, from http://www.eonreality.com/products/eon_studio.htm

Fishwick, P.A. (1995). *Simulation model design and execution: Building digital worlds.* Englewood Cliffs, NJ: Prentice-Hall.

Freeman, R., Steed, A., & Zhou, B. (2005). Rapid scene modeling, registration and specification for mixed reality systems. In Y. Chrysanthou & R. Darken (Ed.), *ACM Symposium on Virtual Reality Software and Technology 2005* (pp. 147-150). Monterey, California: ACM Press.

Gelenbe, E., Kaptan, V., & Hussain, K. (2005). Simulating autonomous agents in augmented reality. *Journal of Systems and Software, 74*, 255-268.

Guvem, S., & Feiner, S. (2003). Authoring 3D hypermedia for wearable augmented and virtual reality. In *7th International Symposium on Wearable Computers* (pp. 118-126). White Plains, NY: IEEE Computer Society.

Hein, C. et al. (2000). *VHDL modeling terminology and taxonomy.* Retrieved February 09, 2006, from http://www.atl.lmco.com/projects/rassp2/taxon/

Isdale, J. (2000). *Augmented reality.* Retrieved February 09, 2006, from http://vr.isdale.com/vrTechReviews/AugmentedReality_Nov2000.html

Lang, U., Wossner, U., & Kieferle, J. (2003). 3D visualization and animation: An introduction. In D. Fritsch (Ed.), *Photogrammetrische Woche 2004*, University of Sttutgart. Retrieved February 9, 2006, from http://www.ifp.uni-stuttgart.de/publications/phowo03/lang.pdf

Ledermann, F., & Schmalstieg, D. (2005). APRIL: A high-level framework for creating augmented reality presentations. In *IEEE Virtual Reality 2005* (pp. 187-194). Bonn, Germany: IEEE Computer Society.

MacIntyre, B. et al. (2003). DART: The designer's augmented reality toolkit. In *The Second IEEE and ACM International Symposium on Mixed and Augmented Reality* (pp. 329-330). Tokyo, Japan: IEEE Computer Society.

Maria, A. (1997). Introduction to modeling and simulation. In S. Andaradóttir et al. (Ed.), *29th Winter Simulation Conference* (pp. 7-13). Atlanta, GA: ACM Press.

Marshall, D. (2001). *What is multimedia?* Retrieved February 09, 2006, from http://www.cs.cf.ac.uk/Dave/Multimedia/node10.html

Milgram, P. et al. (1994). Augmented reality: A class of displays on the reality-virtuality continuum. *Telemanipulator and Telepresence Technologies, SPIE, 2351* (pp. 282-292).

Mixed Reality Lab Singapore. (2006). *MXRToolKit*. Retrieved February 9, 2006, from http://mxrtoolkit.sourceforge.net/

Poupyrev, I. et al. (2001). Tiles: A mixed reality authoring interface. In M. Hirose (Ed.), *Interact 2001 8ᵗʰ IFIP TC.13 Conference on Human Computer Interaction* (pp. 334-341). Tokyo, Japan: IOS Press.

Schmalstieg, D. et al. (2002). The studierstube augmented reality project. *PRESENCE – Teleoperators and Virtual Environments, 11*(1), 32-54.

Sulistio, A., Yeo, C.S., & Buyya, R. (2004). A taxonomy of computer-based simulations and its mapping to parallel and distributed systems simulation tools. *Software - Practice and Experience, 34* (7), 653-673.

Technotecture Labs. (2006). *LibTap: Rapid open reality*. Retrieved February 09, 2006, from http://www.technotecture.com/projects/libTAP/

Virtock Technologies, Inc. (2006). *VizX3D real-time 3D authoring in X3D*. Retrieved February 09, 2006, from http://www.vizx3d.com/

Web3D Consortium. (2006a). *VRML specifications*. Retrieved February 09, 2006, from http://www.web3d.org/x3d/specifications/vrml/

Web3D Consortium. (2006b). *X3D documentation*. Retrieved February 09, 2006, from http://www.web3d.org/x3d/

WorldViz LLC. (2006). *Vizard VR toolkit*. Retrieved February 09, 2006, from http://www.worldviz.com/products/vizard/index.html

Zauner, J., Haller, M., & Brandl, A. (2003). Authoring of a mixed reality assembly instructor for hierarchical structures. In *The Second IEEE and ACM International Symposium on Mixed and Augmented Reality* (pp. 237-246). Tokyo, Japan: IEEE Computer Society.

Chapter XV

Emotional Agent Modeling (EMAM)

Khulood Abu Maria, Arab Academy of Business and Financial Services, Jordan

Raed Abu Zitar, Philadelphia University, Jordan

Abstract

Artificial emotions play an important role at the control level of agent architectures: emotion may lead to reactive or deliberative behaviors, it may intensify an agent's motivations, it can create new goals (and then sub-goals) and it can set new criteria for the selection of the methods and the plans the agent uses to satisfy its motives. Since artificial emotion is a process that operates at the control level of agent architecture, the behavior of the agent will improve if agent's emotion process improves (El-Nasr, Ioerger, & Yen, 1998; El-Nasr & Yen, 1998). In this introductory chapter, our aim is to build agents with the mission "to bring life" several applications, such as: information, transaction, education, tutoring, business, entertainment and e-commerce. Therefore we want to develop artificial mechanisms that can play the role emotion plays in natural life. We call these mechanisms "artificial emotions" (Scheutz, 2004). As Damasio (1994) argues, emotions are necessary for problem solving because when we plan our lives, rather than examining every opinion, some possibilities are emotionally blocked off. We will try to investigate if artificial emotional control can improve performance of the agent in some circumstances. We would like to introduce the readers to our model, which is based on both symbolic and computational relations. Simulations are left for another publication. The space available is barely enough to give an overall picture about our model. The main contributions of this proposal model is to argue that emotion learning is a valid approach to improve the behavior of artificial agents, and to present a systematic view of the kinds of emotion learning that can take place, assuming emotion is a process involving assessment, emotion-signal generation, emotion-response and then emotion learning (LeDoux, 1996). To come across as emotional, an agent needs to incorporate a deeper model of personality, sensitivity, mood, feeling and emotions, and, in particular, directly connect these affective concepts. For agents to be believable, the minds of agents should not be restricted to model reasoning, intelligence and knowledge but also emotions, sensitivity, feeling, mood and personality (Nemani & Allan, 2001). We will propose EMAM (Emotional Agent Model) for this purpose. EMAM generates artificial emotion signals, evaluates and assesses events, takes into account the integration of personality, sensitivity, mood, feeling and motivational states then takes proper action or plans for actions (sequence of actions) (LeDoux, 1996; Gratch, 2000).

Introduction

In this subsection, a list of related works will be briefly described. We feel it is necessary to go through previous studies before we start introducing our work in the next sections. As shown in previous subsections, the topic of emotion was regarded as a very challenging topic, since it was hard to fully understand how we feel and why we do feel that way. Part of the reason for the so-called "mystery of emotions" is due to the fact that most emotions occur at the subconscious level. Moreover, it is still unclear how emotions transition from the subconscious to the conscious brain section. In fact, the complexity in the human mind lies in the complexity of the interaction between both the emotional and the cognitive processes. Searching for a better solution, researchers on agent's technology began working on artificial emotions. However, below is a quick review of some related works:

Magy Seif El-Nasr & John Yen (1998) proposed a model called FLAME – Fuzzy Logic Adaptive Model of Emotions. FLAME was modeled to produce emotions and to simulate the emotional intelligence process. FLAME was built using fuzzy rules to explore the capability of fuzzy logic in modeling the emotional process. Fuzzy logic helped them in capturing the fuzzy and complex nature of emotions. They try to point out the advantages of using fuzzy modeling over conventional models to simulate a better illusion of reality. They concluded that the use of fuzzy logic did improve the believability of the agent simulated. What makes the human mind so complex is the interactions between its emotional and cognitive processes. The cognitive process and the emotional process are not as separate as they use in their model. So they will have to further study the possible ways of interactions between the emotional and the cognitive module.

J. Bates (1994) was building a believable agent (OZ project) using the model described in *The Structure of Emotions* by Ortony, Clore, & Collins (1988). The model only describes basic emotions and innate reactions; however, it presents a good starting point for building computer simulations of emotion. The basic emotions that were simulated in the model are anger, fear, distress/sadness, enjoyment/happiness, disgust, and surprise.

Jean-Marc Fellous (1999) reviewed the experimental evidence showing the involvement of the hypothalamus, the amygdala and the prefrontal cortex in emotion. For each of these structures, he showed the important role of various neuromodulatory systems in mediating emotional behavior. He suggested that behavioral complexity is partly due to the diversity and intensity of neuromodulation and hence depends on emotional contexts. Rooting the emotional state in neuromodulatory phenomena allows for its quantitative and scientific study, and possibly its characterization.

Mannes Poel, Rieks op den Akker, Anton Nijholt, & Aard-Jan van Kesteren (2002) introduced a modular hybrid neural network architecture, called SHAME, for emotion learning. The system learns from annotated data how the emotional state is generated and changes due to internal and external stimuli. Part of the modular architecture is domain independent and part must be adapted to the domain under consideration. The generation and learning of emotions is based on the event appraisal model. The architecture is implemented in a prototype consisting of agents trying to survive in a virtual world. An evaluation of this prototype shows that the architecture is capable of generating natural emotions and, furthermore, that training of the neural network modules in the architecture is computationally feasible. In their paper they introduced a model that makes it possible to talk about an emotional state and emotional state changes because of appraisals of events that the agents perceived in their environment. It is based on the OCC model (Ortony, Clore, & Collins, 1988), a cognitive theory for calculating cognitive aspects of emotions. In their paper they used the OCC model as the basis for the supervised learning approach of emotions they used.

Barry Kort, Rob Reilly, & Rosalind W. Picard proffered a novel model by which they aimed to conceptualize the impact of emotions upon learning, and then built a working computer-based model that could recognize a learner's affective state and respond appropriately to it so that learning could proceed at an optimal pace. They believed that accurately identifying a learner's emotional/cognitive state was a critical indicator of how to assist the learner in achieving an understanding of learning process.

Hirohide Ushida, Yuji Hirayama, & Hiroshi Nakajima (1998) proposed an emotion model for "life-like" agents with emotions and motivations. The model consisted of reactive and deliberative mechanisms. A basic idea of the model came from a psychological theory, called the cognitive appraisal theory. In the model, cognitive and emotional processes interact with each other based on the theory. A multi-module architecture is employed in order to carry out the interactions. The model also has a learning mechanism to diversify behavioral patterns. Their expectation was to build a device that will be capable of "seeing" other facial features, such as eyebrows, lips, and specific facial muscles; tracking them and reacting to them as they occur. They also expected the Learning Companion device to be able to make immediate software-driven evaluations of the emotional state.

Kenji Doya (2002) presented a computational theory on the roles of the ascending neuromodulatory systems from the viewpoint that they mediate the global signals that regulated the distributed learning mechanisms in the brain. Based on the review of experimental data and theoretical models, it is proposed that dopamine signals the error in reward prediction, serotonin controls the time scale of reward prediction, noradrenaline controls the randomness in action selection, and acetylcholine controls the speed of memory update. The possible interactions between those neuromodulators and the environment were predicted on the basis of computational theory of metalearning.

W.F. Clocksin (2004) explored the issues in memory and affect in connection with possible architectures for artificial cognition. The work represented a departure from the traditional ways in which memory and emotion have been considered in AI research, and was informed by two strands of thought emerging from social and developmental psychology. First, there has been an increasing concern with personhood: with persons, agency and action, rather than causes, behavior and objects. Second, there was an emphasis on the self as a social construct, that persons were the results of interactions with significant others, and that the nature of these interactions was in turn shaped by the settings in which these interactions occur.

Jonathan Gratch (2000) discussed an extension to their command and control modeling architecture that addressed how behavioral moderators influenced the command decision-making process. He described one research effort that has addressed how behavior moderators such as stress and emotion could influence military command and control decision-making. The goals were to extend this modeling architecture to support wide variability in how a synthetic commander performed his activities based on the influence certain behavioral moderators (such as stress and emotional state), and to model how certain dispositional factors influenced the level of stress or emotional that arises from a situational context. It differed from some other computational models of emotion in emphasizing the role of plans in the emotional reasoning.

Dylan Evans (2002) claimed that emotions enabled humans to solve the search problem. Emotions prevented us from getting lost in endless explorations of potentially infinite search space by providing us with both the right kind of search strategy for each kind of problem we must solve. The search hypothesis thus offered an account of the relationship between emotions and reason, according to which emotions played a positive role in aiding reason to make good decision.

Nils Nilsson (2001) described architecture for linking perception and action in a robot. It consisted of three "towers" of layered components. The "perception tower" contained rules that created increasingly abstract descriptions of the current environmental situation starting with the primitive predicates produced by the robot's sensory apparatus. These descriptions were deposited in a "model tower" which is continuously kept faithful to the current environmental situation by a "truth maintenance" system.

Emotional Agent Modeling (EMAM)

Model Requirements

To be consistent with the aim of adding artificial emotion to believable agents with different personalities for various application areas, we will introduce a set of requirements for EMAM model:

1. **Cognitive Module:** The module should be flexible and event-action oriented evaluation model. Moreover, the model should have a generic, domain-independent way of appraising events to be useful in a variety of applications.

2. **Quantitative Module:** The module should produce artificial emotions with different intensities. The emotion intensities should also decay over time.

3. **Affects State Module:** Including personality, sensitivity, feeling, mood and other agent's affects state. Personality and other affects state play an important role in determining the emotional behavior and features.

4. **Goals Module:** Goals represent a list of actions (trajectory of actions) one wants to achieve eventually in the future. In this case, this list of actions is initially introduced in a file and it represents the agent's goals.

5. **Learning and Adaptability Module:** The model should be able to learn and adapt to reflect the dynamic status of the environment.

EMAM Conceptual Model

Triple tower model (Nilsson, 2001) is used as a conceptual model for EMAM agents with some extension that deal with reactive and deliberative mechanisms, as shown in *Figure 1*.

Receiver detects stimuli from the behavioral environment to extract objects of events features. Reflex has to direct mapping functions from Receiver to the Executor (reactive

Figure 1. EMAM conceptual model

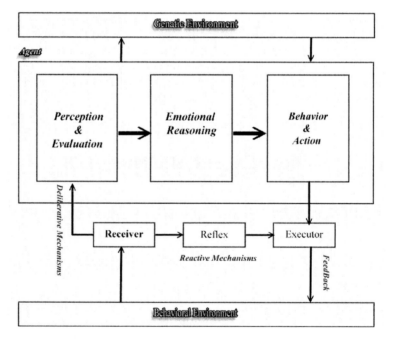

mechanism). Executor has the final action or behavior to feedback to behavioral environment. Receiver and Executor is domain dependent, so the agent designer is the responsible for implementing them.

EMAM conceptual model consists of reactive and deliberative mechanisms. Emotion information path is shown in *Figure 1*. The reactive mechanism (low path) covers direct mapping from Receiver to Executor, while deliberative (high path) mechanism has two processes: the cognitive and emotional processes. The cognitive process executes recognition, evaluation, decision-making and planning. The emotional process generates emotion signals according to the cognitive appraisals results. Agent's appraisal on emotion-inducing situation variables apply to event-based emotions, attribution emotions (agent-based) and attraction emotions (object-based). Emotions, which are generated according to the cognitive appraisals, influence cognitive processes such as empirical goals constructing and conflict dissolving in problem solving. Cognitive and the emotional processes interact with each other. While not of our behaviors are generated under deliberative mechanisms in the case of real life, so reactive mechanisms are required to avoid, for instance, a sudden obstacle and supporting agent survival.

As shown in *Figure 1*, deliberative mechanisms simulate the high information-processing pathways of the mind that lead from emotional stimulus to emotional response; while the reactive mechanisms simulate the low road that provides the quick pathway for our quick reactions (immediate ones). The deliberative mechanisms (high road) leads through the cognitive appraisal process and provides a more accurate representation of the stimulus, but takes a little longer to reach the mind decision-making center then the proper action to be taken by EMAM agent.

EMAM Components

Agent

Agent is an entity that is constantly changing. So, when we speak about an agent we always refer to it relative to a time. The moment that the agent starts existing is defined by the initial time setting. An agent has *personality, emotional activation level (sensitivity), mood (feeling), motivation (affective concerns)* and *emotional state*.

* Agent should be able to:
    o Represent a wide range of artificial emotions, not just the few basic ones that we can clearly differentiate, but also a lot of more complex, secondary, memory and learned emotions.
    o Make emotional judgments about incoming data as well as external perceptions.
    o Remember and learn from past emotional experiences.

- An agent has two classes of attributes:
  - o Cognitive Attributes:
    - Beliefs
    - Standard
    - Goal (to achieve)
      - o Main goals
      - o Innate goals (built in goals)
      - o Empirical goals
  - o Emotional Attributes:
    - Agent emotional state.
    - Emotional activation level
    - Agent mood
    - Personality
    - Previous experience (memory components)
    - Agent's kind: Male and female are different in their reaction to the events and the resulted emotions. Females are more emotional than males.
- Desired Value or Set Points:
  - o Drive releasers represent control systems that maintain a controlled variable within a certain range. This variable is measured through some of the agent's receivers (sensors) and compared to a desired value or *set point*.
    - If its value does not match the set point, an error signal is produced and alerts the agent's current process.
    - This error signal is fed to the appropriate drive (innate goals), in which it can be combined with error signals from other relevant control systems.

Agent Affects State

Personality

Personality is an important aspect for modeling emotional agent. Personality is static part of agent's being. Personality should be imported from the agent profile *(at the Initialization Stage)*. There exist many personality models, each of them consisting of a set of dimensions, where every dimension is a specific property of the personality. There exist many personality models, each of them consisting of a set of dimensions, where every dimension is a specific property of the personality. Take for example personality five dimensions FFM (openness, conscientiousness, extraversion, agreeableness, and neuroticism). Generalizing from personality models, we assume that a personality has *x-dimensions*, where each dimension represented by a value in the define interval **[a, b]** as a representation for personality dimensions, where the value of **a** corresponds to an absence of the dimension in the personality; while the value of **b** corresponds to a maximum presence of the dimension in the personality.

$$PD = [p1..pn], \forall i \in [1..x]: pi \in [a,b]$$

Personality is modeled through the different values that parameters (e.g., thresholds, gains, and decay rates) within each emotional system can have. For instance, if we want to model a "grumpy" agent, we might lower the emotion activation threshold and emotion decay rate for the *negative* emotion as well as increasing those for the *positive* emotion.

The relationship between personality and emotions remains problematic model. Some map emotion to behaviors in a personality specific way. Others treat personality as a variable that determines the intensity of a certain emotion, and some define a relationship between every personality factor and the goals, standards and attitudes.

In EMAM model, we will consider that every emotion is influenced by one or more dimensions of the personality. We need to define the relationship between agent personality and emotion state. However, we will provide the possibility of having multiple personality dimensions influence emotion state.

Emotions

Emotions are the dynamic part of agent being. The starting values of emotions should be reset to 0. For example, agent will portray dynamic emotions (that change over time) based on what happens, but how the agent obtains these dynamic emotions and the behavior that results from it depends on the agent's personality, emotion activation threshold and mood.

Dynamic Nature of Emotions

•   The role of time is an important factor in modeling artificial emotion. Agent's emotional state can be changed over time. It may be represented by a set of dynamic emotions whose intensities are changing continually.

Emotion Families

•   Emotions can be grouped into families. Each member of emotion family shares certain characteristics and mechanisms, including the similarities in previous events (triggered event), expression, as emotional behavioral response, and physiological patterns (somatic markers). These characteristics differ between emotion families, distinguishing one from another. Emotions can be grouped as:

   o   *Primary Emotions*: are activated, via natural releasers (innate), of one particular emotional system, such as *disgust* or *fear*. These primary emotions play an essential role in the preparation of appropriate emotional behavioral responses that are adaptive for the agent. The primary emotions can be anger, fear, distress/sadness, enjoyment/happiness, disgust and surprise.

   o   *Secondary Emotions (Emotional Memories, Previous Experience)*: are emotional acquiring learned releasers, which correspond to stimuli that tend to be associated with and predictive of natural releasers. The activation of emotional systems via these learned releasers correspond to what are referred to by some researchers as *emotional memories* or *secondary emotions*. These secondary emotions occur only after an agent tries to make orderly associations between *objects, events* and *situations,* and *primary emotions.* They require more complex processing for the retrieval of emotional memories of the previous experiences. Secondary emotions also play an adaptive role in dealing with situations that have occurred over and over throughout evolution, such as escaping danger, finding food, and mating (support survival for agent).

Emotions Interactions

- Different event-appraisals act as an elicitor that excite or inhibit different emotional states, and decay over time.

Emotions Definition

Agent's emotional state (AES) is a set of emotions with specific intensity. We will define emotional state as a y-dimensional vector, where emotion intensities are represented by a value in specific interval **[a, b]**, where the value of **a** corresponds to an absence of the emotion; a value of **b** corresponds to a maximum intensity of the emotion.

$$AES = \begin{cases} [e1..ey], \forall i \in [1, y] : ei \in [a,b] & if \quad t > 0 \\ 0 & if \quad t = 0 \end{cases}$$

Emotions Activation Triggering Events

Triggering event mechanisms are perceived events that stop and interrupt current agent processes. Examples for primary emotions mechanism:

- **Anger:** A mechanism to block the influences from the environment by suddenly stopping the current process. Anger triggering event is the fact that the accomplishment of a goal is undone.

- **Boredom:** A mechanism to stop repetitive action/behavior that does not contribute to satisfy the agent's needs.

- **Fear:** A defense mechanism against external threats or agents. Its triggering event is the presence of agent's enemies.

- **Happiness**: A mechanism for the agent goals achievement.

Emotions Functions

- Emotion with a dynamic nature means that once it becomes active, unless there are some continuous eliciting stimuli, its intensity decays according to some specific function until it becomes inactive again. For modeling the dynamic nature of emotions we need two functions:

o   Activation function for each emotion class.

   •   It depends on activation factors (drives or cognitive factors).

o   Decay function for each active class of emotions.

Emotions Attributes

1.   Emotion Intensity: Emotion intensity is the strength at which emotion fires. Intensity is a measure of how well a feature matches or does not match an expected pattern

2.   Emotion Activation Level (*as a threshold*): Activation level (we may call it agent's sensitivity) is considered as the threshold of the emergence of the emotion according to agent's personality, kind and age. Emotion activation threshold represents the minimum value that the intensity of every emotion should have to be considered.

3.   Saturation Point (*as a threshold*): There is a limit to how strong an emotion can be. When a particular emotion fires numerous times in succession it will become stronger until it reaches its *saturation* point.

4.   Polarity: It is the positive/negative metaphor as an approximation of emotion's valence. Polarity is the category of feeling associated with an emotion. These can possibly be separated into *positive* and *negative* emotions.

5.   Emotion Life Time *(Duration)*: Is the lifetime an emotion can be active. There is differences in considering both the long-term and short-term effect on emotions according to personality, age and experience. The older the person is, the less response to the emotional event and less the effect of emotion on his behavior. So the sensitivity, emotion intensity and life time will change through time.

6.   Decay Confident: Once emotions are generated, they do not remain active forever. After some period of time they disappears.

Mood

Some researchers differentiated mood from emotions based on specificity of the targets (e.g., emotions are specific and intense and are a reaction to an appraisal event, whereas mood is unfocused). Others differentiated them according to time (e.g., emotions are caused by something more immediate in time than moods). Others differentiated mood and emotion based on functional differences.

While emotions consist of high tonic levels of arousal, moods may be explained as low tonic levels of arousal within the same agent. Mood has a rather static state of being; that is, less static than personality and less fluent (dynamic) than emotions. Mood can act as a filter for the emergence of artificial emotions. A person in a positive mood tends to have a positive interpretation of the situation, which moderates the emotions a person feels; whereas, a person in a negative mood has a negative interpretation of the situation.

Mood can be represented as a one-dimensional vector (being in a good or bad mood) or perhaps multi-dimensional (for example, feeling in love). We will define a mood state as one-dimensional, where mood value is represented as a value in the interval **[-a, +a]**. While **+a** refers to good mood, **-a** refers to a bad mood. We may need to store the history of both emotional and mood states to use them for studding agent affective status through certain period (simulation time) or as an indication for the evolution of the affective state for our running agent.

EMAM Symbol-Processing Requirements

In EMAM we will try to employ computers in order to represent artificial emotions. EMAM will use *rule-based production systems*. It is also symbol-processing systems. A rule-based production system has as minimum requirements (set of standard components):

1. **Knowledge Base:** that contains the processing rules of agent's application.
2. **Global Database:** that represents the main memory of the application.
3. **Control Structure:** that analyzes the contents of this *global database* and decides which processing rules of the *knowledge base* are to be implemented.

So EMAM needs several databases for its functioning, to which it must have access to at any time:

1. **A database for the primary emotion:** Special *emotion eliciting conditions* are assigned to each of these emotion types.
2. **A database for the goals and cognitive factors.**
3. **A database for the reaction patterns:** which depend upon type of emotion.

EMAM Emotions

Artificial emotion in EMAM is a sequential possibly iterative process that involves emotional evaluation, emotional triggers used to regulate the agent's behavior, and finally emotional responses. In the first stage, a set of emotion eliciting structures evaluates the global state of the agent (internal state plus external environment). If certain conditions hold, an emotion trigger is generated informing the agent of the result of the evaluation stage. Emotional triggers generated during the stage of emotion eliciting are sent to the behavioral module, possibly giving rise to an emotional response. After an emotion trigger has been generated and an emotion response has been performed, the global state of the agent changes and a new assessment takes place, possibly generating a new emotion.

EMAM Emotions Algorithm

1. Receive event.

2. Interpret event for personality.

3. Factor the event into memory.

4. Compare event impact with activation emotion threshold.

5. Depending upon the current affects state (e.g., mood) of the agent, filter/modify the emotion.

6. Use emotion life time, saturation point and decay factor according to emotion family.

7. Agent final interpreted event, which could be classed as the reaction, is then used as feedback to modify some of the personality aspects, especially mood.

8. This final reaction is then returned to the event appraisal module and used to decide what is the emotional response.

EMAM Architecture

The mind consists of many small components called agencies, which have a very simple function, but the interaction among them leads to complex behaviors (Minsky, 1986). We used this idea for building EMAM architecture. The configuration of EMAM architecture is mention in *Figure 2*. EMAM consists of three categories of modules, which are functional module, supportive module and memory module.

Memory Components

There are four kinds of memory components in EMAM:

- **Agent Profile:** From the profile EMAM receives initial states of its supportive parameters for behaving, such as primary emotions, emotions activation threshold, emotion lifetime, emotion saturation point, personality and others.

- **Agent Goals:** Goals represent a list of actions one wants to achieve eventually in the future. In this case, this list of goals is initially introduced in a file and it represents the agent's initial goals. During run-time, the system will update this list. The order of EMAM goals is set according to the priority value each one has attached. Goals with higher priority will be evaluated/implemented first.

  o **Innate Goals:** Are built-in goals and assumed to correspond to the instinct of agent-preservation.

  o **Main Goals:** Each agent has one or more desire goals.

Figure 2. EMAM architecture

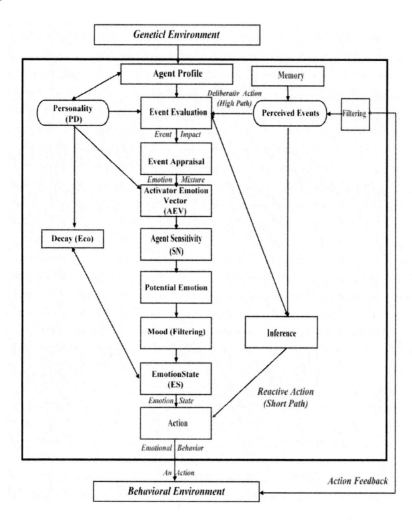

The goals creation depends on the situation and experience in interaction with the environment. The agent learns causal relations between stimuli and goals. For example, an agent empirically finds objects that contribute to satisfying hunger, and stores the knowledge in memory.

- **Buffers:** The interaction between the agent mind and the receiver/executor functions is mediated by the input and output buffers.

- **Long Term Memory:** Here an agent stores knowledge, which is captured via interaction with the environment and agent motivations. The knowledge of the agent

Table 1. EMAM goals tables

Attribute	Description
Goal Type	Goal kind that the user wants to achieve. • Main Goal • Sub-Goal • Innate Goal
Time Stamp	The time when the fact occurred
Decay Coefficient Value	Is used for deciding whether a goal should be memorized or not. It increases when the agent is acting with goals, and decrease as time elapses. If it becomes lower than zero, the memory about the goal is deleted. This helps to save the memory capacity.
Desire Level	The level of innate goals increases according to time series and decreases when these goals satisfied.
Priority	Determines the order of the goals in the list

is represented by a list of facts describing the objects, agents or events previously or presently observed in a given situation; and the previous emotional reactions to agents, objects and events.

Table 2. EMAM knowledge

Attribute	Description
Type	Type of fact. It can be a reaction or a description. Is it internal needs for agent or external from the interaction with the environment or other agents?
Target	The agent, event or object that is referred in the fact
Object Attributes	Are physical features of the percept object, such as colors, size, shapes and others. They are used to recognize objects and stimuli.
Novelty Value	Is set to the heights value at first and decreases according to the number of contact times. It helps an agent to express interest emotion to invest age such an object.
Contribution Degree Value	Represents how much an object contributes to each innate agent goals.
Decay Coefficient Value	Is used for deciding whether an object should be memorized or not. It increases when the agent contacts an object, and decreases as time elapses. If it becomes lower than zero, the memory about the object is deleted. This helps to save the memory capacity and diversify behaviors.
Time Stamp	The time when the fact occurred

The knowledge includes the following attributes for each event, object and others agent:

Functional Components

1. **Filtering:** Filtering process is responsible for determining the relevant information about the behavioral environment for the emotional response process. The filtering mechanism is necessary in agents, so as to increase its performance.

2. **Perception:** The perceptions contain the description of the following attributes.

3. **Evaluation:** The evaluation unit appraised events with regard to their emotional significance to the agent. According to OCC (7) each agent should hold a variety of different cognitive factors (attitude, goals, standards and beliefs) with which this evaluation proceeds. In this process the reactive or deliberative action (cycle) is selected according to the event if it is motivation action or external action.

4. **Emotional Reasoning:** It takes the event, activation threshold, mood and personality as inputs to produce the active emotion as output. The emotion reasoning component is in charge of generating emotions when external events or innate goals trigger it. This is done based on a specific emotional rule base system. Emotions are generated by evaluating events relating to goals. For evaluating the significance of an event to the agent's goal, we store event trajectory to reach the goal in the memory component. The aim is to find the optimal empirical goals of achieving main goals.

5. **Inference:** This component receives information from perceptions, goals, knowledge, and behavior components. The inference process will then infer the new current emotional state for the agent, given the knowledge of such information. This module is related to the evaluation component. It has a set of if-then rules that will activate an emotion when a given situation is verified.

6. **Behavior:** Is performing actions toward the behavioral environment and those actions, in time, produce the actions trajectory. The behavioral environment can be changed due to the agent's behavior. The agent can learn to perform in the behavioral environment, even if it is changed.

Table 3. Perceptions

Attribute	Description
Type	Type of the perception: • Drive • Event • Agent • Object
Description	Contains the features of the event, agent or object
Information Path	• Long Path (Deliberative Process) • Short Path (Reactive Process)
Time Stamp	The time when the perception occurred

Supportive Components

1. **Environments:** Our model consists of two types of environments:

    •   Genetic Environment: The agent receives initial states from its memory and other initial parameters. EMAM may be able to export *genomes*; data structures reflecting the adaptation of the agent in the behavioral environment to its child so the *new generation* of agents can be adapted to the same behavioral environment (we may use a genetic algorithm to improve EMAM model in future).

    •   Behavioral Environment: It can be changed due to the agent's behavior or to some other influences (simulation or user interaction).

2. **Motivation:** We consider for example hunger, fatigue, thirst, and pain motivational states. It will influence emotions by changing the activation threshold for each emotion. When the level of *fatigue* gets higher, the activation thresholds for negative emotions decrease and the activation thresholds for positive emotions increase.

3. **Personality:** Personality gives some specificity of the agent among other agents of the same kind in the agent's environment. The personality system is initialized by the agent's profile.

Principle of Emotional Reasoning

Emotion reasoning is aimed at determining whether or not an event means a specific situation; that is, does it contain emotion arousal features? Since EMAM framework is based around the concept of artificial emotion, it is necessary to provide a definition for artificial emotion features frame (see Table 4).

Table 4. Emotion features frame

Attribute	Description
Category	The ID of the artificial emotion family being experienced
Type	The ID of the artificial emotion class being experienced
State Comparison	After-state with before-state
Physiological State	Somatic markers related to specific emotion
Mental State	Goal achievement: achievement or not
	Planning: plan creation or not
	Result of execution: plan's effectiveness or not

Table 4. continued

Attribute	Description
Goal Parameters	• Goal importance: Low/High • Goal likelihood: Low/High
Cognitive Factors	• Standards • Beliefs • Attributes • Trust
Desirability	Event, object, goal and other desirability
Personality	Agent personality type (for example, FFM) (personality determines emotion type intensity)
Mood	Current agent affective state (bad, good feeling) (mood affects emotion)
Responsibility	Situation responsibility
Intensity	The intensity of the emotion.
Activation Level	Minimum vale for emotion emergent to the system (*as a threshold*)
Saturation	Emotion saturation level (*as a threshold*)
Decay	Emotion decay nature (decay coefficient) (is it linear, exponential or other?)
Time Stamp	The moment in time when the emotion was felt

According to the OCC theory (Ortony, Clore, & Collins, 1988), the emotions are organized hierarchically. This hierarchy is in sum defined by three types of reactions, depending on three types of aspects of the real world: events, agents and objects. These aspects are responsible for causing emotional reactions.

Emotional Process

In order for an agent to act emotional, the agents need an *object domain* within which situations occur that can lead to emotions and response. From perceptive input, an evaluation module will obtain emotional information (event impact on agent emotional state). This information is then used to update agent affective states. Event impact information will be represented as active emotions that include the information about the desired changes in emotion intensity and the event impact for each emotion. An agent receives its knowledge over emotions. In order to integrate this learning process into the *EMAM*, the learning component permits the agent to draw conclusions about emotions.

First of all, the observed emotional reaction is compared with a database of emotional reactions in order to define the underlying emotion. Then the observed event is filtered through the agent's previous experiences for the observed agent in order to determine whether this reaction is already registered. If this is the case, it can be assumed that the memory contains a correct representation of the emotion-triggering situation.

Emotion-Action Process

Step (1) Evaluation Stage

Determining of the Event Emotional Impact:

a.   After event occurs, an active emotion is generated by appraisal module that evaluates event's importance, disability, likelihood, desirability, novelty and contribution degree to agent's goals using the certain emotional appraisal rules base.

EMAM will use the Bates & Reilly model (Bates, 1994) of emotions, Em. *Em* contains an emotion system, which is based in the model of OCC, and we add our appraisal rules that fire each emotion kind. Emotions that can be generated by *EMAM* on the basis shown in *Table 5*.

Step (2) Emotional Activation Threshold Phase

Determining the ability of active emotion to influence the system:

a.   At the emotional module, every active emotion is compared with the agent's emotional activation threshold (this threshold is related to agent personality) to decide if an emotion may emerge or may not to an agent brain and affect its behavior and affective state.

b.   If the condition is accepted; the emotion module will calculate the agent's emotion state.

c.   Else the agent will continue to express its emotion state with some decay (using emotion decay coefficient **(Eco)**), as the following formula:

$$\text{Intensity}_{em\,t} = Intensity_{em\,t-1} \times Eco_t$$

Then return back to step 1 (to the beginning of the loop).

Table 5. Emotion types and their generation

Emotion Type	Emotion Cause	EMAM Emotional Rule-Based
Distress	Goal *fails* or becomes more likely to fail and it is important to the agent that the goal not fail.	IF (Current-Goal = 'Fail' Or Likehood(Current_Goal)="Low") And Important(Goal)="High" And Desirability="High" And Target="Event" And Responsibility="Self_Event"
Joy	Goal *succeeds* or becomes more likely to succeed and it is important to the agent that the goal succeeds.	IF (Current-Goal = 'Success' Or Likehood(Current_Goal)="High") And (Important(Goal)="High" And Desirability="High") And Target="Event" And Responsibility ="Self_Event"
Fear	Agent believes a goal is likely to fail and it is important to the agent that the goal not fail.	IF Likehood(Current_Goal)="Low" And Important(Goal)="High" And Desirability="High" And Target="Event" And Responsibility ="Self_Event"
Hope	Agent believes a goal is likely to succeed and it is important to the agent that the goal succeeds.	IF Likehood(Current_Goal)="High" And Important(Goal)="High" And Desirability="High" And Target="Event" And Responsibility ="Self_Event"
Like	Agent is near or thinking about a liked object or agent.	IF Desirability="High" And (Target="Agent" Or Target="Object")
Dislike	Agent is near or thinking about a disliked object or agent.	IF Desirability="Low" And (Target="Agent" Or Target="Object")

Table 5. continued

Emotion Type	Emotion Cause	EMAM Emotional Rule-Based
Anger	Another agent is responsible for a goal failing or becoming more likely to fail and it is important that the goal not fail.	IF Current_Goal="Fail" And Likehood="High" And Desirability="Low" And (Target="Object" Or Target="Agent") And (Responsibility ="Self" Or Responsibility ="Others")

Step (3) Mood Affect

Computing Mood Affect on Emotion State

a. Mood acts as a filter for the emergence of emotions. Agent in a positive mood tends to have a positive interpretation of the situation. Agent in a negative mood has a negative interpretation of the situation. In EMAM emotional model, we consider negative mood as an agent in a bad mood while positive mood value is considered as a good mood for an agent. If the agent's mood is positive then it will exhibit positive emotions with a degree of β more than the negative emotions and vice versa.

Step (4): Updating the Emotional State and the Mood

a. The impact of each emotion felt on the mood will be calculated proportionally to its intensity. So mood will be calculated according to the average of emotions felt by agent:

Mood (emotion) = Average (intensity of emotion)

Step (5): Calculate Emotion Decay Coefficient

a. The emotional model takes into account the fact that the intensity of emotions decreases over time. The decay of emotion differs according to the type of emotion and the personality of the agent. As a result, in the absence of any emotional stimulus the emotional arousal will converge towards zero. We will define emotion coefficient **(Eco)** using hyperbolic tangent function **(tanh ())**. The emotional intensity coefficient is a real value that can be normalized and multiplied by the intensity to produce the proper output. The function need specific variables that described the nature of the active emotion such as:

o   Current time value **(t)**.

o   Emotion intensity saturation value **(b)**.

o   Emotion intensity rate of degradation **(d)**.

o   The latency of degradation **(l)**.

$$Eco~(t) = -a~tanh~(t-c) +b$$

Step (7): Take Action and Return to Step 1

a.   Action will be taken according to emotion factor and influence the behavioral environment. Then the process will return to step 1.

Discussions and Applications

We will use EMAM as decision-making agent in dynamic system models. Decision-makers' agent controls the system variables. The decision is made based on the information of the up- and down-stream levels of each variable received at the decision points. In dynamic systems it is common practice to assume that decisions are made according to a rule base or a guidance table or graph. This deterministic approach is hardly able to model systems in which decisions are taken by people. People may decide differently in the same conditions, not because they are rational, but because they sometimes decide emotionally. The decision-maker is always trying to maximize her/his explicit profits by taking decisions that are known to the modeler. Emotionality is very personal and often leads to un-ustifiable decisions. A rational decision-maker may be replaced by a rational-emotional one.

EMAM will receive information from the environment and decide in-line with its personality. The environment is being changed by the decisions made, so EMAM faces a new condition to decide in. The environment also rewards or punishes the agent by the result of the decisions taken. Therefore, the personality of EMAM is a set of dynamic levels under the influences of the environment.

The business application in which EMAM will be implemented is material requirements planning (MRP). MRP is one of the most importing manufacturing models that are a form of backward scheduling. MRP is an information management system that ensures that material is not ordered for production until there exists end item for it to produce: MRP starts with the finished product and works backward to plan the requirements of all neces-

Figure 3. Manufacturing planning and control system

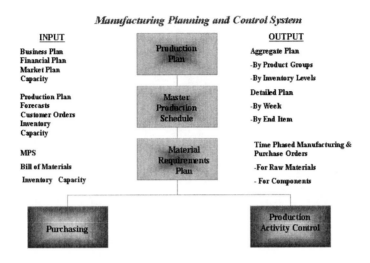

sary raw materials using Bell of Material (BOM) issue. The major goal of MRP is to keep stock value to minimum. MRP is applicable to discrete and dependent demand systems, allocating items in inventory when they are required. The system must be time phased so that predetermined materials arrive at the point in time when they are needed. Time phasing results in reduced inventory levels, since holding costs are a factor in determining ordered raw material or work-in-progress (WIP).

Agent simulation aims to realize an integrated system that has rapid response to changing requirements and the capability to integrate heterogeneous manufacturing activities with EMAM agent supervisory. EMAM will be the MRP agent (with supervisory & decision-making role) who is responsible for the effectiveness of the application. At first, the agent accepts the customer's order (through sales module) and performs the material requirement planning process, which results is material scheduling and deliveries dates. Then agent will monitor daily planned order to production model that carries out real-time scheduling and production. At the same time, EMAM will send an order to supplier for purchasing raw materials (shortage). After completing production task, production model returns the relevant information about finished parts. Once the customers add or modify orders, MRP agent will update data automatically to respond to the changes of customer requirements rapidly.

MRP is a computer modeling technique that allows for demand-driven production plans to be made. It determines what to produce, when to produce and how much to produce. We will design a MRP system that considers the process that begins when an order is placed by a customer and ends with the production of the corresponding item.

References

Bates, J. (1994). The role of emotion in believable agents. *Communications of the ACM*, *37*(7), 122-125.

Clocksin, W.F. (2004). Memory and emotion in the cognitive architecture. In D. Davis (Ed.), *Visions of mind* (pp. 122-139). Hershey, PA: Idea Group Publishing.

Doya, K. (2002). Metalearning and neuromodulation. *Neural Networks*, *15*(4-6), 495-506.

El-Nasr, M.S., Ioerger, T.R., & Yen, J. (1998). Learning and emotional intelligence in agents. In Proceedings of AAAI Fall Symposium.

El-Nasr, M.S., & Yen, J. (1998). *Agents, emotional intelligence and fuzzy logic*. Retrieved June 2005, from http://www.citeseer.ist.psu.edu/218095.html

Evans, D. (2002). The search hypothesis of emotion. *The British Journal for the Philosophy of Science*, *53*(4), 497-509.

Fellous, J-M. (1999). The neuromodulatory basis of emotion. *The Neuroscientist*, *5*(5), 283-294.

Gratch, J. (2000, May). Modeling the interplay between emotion and decision making. In *Proceedings of the 9th Conference on Computer Generated Forces and Behavioral Representation*.

Kort, B., Reilly, R., Rosalind, W., & Picard (2001). An affective model of interplay between emotions and learning: Reengineering educational pedagogy. In *Building a Learning Companion* (pp. 43-48). ICALT.

LeDoux, J.E. (1996). *The emotional brain*. New York: Simon & Schuster.

Nilsson, N. (2001). Teleo-reactive programs and the triple-tower architecture. In *Electronic Transactions on Artificial Intelligence* (Vol. 5, Section B) (pp. 99-110).

Nemani, S.S., & Allan, V.H. (2001). Agents and the algebra of emotion. In *Second International Joint Conference on Autonomous Agents and Multi-agent Systems* (AAMAS 2003).

Ortony, A., Clore, G.L., & Collins, A. (1988). *The cognitive structure of emotions*. Cambridge, UK: Cambridge University Press.

Poel, M., op den Akker, R., Nijholt, A., & van Kesteren, A-J. (2002). Learning emotions in virtual environments. In R. Trappl (Ed.), *Proceedings of the Sixteenth European Meeting on Cybernetics and System Research* (Vol. 2) (pp. 751-756).

Ruebenstrunk, G. (1998, November). *Computer models of emotions and their meaning for emotion-psychological research*.

Scheutz, M. (2004). Useful roles of emotions in artificial agents: A case study from artificial life. *In Proceedings of AAAI*. AAAI Press.

Ushida, H., Hirayama, Y., & Nakajima, H. (1998). Emotion model for life-like agent and its evaluation. *AAAI/IAAI*, pp. 62-69.

About the Contributors

EDITORS:

Asim Abdel Rahman El Sheikh is dean of information technology in the Arab Academy for Banking and Financial Sciences (AABFS). He supervised a number of theses in simulation and software engineering. He earned his PhD and MSc from London School of Economics & Political Science, and his BSc from University of Khartoum. He was a researcher in the Computer-Aided Simulation Modelling (CASM) Research Group. He worked as programmer, system analyst & designer in many organizations. He has authored two books and many articles. His research interest areas include SW piracy, software outsourcing, simulation modeling, and SW engineering.

Evon M. Abu-Taieh is a PhD holder and assistant professor in the Arab Academy for Banking and Financial Sciences (AABFS). She is also assistant dean in the Information Systems College and director of the London School of Economics program in the AABFS. She earned her PhD from AABFS in 2005 in simulation. She received her master's degree in computer science from Pacific Lutheran University, and her BSc from St. Martin's College, both in Washington State, USA. She has published many research papers in many topics, such as GIS, RSA, neural networks, simulation, and data mining in airline reservations. She was appointed in many conferences as reviewer, track chair, or track co-chair. Dr. Abu-Taieh worked in the field of computers for almost 17 years as system analyst, software engineer, and head of IT departments in many organizations: Mutah University, Ministry of Transport, Baccalaureate School.

Abid Al-Ajeeli received his BSc from University of London (Queen Mary College), MSc from University of Southampton, and PhD from University of Keele, UK. He worked in the oil industry for several years. Currently he is an associate professor in the College of Information Technology and the chairman of Information System Department at the University of Bahrain. His main research interests are software engineering, computerized manufacturing, and natural language processing.

AUTHORS:

Abdullah Abdali Rashed is an assistant professor of information systems security. He holds a PhD and an MSc in computer information systems from the Arab Academy for Banking and Financial Sciences and a Bachelor of Science in computer science from the Applied Science University, Amman, Jordan. Dr. Abdali's research interests include software piracy, cryptography, and computer programming languages. Prior to entering academia, he worked as a programmer and system analyst in Amman, Jordan.

Jeihan M. Auda Abu-Tayehwas born in 1978 as the youngest daughter to Mohammad Pasha Abu-Tayeh, the son of Sheikh Auda Abu-Tayeh. Raised in the badia region, she managed to attend school at the Rosary School in Amman, and then she acquired her bachelor's degree in pharmaceutical science and management from Al-Ahlyya Amman University. Furthermore, in 2002, she got her MBA with emphasis on "International Marketing & Negotiations Technique," with a GPA of 3.87 out of 4 (Hons.) from Saint Martin's College, in Washington State, USA. Currently, Abu-Tayeh is a head of the International Agencies & Commissions Division at the Jordanian Ministry of Planning and International Cooperation. In her capacity, she has the opportunity to maintain sound cooperation relations with the World Bank Group, as well as the UN Agencies, in order to extend to Jordan financial and technical support for developmental projects through setting appropriate programs and plans, and building and improving relations with those organizations. This is achieved through loans and aid programs, by means of designing project proposals, and conducting problem & needs assessment for the concerned governmental and non-governmental Jordanian entities, followed by active participation in extensive evaluation processes, conducted by either the UN Country Team or the World Bank Group Country Team.

David Al-Dabass holds the chair of intelligent systems in the School of Computing & Informatics, Nottingham Trent University. He is a graduate of Imperial College, holds a PhD and has held post-doctoral and advanced research fellowships at the Control Systems Centre, UMIST. He is fellow of the IET, IMA and BCS and serves as editor-in-chief of the *International Journal of Simulation: Systems, Science and Technology*; is chairman of the UK Simulation Society and Treasurer of the European Council for Modelling and Simulation. He has authored or co-authored over 170 scientific publications in modelling and simulation. For more details see his Web site: http://ducati.doc.ntu.ac.uk/uksim/dad/webpage.htm.

Jens Alfredson is since 2001 a researcher at the Department of man-system-interaction, Swedish Defense Research Agency in Linköping. He received an MSc in industrial ergonomics from Luleå University of Technology in 1995. He received a Licentiate of Technology in human-machine interaction from Linköping University of Technology in 2001. Since 1999, he is certificated as an Authorized European Ergonomist (CREE). He has previously worked at Saab as a senior research engineer, developing and evaluating novel presentations for fighter aircraft displays.

Firas M. Alkhaldi is an assistant professor of knowledge management. He holds a BA and MA in applied economics from WMU, U.S., and a PhD in knowledge management from Huddersfield University, UK. He is a certified e-business consultant and a KM professional. He is the dean of scientific research at the Arab Academy for Banking and Financial

Sciences and a professor in the Faculty of Information Systems and Technology, AABFS. His research interests are in knowledge conversion and transfer, organizational knowledge theory, knowledge culture, business process management, innovative work environment, and human and social implications of enterprise systems (ERP, CRM, and SCM). His work appears in a number of international journals and conferences

Torbjörn Alm is head of the VR & Simulation Lab at the Division of Industrial Ergonomics at Linköping University. He has a long industrial career with systems integration and cockpit design experience from the Swedish aerospace industry preceded by education and service as an officer in the Swedish Air Force. Later he entered the IT industry with focus on user interaction design. Retired from industry, he started his employment at Linköping University in 1996. In parallel he studied at the National Graduate School of Human-Machine Interaction and earned his PhD in early 2007.

Arijit Bhattacharya received a bachelor's degree of mechanical engineering, a master's degree in production engineering, and a PhD (engineering) degrees from Jadavpur University, Kolkata, India. Currently he is working as an examiner of patents & designs at the Patent Office, India. He was a senior research fellow at the Production Engineering Department of Jadavpur University. He served for a short period for construction and manufacturing sectors. His active engagement in research includes application of optimization techniques, multi-criteria decision-making theories in strategic management, operations research techniques, soft computing techniques, theory of uncertainties, and industrial engineering. He has about 36 publications of his research works in various journals, conferences/symposia and book chapters. He was conferred with the best paper award for the year 2002–2003 from the Indian Institution of Industrial Engineering. He is an Associate of the Institution of Engineers (India), member of the EURO Working Group on Multicriteria Aid for Decisions (EWG-MCDA), member of international society on MCDM, and member of the Industrial Applications Technical Committee of the IEEE Systems, Man & Cybernetics Society. He has served as anonymous reviewer for several international journals. He is actively engaged with the activities relating to commercialization of technology.

Peter Bollen received the MSc in industrial engineering and management science from Eindhoven University of Technology (The Netherlands) and a PhD in management information systems from Groningen University (The Netherlands). He is a senior lecturer in organization and strategy in the Faculty of Economics and Business Administration at Maastricht University (The Netherlands). His research interests include organizational engineering, conceptual modeling, business rules and business process design and modeling, and business simulation.

Mário M. Freire is an associate professor at the Department of Informatics of the University of Beira Interior, Covilhã, Portugal, where he is the head of the Department and the director of the MSc Programme in Informatics Engineering. He is also the leader of the Networks and Multimedia Computing Group at his department. His main research interests include optical Internet, high-speed networks, network security, and Web technologies and applications. He has been the editor of two books and has authored or co-authored over 90 papers in international refereed journals and conferences. He is member of the EU IST FP6 Network of Excellence EuroNGI (Design and Engineering of the Next Generation Internet). He is or was a member of Technical Program Committee of several IEEE and IASTED conferences. He was the general chair of HSNMC 2003, co-chair of ECUMN 2004, program chair of ICN 2005, and TPC co-chair of ICIW 2006. He is a licensed professional engineer and he is a member of IEEE Computer Society and IEEE Communications Society, a member of the ACM SIGCOMM and ACM SIGSAC, and a member of the Internet Society. He is also the Chair of the IEEE Portugal Section – Computer Society Chapter.

Jennie J. Gallimore is a professor in the Department of Biomedical, Industrial, and Human Factors Engineering at Wright State University. She received her PhD in industrial engineering and operations research from Virginia State and Polytechnic University in 1989. Dr. Gallimore applies human factors engineering principles to the design of complex systems. She conducts research in the areas of aviation spatial orientation and investigation of pilot spatial sensory reflexes, design of displays for advanced cockpits, design of displays for interactive semi-autonomous remotely operated vehicles including uninhabited combat aerial vehicles, human performance in virtual environments, and human factors issues in medical systems.

Subhashini Ganapathy (ganapathy.2@cs.wright.edu) is a software engineer at Intel Corporation, Chandler, AZ, USA. She holds a PhD in engineering with a focus in humans in complex systems from Wright State University. Her research interests include predictive analysis in model-based information technology systems, design optimization, and simulation and modeling. For additional information, visit www.wright.edu/~ganapathy.2.

Raymond R. Hill (ray.hill@wright.edu) is a professor of industrial and human factors engineering with the Department of Biomedical, Industrial & Human Factors Engineering at Wright State University. He received his PhD from The Ohio State University in 1996 and has research interests in heuristic optimization analysis, applied optimization, discrete-event and agent-based simulation modeling and decision supporting technologies for military-focused applications.

Claudio Kirner is an associate professor in computer science at the Methodist University of Piracicaba in Brazil. He received his PhD in systems engineering and computing from Federal University of Rio de Janeiro, in Brazil in 1986. He also spent two years as a visiting research scholar at the University of Colorado at Colorado Springs, USA. Dr. Kirner was chair of the First Symposium on Virtual Reality (1997) and First Workshop on Augmented Reality (2004) in Brazil and has published over 60 papers at international conferences and journals. His research interests include virtual and augmented reality, simulation and distributed systems.

Tereza G. Kirner is an associate professor in computer science at the Methodist University of Piracicaba in São Paulo state, Brazil. She earned an MS in information systems from the Federal University of Rio de Janeiro and a PhD in software engineering from the University of São Paulo, in Brazil. She also spent two years as a visiting research scholar at the University of Colorado at Colorado Springs, USA. Dr. Kirner has published over 50 papers at international conferences and journals. Her research interests include software engineering, simulation, and development of virtual reality applications.

Robert Macredie has over 15 years of research experience in working with a range of organizations, ranging from large, blue-chip companies, through small businesses, to government agencies. Macredie's key interest lies in the way in which people and organizations use technology, and his research aims to determine how work can be more effectively undertaken by improving the way that we understand how people and technology interact in organizational (and social) settings. He is professor of interactive systems and pro-vice-chancellor, Brunel University. He has undertaken work on a range of issues associated with people, technology and organizations and has over 180 published research contributions in these areas.

R. Manjunath is a research scholar from the University Visveswaraiah College of Engineering, Bangalore University, India. He was born in 1971 in Kolar, India. His doctoral thesis spans signal processing, neural networks and simulation of data transfer over the network. He has published about 55 papers in international conferences and journals in diverse areas involving the applications of signal processing. He has chaired many international conferences. His research interests include networking, signal processing, supply chain, validation methodologies, and so forth. He has industrial and academic experience over 11 years in various fields including signal processing, data transfers, validation strategies and neural networks.

Khulood Abu Maria is a PhD student and was born in Al Zarqa-Jordan in 1971. She earned her BS in computer science from Mut'a University in 1992, and a master's degree in information technology from Al Neelain University in 2002. She spent 14 years as a programmer, system analyst and an IT manager in a big international industrial company. She faced a good experience in the practical part of computer science and IT section. She is currently working on her PhD thesis. The thesis research interest is on artificial emotion and its application.

Roberto Mosca is an industrial plants management full professor and a DIP (Department of Production Engineering) director. He has fulfilled every academic career step beginning in 1972. He has served as logistic and production eng., CCDU president and management eng., and as CCL president until 1997. With courses started up in University of Genova Polo of Savona; he planned and cooperated to realize this Polo where he directs the Discrete Simulation and the Automated Industry Laboratories. Mosca is the author of more than 130 works, published in international congress acts or refereed papers. He works on industrial plants design and management modelling. Particular attention has been given to the discrete and stochastic simulators development for complex systems and to the application of original and innovative techniques for simulation experiment design and optimization (from new methodologies creation for run duration determination and the experimental error evolution control until the independent variables effects analysis on dependent variables). Particularly interesting is the design of online simulators for the real-time production management, with incorporated automated decision rules, a technique later used by many other researchers. Since 1990, he has worked on combined utilization of simulation and AI techniques and on the applicability conditions and statistical reliability of special DOE techniques to simulation problems involving an elevated number of independent variables. He is member of important papers' scientific committees (*International Journal of Modeling & Simulation, Impiantistica Italiana, Production Planning and Control*) and national and international congresses (ANIMP, IASTED, ESS, etc.). He is also a member of ANIMP, SCS, AIRO, PM, and AIIG. Titular in

1998 of FESR funds about 1 billion ITL for three applied research projects. He increased the industrial research contract amount as DIP Director from 250,000 USD/year in 1998 to over 1 million dollar/year in 2001 working with major international/national companies and agencies (Marconi, Elsag, COOP, NATO, Italian Navy, Fincantieri, PSTL, etc.)

S. Narayanan (PhD, PE) is a professor and chair of biomedical, industrial and human factors engineering at Wright State University in Dayton, Ohio, USA, where he directs the interactive systems modeling and simulation laboratory. He received a PhD in industrial and systems engineering from the Georgia Institute of Technology in 1994. His research interests are in interactive systems modeling and simulation, cognitive systems engineering, and human decision aiding in complex systems. He is a member of IIE, SCS, IEEE, IEEE Systems, Man & Cybernetics, HFES, and INFORMS. He is a registered professional engineer in the state of Ohio.

Kjell Ohlsson is a professor of human-machine interaction at the Division of Industrial Ergonomics at Linköping University. He received a BSc in 1974 at Uppsala University and a PhD in experimental psychology 1982 at Umeå University. Between 1992-1998 he held a position as professor of engineering psychology at Luleå University of Technology. He has experience in human factors simulator-based research in different simulator environments. Since 1998, he is appointed as program director of Graduate School for Human Machine Interaction and research director of the Swedish Network for Human Factors.

Mohammad Olaimat holds his bachelor's degree in mechanical engineering from Jordan University for Science and Technology, Jordan, and he received a master's degree in computer information systems from Arab Academy for Banking and Financial Sciences in 2006. His research interest is in knowledge representation, knowledge management, supply chain management, software engineering industry, business process reengineering, and secured organization. He has five published papers.

Ray J. Paul is a professor of simulation modeling and director of the Centre for Applied Simulation Modeling at Brunel University, UK. He received a BSc in mathematics, an MSc and PhD in operational research from Hull University. He has published widely, in books, journals and conference papers, many in the area of simulation modeling and software development. He has acted as a consultant for a variety of United Kingdom government departments, software companies, and commercial companies in the tobacco and oil industries. He is the editor of the Springer Verlag Practitioner book series. His research interests are in methods of automating the process of modeling, and the general applicability of such methods and their extensions to the wider arena of information systems. He is currently working on wider aspects of simulation, in particular in Web-based simulation and the new Grab-and-Glue modeling technique

Sasanka Prabhala is a PhD candidate in the Department of Biomedical, Industrial, and Human Factors Engineering at Wright State University. His research interests are in usability testing, developing advanced user interface designs, modeling human-machine interactions in complex environments, affective computing, and decision-making.

Roberto Revetria earned his degree in mechanical engineering at the University of Genoa and he completed his master's thesis in Genoa Mass Transportation Company developing an automatic system integrating ANN (artificial neural networks) and simulation with the ERP (enterprise resource planning) for supporting purchasing activities. He had consulting experience in modeling applied to environmental management for the new Bosch plant facility TDI Common Rail Technology in construction near Bari. During his service in the Navy as officer, he was involved in the development of WSS&S (Weapon System Simulation & Service) Project. He completed his PhD in mechanical engineering in 2001, defending his doctoral thesis on "Advances in Industrial Plant Management" by applying artificial intelligence and distributed simulation to several industrial cases. Since 1998, he is active in distributed simulation by moving U.S. Department of Defense HLA (high level architecture) Paradigm from military to industrial application. In 2000 he successfully led a research group first demonstrating practical application of HLA in not dedicated network involving an eight international university group. He is currently involved, as researcher, in the DIP of Genoa University, working on advanced modeling projects for simulation/ERP integration and DSS/maintenance planning applied to industrial case studies (contracting & engineering and retail companies). He is active in developing projects involving simulation with special attention to distributed discrete event and agent-based continuous simulation (Swarm Simulation Agents). He is teaching modeling & simulation, VV&A, distributed simulation (HLA), and project management in master's courses worldwide; and he is teaching industrial plants design in the University of Genoa master's courses. He is a member of SCS, IASTED, ACM, ANIMP, AICE, MIMOS and Liophant Simulation Club. He is associated professor in mechanical engineering.

Joel J. P. C. Rodrigues is a professor at the Department of Informatics of the University of Beira Interior, Covilhã, Portugal, and researcher at the Institute of Telecommunications, Portugal. He received a PhD in informatics engineering, an MSc from the University of Beira Interior, Portugal, and a 5-year BS degree (licentiate) in informatics engineering from the University of Coimbra, Portugal. His research interests include optical Internet, optical burst switching networks, high-speed networks, ubiquitous systems, and knowledge networks. He is member of the EU IST FP6 Network of Excellence – EuroNGI. He is member of many international program committees and several editorial review boards, and he has served as a guest editor for a number of journals including the *Journal of Communications Software and Systems*. He chaired several technical sessions and gave tutorials at major international conferences. He has authored or co-authored over 30 papers in refereed international journals and conferences. He is a licensed professional engineer and he is member IEEE Computer Society and IEEE Communications Society, a member of the ACM SIGCOMM, and a member of the Internet Society.

Sani Susanto received a bachelor's degree in mathematics and a master's degree in industrial engineering and engineering management (IE&EM) from Bandung Institute of Technology, Indonesia, and a PhD in IE&EM from Monash University, Australia. Currently, he is a senior lecturer at the Department of Industrial Engineering, Parahyangan Catholic University, Bandung, Indonesia. His active research includes application of optimization techniques, fuzzy logic, data mining, natural and soft computing techniques, and industrial engineering. He has published about 25 research papers in various international/national journals, conference proceedings, and several national conference papers. He is a member of UNESCO International Centre for Engineering Education (UICEE). Dr. Susanto was awarded the UICEE Silver Badge of Honor for distinguished contributions to engineering education, outstanding achievements in the globalisation of engineering education through the activities of the Centre, and, in particular, for remarkable service to the UICEE. He has served as anonymous reviewer for international journals, book chapters, and international conferences.

Pandian M. Vasant is a lecturer and coordinator of engineering mathematics for Electrical & Electronics Engineering Program at University Teknologi Petornas in Tronoh, Perak, Malaysia. He obtained a BSc (Hons.) in mathematics from the University of Malaya, Kuala Lumpur, and obtained a Diploma in English for business from Cambridge Tutorial College, Cambridge, UK. He received an MSc in engineering mathematics from School of Engineering & Information Technology at University Malaysia, Sabah. Currently he is a PhD candidate at University Putra Malaysia. During 1996-2003 he became the lecturer in advanced calculus and engineering mathematics at Mara University of Technology. He took the position of senior lecturer of engineering mathematics for the American Degree Program at Nilai International College during 2003-2004. His main research interests are in the areas of optimization methods and applications to decision and management, fuzzy optimization, soft computing, computational intelligence, and industrial production planning. Vasant has published more than 30 research papers in various national and international journals, and more than thirty papers in national and international conference proceedings. He is a reviewer for several international journals, book chapters and conference proceedings.

Raed Abu Zitar, an associate professor, was born in Gaza in 1966. He earned his BS in electrical engineering from University of Jordan in 1988, a master's degree in computer engineering from North Carolina A&T State University, Greensboro, in 1989, and his PhD in computer engineering from Wayne State University in 1993. He is currently the dean of College of Information Technology, Philadelphia University, Jordan. He has more than 40 publications in international journals and conferences; his research interests are machine learning, simulations, modeling, pattern recognition, and evolutionary algorithms with applications.

Index